for jenny

AMERICAN MATHEMATICIANS AS EDUCATORS, 1893 ★ 1923

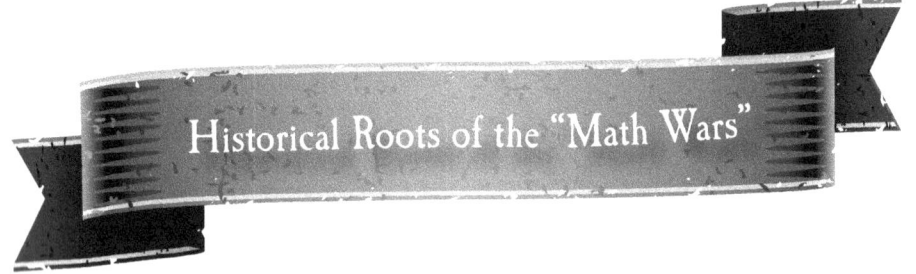

Historical Roots of the "Math Wars"

DAVID LINDSAY ROBERTS

Docent Press

DOCENT PRESS
Boston, Massachusetts, USA
www.docentpress.com

Docent Press publishes monographs and translations in the history of mathematics for thoughtful reading by professionals, amateurs and the public.

David Lindsay Roberts has an M.A. in mathematics and an M.S. in industrial engineering from the University of Wisconsin-Madison. His Ph.D. is from the Johns Hopkins University, in the history of science. He is currently chair of the Americas Section of the International Study Group on the Relations between the History and Pedagogy of Mathematics, and an adjunct professor of mathematics at Prince Georges Community College, in Largo, Maryland.

Cover design by Brenda Riddell, Graphic Details, Portsmouth, New Hampshire.

The cover photograph on the left is of Simon Newcomb and the photograph on the right is of Eliakim Hastings Moore.

© David Lindsay Roberts 1997, 2012

All rights reserved. No part of this book may be reproduced or utilized in any form or by any means, electronic or mechanical, including photocopying and recording, or by any information storage and retrieval system, without permission in writing from the author.

Contents

List of Tables ... ix

Acknowledgements ... xi

Chapter 1. Introduction ... 1

Chapter 2. Mathematics Education in Nineteenth-Century America ... 9
  2.1. Early Century Justifications for Studying Mathematics ... 10
  2.2. Charles W. Eliot: A Harvard Man Confronts Mathematics ... 18
  2.3. Emerging Challenges to Mathematics in the Curriculum ... 28
  2.4. Mathematics in Nineteenth-Century Educational Institutions ... 37
  2.5. Establishment of Mathematics at the Johns Hopkins University ... 41
  2.6. An 1890 Survey of the Educational Scene in Mathematics ... 47

Chapter 3. Simon Newcomb: One Mathematician's Educational Theory and Practice ... 55
  3.1. Newcomb's Educational Roots ... 56
  3.2. Newcomb at Johns Hopkins ... 62
  3.3. Newcomb as Textbook Writer and Educational Theorist ... 67

Chapter 4. The Committee of Ten and Its Mathematics Conference ... 83
  4.1. Immediate Origins and Basic Facts ... 84
  4.2. The Mathematics Conference and Its Report ... 86
  4.3. Near Term Response to the Report ... 95
  4.4. Long Term Response to the Report ... 106

Chapter 5. E. H. Moore: Leader of a New Generation of American Mathematicians ... 111
  5.1. Elite American Mathematical Activity from Newcomb to Moore ... 112

5.2.	Moore's Education and Early Career	119
5.3.	Moore, Harper, and the University of Chicago	122
5.4.	Moore's Initiatives for Enhancing American Mathematical Research	134

Chapter 6. The Development of
E. H. Moore's Pedagogic Program — 143
- 6.1. The Place of Pedagogy in Moore's Department of Mathematics — 144
- 6.2. Initial Instances of Moore's Interest in Pedagogy — 154
- 6.3. Moore's Championship of the Laboratory Method of Instruction — 164

Chapter 7. Moore's Pedagogy in Relation to
Contemporary Educational Thinkers — 179
- 7.1. William Rainey Harper — 180
- 7.2. John Dewey — 182
- 7.3. John Perry — 192
- 7.4. Felix Klein — 195
- 7.5. Simon Newcomb and the Committee of Ten — 202

Chapter 8. The Reception of Moore's Program — 207
- 8.1. Promotion of Moore's Ideas at the University of Chicago — 208
- 8.2. National Reaction to Moore's Proposals and Vision — 228

Chapter 9. School Mathematics on the Defensive — 241
- 9.1. Questioning the Place of Mathematics in the School Curriculum — 242
- 9.2. Organized Responses By Mathematical Educators — 260
- 9.3. The *Reorganization* Report of 1923 — 272

Chapter 10. Conclusion — 283

Appendix A. Acronyms — 293

Appendix B. Bibliographical Note — 297
- B.1. Introduction — 303
- B.2. Mathematics Education in Nineteenth Century America — 303
- B.3. Simon Newcomb: One Mathematician's Educational Theory and Practice — 307
- B.4. The Committee of Ten and Its Mathematics Conference — 309
- B.5. E. H. Moore: Leader of a New Generation of American Mathematicians — 310

B.6.	The Development of E. H. Moore's Pedagogic Program	312
B.7.	Moore's Pedagogy in Relation to Contemporary Educational Thinkers	312
B.8.	The Reception of Moore's Program	313
B.9.	School Mathematics on the Defensive	314
B.10.	Conclusion	315
Index		317

List of Tables

4.1 The Mathematics Conference of the Committee of Ten	87
A.1 Organizations and Committees	294
A.2 Serials	295
A.3 Encyclopedias	295
A.4 Archival Sources	296

Acknowledgements

The present book grew out of a doctoral dissertation in the history of science written at the Johns Hopkins University under the insightful supervision of Bob Kargon. It was in his seminar that I first realized the importance of the early University of Chicago for the evolution of American mathematics and mathematics education.

The Hopkins environment, to which I came after an extended sojourn as an operation research analyst in the defense consulting field, was delightfully stimulating. I received great benefit from informal discussions on miscellaneous topics in the history of science with my fellow students, in particular Takehiko Hashimoto, Barbara Becker, Louie Carlat, Larry Principe, Susan Morris, Elizabeth Melia, Ian Corbyn, Jennifer Tucker, Buhm Soon Park, and Carl-Henry Geschwind. In addition to Bob Kargon, other Hopkins faculty members from whom I received valuable encouragement and advice include Bill Leslie, Sharon Kingsland, Lou Galambos, and Robert Smith.

It was Robert who expertly guided me to a pre-doctoral fellowship at the Smithsonian's National Museum of American History. This proved to be immensely fruitful for my scholarly career, well beyond this book, and the person I have most to thank for this is Peggy Kidwell, the museum's curator of mathematics. Peggy then mentored me again during a postdoctoral fellowship at the Smithsonian. She has been a source of ideas and a discerning critic of my writing, and has introduced me to many other scholars in the history of mathematics.

Two fellowships from the Spencer Foundation (pre-doctoral and postdoctoral) deepened my research skills and introduced me to a wider community of scholars, the most fascinating of whom being Tomás Mario Kalmar, who never fails to make some revealing observation whenever we communicate.

The history of mathematics education after 1923 is described only fleetingly in this book. Nevertheless, I hope that my understanding of this later period (which may yet result in another book) informs the current book's commentary on earlier times. Said understanding has been greatly improved by my association with the following: the Oral History Task Force of the National Council of Teachers of Mathematics (supported by Harry Lucas of the Educational Advancement Foundation); the University of Maryland Department of Curriculum and Instruction; and the Americas Section of the International Study Group on the Relations Between the History and Pedagogy of Mathematics.

A reader for a publisher who declined an earlier version of the manuscript entirely failed to appreciate my sense of humor, but nevertheless made several useful suggestions.

I thank Scott Guthery of Docent Press for his cheerful, and wonderfully efficient, editorial guidance.

Lastly I am grateful for the continuing love and support of my wife, Jenny Scott, without whom I might still be laboring in far less pleasant fields.

CHAPTER 1

Introduction

From the 1890s to the 1920s American mathematical research grew substantially in quantity and quality. However, few mathematicians were able to pursue research exclusively; most mathematicians, whether active researchers or not, were employed as teachers by colleges and universities. Their degree of interest in the teaching role varied greatly: while some shunned instructional issues, a few mathematicians became deeply involved not only in teaching undergraduates but in attempting to influence mathematics education in the secondary and even the elementary schools. This book seeks to understand these divisions among the mathematicians regarding pedagogy, to explore alliances and conflicts between mathematicians and other educators, and to explain the resulting effects on educational institutions and on the mathematicians themselves.

In essence this book is an extended excursion in the social history of mathematics, using education as the organizing focus. Other aspects of mathematics that many would consider more important are touched on only peripherally, but education, an area of wide concern even to those with no interest in esoteric mathematics, provides an especially accessible demonstration that mathematicians are indeed social actors, a fact that has often proved easy to forget or ignore.

This book emerged from the perception that two events of 1893 are significant for understanding the developing relationship between university mathematicians and school education in the United States during the early twentieth century. The first event was the publication of the *Report of the Committee on Secondary School Studies*, popularly known as the Committee of Ten Report. This report, overseen by Harvard president Charles W. Eliot and written under the auspices of the National Educational Association, included sections written by subcommittees of specialists, one such devoted to mathematics. The second event was the International Mathematical Congress held in conjunction with the World's Columbian Exposition in Chicago. Unlike the Committee of Ten

Report, this event was not overtly concerned with secondary school instruction, but rather with the leading edge of mathematical research. Nevertheless, by advancing the interests of certain mathematicians the Congress had implications for school mathematics. The present work follows the consequences of these two events forward into twentieth century mathematics education, and culminates in 1923, the date of publication of another major curricular study, *The Reorganization of Mathematics in Secondary Education*, written under the auspices of the Mathematical Association of America.

The Mathematics Conference of the Committee of Ten was headed by the noted mathematical astronomer Simon Newcomb. Newcomb and his subcommittee, and the Committee of Ten generally, were reacting to the early stages of major ideological and demographic challenges that were overtaking American educational institutions at the end of the nineteenth century. Under scrutiny were such issues as the relationship between the secondary schools and the colleges, and the appropriate composition of the curriculum. A basic question at issue was, why study mathematics? Justification in terms of the mental discipline that mathematics allegedly conferred on students had been commanding decreasing respect since the middle of the nineteenth century. Challenges to mental discipline often came from advocates of the natural sciences, who were agitating for more time in the school curriculum for these studies. An important effect of the Committee of Ten was to endorse the notion that the sciences were indeed worthy to stand with the subjects that had previously formed the basis of the classical curriculum: Greek, Latin, and mathematics.

The report of Newcomb's subcommittee was in part an attempt to cope with the rise of science as a school subject. The virtues of mathematics as a source of mental discipline were not denied but were played down. Likewise, the deductive character of mathematics was given less emphasis. Instruction was to be recast in terms of induction and experiment, and abstract concepts were to be thoroughly motivated by concrete examples.

Newcomb's report by no means disowned abstraction, however. Indeed we can observe in this report a good deal of deference paid to the abstract core of mathematics, and a concomitant disdain for some traditional topics of school mathematics, most notably commercial arithmetic. The mathematicians on the committee, cognizant of the nineteenth-century triumphs of abstraction in mathematical and astronomical research, objected to laborious treatment of special cases in arithmetic which could be subsumed under general algebraic procedures.

The depreciation of arithmetic and the promotion of algebra were in some respects agreeable to general educators looking to squeeze more subjects into the curriculum and to make schooling more efficient, but this accord between mathematicians and other educators was short-lived. When in the years after 1893 the secondary schools experienced a great influx of students, most of whom were not bound for college, more radical advocates of efficiency and social utility were emboldened to call for deep reductions in the time devoted to mathematics in the schools. Neither the Committee of Ten, nor Newcomb's subcommittee, was prepared for this great increase in numbers of students, or for the resulting pressure on mathematics in the schools.

The generation of mathematicians who followed Newcomb was somewhat more attuned to the new conditions, but at the same time the professional evolution of these mathematicians was leading many of them to be more resistant to significant involvement with secondary school education. The Mathematical Congress of 1893 indicated this trend. The Congress was dominated by pure mathematicians, who would soon dominate the universities and the national professional association of mathematicians, the American Mathematical Society (AMS). Mathematicians of the older generation, typified by Newcomb, would feel somewhat out of place under this new regime. One young pure mathematician played a central role in organizing the Congress, and skillfully used it to advance his career and his objectives for American mathematics: Eliakim Hastings Moore of the University of Chicago.

Moore functioned primarily as a researcher in pure mathematics all his life, but within a few years of the Congress he began to exhibit a special interest in mathematical pedagogy, at both the school and college level. His pedagogic program in some ways was a natural extension of the ideas of Simon Newcomb and his subcommittee of the Committee of Ten. Like Newcomb, Moore called for greater emphasis on developing abstract concepts by means of concrete examples. This was part of what Moore labeled his "laboratory method." Newcomb had defended his pedagogic proposals as being in accord with the psychology of learning. Moore, in addition, explicitly designed his pedagogic proposals as stratagems for the jurisdictional competition in which he saw mathematics. Mathematicians needed to protect their own domain while promoting their usefulness to scientists and engineers. Moore's advocacy of graphs, models, and other "laboratory" techniques was founded on his vision of a unified mathematics curriculum which would smoothly serve the needs of those using mathematics as a tool, while at the same time yielding a steady production of mathematical educators at all levels, from school teachers to pure mathematical investigators.

But despite Moore's high place in the mathematical world, and the vigor with which he and his supporters pushed his program, mathematics in the United States did not develop according to his vision, except in minor respects. This book emphasizes two major factors that contributed to this outcome: Moore's program was at odds both with the internal dynamics of the mathematics profession and with the external development of the schools. On the one hand, mathematicians desired to cultivate the abstract core of their subject, unconstrained by problems of pedagogy at any level. On the other hand, educators trying to cope with the vast increase of students in the secondary schools were more and more susceptible to proposals to decrease mathematics instruction because of its low contribution to social efficiency and utility. Moore's program, which had largely ignored the old justification of mathematical training as mental discipline in favor of stressing the utility of mathematics for science and engineering, was ill-equipped to respond to the new anti-mathematical sentiments, since only a small proportion of students were training to become scientists or engineers.

The book concludes with an examination of what can be justly considered the final phase of the Moore program, *The Reorganization of Mathematics in Secondary Education* of 1923. Several features of this massive report, written with Moore's participation, illuminate the attempt of the mathematicians to cope with the changed educational landscape. First, the report was written by mathematical educators independently of other disciplines, unlike the Committee of Ten Report. Second, it was written under the auspices of a new organization, the Mathematical Association of America (MAA), which had branched from the AMS specifically in order to absolve the latter from concern with pedagogy. Third, Moore's original laboratory method proposals were much watered-down, while mental discipline was resurrected. Last, the report made an ambivalent effort to introduce into the secondary school curriculum one abstraction that had proved to be of major importance in research: the function concept.

The *Reorganization* report failed to stem the tide of anti-mathematical sentiment in the secondary schools, although for some time it offered a convenient standard around which to rally the partisans of mathematics. Meanwhile, although Moore had expressed the belief that the health of university mathematics was ultimately dependent on the vitality of mathematics at lower educational levels, pure mathematical research survived and even prospered in the universities, suffering little apparent ill effects from the "crisis" of mathematics in the secondary schools.

This book is therefore in large part a story of fragmentation. Although many of the actors claimed to see overriding unity, the events recounted here suggest fundamental disagreements regarding the aims of mathematics education, disagreements that have persisted to the present time. How should we allocate resources between educating the most promising students to high excellence versus raising a larger population of students to competency? Where is the appropriate balance between teaching abstract concepts and teaching concrete applications? How do we encourage students to become resourceful and original problem solvers and at the same time ensure that they learn and retain basic skills? Such questions, all the subject of current debate, are already visible in the period 1893–1923 and even before. Those who would answer these questions once and for all by clinging relentlessly to some simple version of progressive or classical education will find little support herein.

The narrative proceeds essentially chronologically. Chapter 2 provides background on nineteenth century education necessary to appreciate the significance of the Committee of Ten. Mental discipline and other justifications for the place of mathematics in the curriculum are surveyed, and the rise of challenges to mathematics is described. The chairman of the Committee of Ten's mathematics subcommittee, Simon Newcomb, is the subject of Chapter 3: his educational views, and how they grew out of his own experiences, are treated in detail. Chapter 4 then discusses the 1893 report produced by Newcomb's subcommittee, and the reception of that report.

The generational transition from Newcomb to E. H. Moore is described in Chapter 5, with special attention to Moore's institution, the new University of Chicago. Chapter 6 walks through Moore's pedagogic proposals at length, especially his 1902 address before the AMS. Chapter 7 interprets Moore's ideas in relation to those of other contemporary educators, both American and European, including Newcomb and the Committee of Ten. Chapter 8 describes the most important responses to Moore's program. In particular, it explores the conditions giving rise to the formation of the MAA in 1915. The more hostile environment for mathematics education in the 1910s and 1920s is sketched in Chapter 9, and significant responses by school teachers and mathematicians are examined, culminating in the 1923 report by the MAA's National Committee on Mathematical Requirements. Chapter 10 offers conclusions and looks briefly ahead to developments later in the twentieth century.

Mathematics has continued to be divisive in many respects. Most people encounter it primarily as an unavoidable school subject, and often find this a weary slog, if not worse. Meanwhile, a small minority holds mathematics to be among the most glorious of all human activities, perhaps supreme, yet even these partisans are not united as to its chief merits. Some tout mathematics for its aesthetic pleasures in the most vivid terms; indeed, one notable twentieth-century mathematician remarked of the "lucid exaltation" produced by mathematical discovery that, "unlike sexual pleasure, this feeling may last for hours at a time, even for days."[1] Surely this is a comparison baffling to the great majority, and even many skilled mathematicians would rather laud mathematics primarily on the basis of its utility in science and engineering. This latter defense seems widely accepted, although in most cases with only the vaguest comprehension of how the asserted utility is accomplished in practice.

The difficulty of clearly demonstrating the practical value of anything beyond the most basic mathematics has meant that some version of what earlier educators referred to as mental discipline is often not far from the surface of any argument about the desirability of mass education beyond arithmetic. Even among those who regard mathematics with distaste, few are willing to openly challenge its prominent and pervasive place in school education, but this does not entail any generally agreed upon justification for mathematics in the curriculum. Specific claims regarding what all students should know about mathematics, or how all students should be taught, can elicit sharp disputes. Do all students need to know algebra? Why or why not? Many readers may know of the "math wars" that erupted in the late twentieth century and have continued into the twenty-first, and some with longer memories may recall the "new math" controversies of the 1960s.

It is a major aim of the present book to demonstrate, for the case of the United States, that disagreements such as those just alluded to do not represent a recent fall from grace, but have roots well back in the nineteenth century. Indeed, modern-seeming disputes about mathematics education became notably visible as soon as basic features of our present educational structure began to emerge: mass schooling, an educational ladder from kindergarten to graduate school, and communities of specialists in both research mathematics and education. It is the role of the mathematicians in these educational arguments

---

[1] André Weil, *The Apprenticeship of a Mathematician* (Basel: Birkhauser, 1992), 91. For a nice evocation of the rollicking good time had by some research mathematicians, see Paul Nevai's review of Barry Simon's *Orthogonal Polynomials on the Unit Circle*, in *BAMS* 44 (July 2007): 447–470.

that has received the least attention heretofore and that will be central to the pages that follow. Examining mathematicians as educational actors during this formative period provides indispensable background for understanding more recent forays by mathematicians into school mathematics.

I also like to think that today's mathematicians, and others, who are interested in educational reform may benefit from the history of previous efforts as recounted in this book; not in explicit guidance to action, but rather in an awareness that seemingly obvious assumptions about mathematics education are not so obvious after all, and require attention and analysis again and again. Hyman Bass, a notable present-day research mathematician who has been active in education issues, has written of research mathematicians as having a "tradition of concern for pre-college mathematics education," and moreover has found this tradition "both edifying and inspiring."[2] I am less convinced than Bass that tradition is the appropriate word to apply; I see a more convoluted picture. I am also doubtful that many will find the present account inspiring. But I do have modest hopes for edifying.

---

[2]Hyman Bass, "Mathematics, Mathematicians, and Mathematics Education," *BAMS* 42 (2005): 417. In supporting his "tradition" claim Bass emphasizes the educational activities of two European mathematicians: one, Felix Klein, who does figure in this book; and one, Hans Freudenthal, who does not. He does not discuss the Americans central to this book.

CHAPTER 2

Mathematics Education in Nineteenth-Century America

The ideas of the educational commentators central to this book, from the 1890s to the 1920s, were built upon institutional and ideological foundations gradually established during the course of the nineteenth century. That century saw the emergence of now familiar educational institutions, such as the public high school and the research university, blending with preexisting institutions to form an overall structure by the century's end which has changed little in all the decades since. Mathematics instruction, in some form, pervaded most of this educational system as it developed, but not without comment, criticism, and shifting emphases. In examining nineteenth-century justifications for the position of mathematics in the curriculum, and the challenges to this position that emerged, we will encounter much of the vocabulary that will be central to later chapters, including mental discipline, induction, laboratory methods, student interest, and efficiency.

In the United States in the early nineteenth century it was considered useful to know portions of mathematics for specific practical purposes, especially in business and commerce. There were also those who claimed that the logical and well-defined nature of the subject helped train a student's mind in general, so that the student would be able to solve a range of problems, not necessarily explicitly mathematical. We will find this mental discipline thesis, in a moderate form, given especially influential currency in the Yale Report of 1828, largely written by Jeremiah Day, Yale president and mathematics professor. Day would probably not have seen the disciplinary value of mathematics as fundamentally in conflict with its utilitarian value, but as the century advanced educational commentators came more and more to emphasize one pole over the other. We will also find disagreements as to the best teaching method by which to impart mathematics to students, with some emphasizing memorization of rules and others advocating induction of rules from examples. These camps did not necessarily align in a clear way with the split between the mental disciplinarians and the utilitarians.

These arguments about educational philosophy occurred against a background of institutional change of which the colleges were only one part. Secondary schools, thought of either as preparing students for college or as a culminating educational experience, began to grow in number. Mathematics education in these schools, whether in the academies that had survived from the eighteenth century or in the public high schools which would proliferate in the twentieth, initially held a position of relatively unexamined high regard. Over the course of the nineteenth century the position of mathematics would become more embattled but still central, as these schools grew in number and importance and as other subjects, notably the sciences, began to compete for space in the curriculum. Meanwhile, in the last quarter of the nineteenth century, the first significant instances of graduate level mathematical instruction and a self-replicating research community began to blossom, beginning with the Johns Hopkins University. Thus the traditional colleges found themselves having to react to new dynamism in both the secondary schools below and the graduate schools above. Some ambitious college presidents attempted to exert influence on this situation, most prominently Harvard's Charles Eliot, by many measures the most powerful educational leader of his time. Both Hopkins and Eliot will prove to be useful lenses through which to view American mathematics education, within this chapter and also later in the book.

The chapter concludes by examining a survey published in 1890 which covers mathematics teaching in the United States at a wide variety of institutions, from secondary schools to graduate schools. This will provide a helpful benchmark for our eventual discussion of developments in the last decade of the nineteenth century and beyond.

2.1. Early Century Justifications for Studying Mathematics

Mathematics was given little educational attention in early colonial America, but when it did start to attract more attention, in the middle of the eighteenth century, it was for two seemingly disparate reasons, a circumstance highly favorable to securing a solid foothold in the educational system. On the one hand mathematics began to take its place beside Latin and Greek as a recognized part of the upbringing of the male colonial elite, gradually vanquishing suspicion of the subject as being vulgar. The rising prestige of Newtonian science amid the more general stirrings of the Enlightenment likely played a role here. By the early nineteenth century mathematics in some form had become an established part of the classical curriculum in the Latin grammar schools and academies in the more populous parts of the Northeast, and in the colleges.[1]

At nearly the same time, the commercial utility of portions of arithmetic began to be more valued, so that the rudiments of this subject were also an expected acquirement of even those with access only to the more limited facilities of the rural areas.[2] Abraham Lincoln (1809–1865) testified on this point:

> There were some schools, so called; but no qualification was ever required of a teacher, beyond *"readin, writin, and cipherin,"* to the Rule of Three. If a straggler supposed to understand latin, happened to sojourn in the neighborhood, he was looked upon as a wizard. There was absolutely nothing to excite ambition for education. Of course when I came of age I did not know much. Still somehow, I could read, write, and cipher to the Rule of Three; but that was all.[3]

There was thus Latin, Greek, and mathematics for the few, and "readin, writin, and cipherin" for the many.

Arithmetic instruction in the early nineteenth-century United States had inherited the tradition established in the English colonies in the eighteenth century of relying on memorization of rules, with a minimum of explanation and no symbolic aids.[4] The so-called Rule of Three referred to by Lincoln was one such rule extensively belabored by eighteenth and early nineteenth century textbooks. The goal of the Rule of Three was to "Teacheth, by having three numbers given, to find a fourth, that shall have the same proportion to the third, as the second hath to the first." The student was directed to solve such problems by proceeding as follows:

> 1. State the question by making that number, which asks the question, the third term, or putting it in the third place; that, which is of the same name or quality as the demand,

---

[1] Patricia Cline Cohen, *A Calculating People: The Spread of Numeracy in Early America* (Chicago: University of Chicago Press, 1982), 118–119. Frederick Rudolph, *The American College and University: A History* (New York: Knopf, 1962), 29–30.

[2] Cohen, 130–132.

[3] Abraham Lincoln, *Speeches and Writings 1859–1865* (New York: Library of America, 1989), 107. Spelling, punctuation, and italics as in original.

[4] Cohen, 120–122.

> the first term; and that, which is of the same name or quality with the answer required, the second term.
> 2. Multiply the second and third numbers together; divide the product by the first, and the quotient will be the answer to the question.[5]

This rule was applicable to problems involving questions of proportion, frequently encountered in commercial transactions, such as the following:

> My correspondent in Maryland purchased a cargo of flour for me, for £437 that currency, how much Massachusetts money must I remit him, £125 Maryland being equal £100 Massachusetts?[6]

The ease with which such problems are handled with a smidgen of algebra, notation dissipating the fog of verbiage and taking away most of the burden of memorization, may make it difficult for the modern reader to credit the intensity of instructional attention devoted to them in the eighteenth and nineteenth centuries.

It was not until the early nineteenth century that the dominance of rote learning in the case of arithmetic was seriously challenged. Patricia Cline Cohen has suggested that there was a new climate in the United States which increased the prestige of arithmetic and changed the value attached to its instruction. The burgeoning market economy meant that a flexible command of figures was more and more useful for a wide variety of daily transactions. Further, those concerned with the maintenance of the republican form of government saw in arithmetic an excellent means of training the reasoning powers of the citizenry. The consequences, according to Cohen, included the excision of useless drill and the abandonment of memorizing rules in favor of a more logical motivation of methods by leading the student through carefully selected problems. Warren Colburn was the leading figure in this new approach, publishing texts based on "the inductive method of instruction," in the 1820s.[7] Nevertheless, drill, memorization, and rules would remain central to much mathematics instruction through the nineteenth century. We will find reform-minded educators still attacking these features at the end of the century, and promoting "inductive methods " with renewed fervor.

---

[5]Nicholas Pike, *A New and Complete System of Arithmetic Composed for the Use of Citizens of the United States* (Newburyport: J. Mycall, 1788), 126.

[6]Ibid., 134.

[7]Cohen, 117, 130–138.

It is worth remarking on one part of the evidence for Cohen's argument: the idea that studying mathematics, arithmetic in particular, was valuable as general mental training, apart from the specific knowledge conveyed, had not descended in an unbroken tradition from time immemorial; rather, it was only in the early nineteenth century that it was firmly established in the United States. By the close of that century this genealogy was little remembered by many educators. The justification for mathematics instruction on the grounds of "mental discipline " (as the notion came to be called) would by that time no longer be in close alliance with the justification on the grounds of practical utility. Mental discipline would become an increasingly embattled conservative position, not only in mathematics, while utility would become, in some guises, a very radical position.

Before considering the elaborate intellectual justifications that were advanced on behalf of the classical curriculum, it is well to note strong material factors in its favor: it was simpler and cheaper than the alternatives. One late nineteenth-century educator affirmed its essential simplicity as follows:

> Grammatical and mathematical studies are the easiest to teach. They become powerful pedagogical instruments of mind-training, even with poor teaching. The reason for this is that they are perfectly definite ... A lesson in Latin or Greek has so many sentences to translate, so many expressions to be noted. A lesson in mathematics has so many problems to solve.[8]

Another educator proclaimed mathematics "one of the virile studies":

> One peculiar advantage of right mathematical work lies in the completeness and accuracy of the results attainable: Most kinds of study are like woman's work—they are never done.[9]

No laboratories were needed for mathematics, and few books. In elementary mathematics, in particular, a single text well stocked with problems could keep a class busy for years; surely this goes far to explain the long dominance of arithmetic in the schools. Frederick Rudolph suggests that many educators were sensitive to the economic advantages of the classical curriculum from

---

[8]Charles DeGarmo, "Formal Vs. Concrete Studies in the College," *School Review* 2 (1894): 22.

[9]Frank A. Hill, "The Educational Value of Mathematics," *Educational Review* 9 (Apr. 1895): 355–356.

early in the century,[10] but it seems that the low capital requirements were not often explicitly cited until the dawn of far more substantial support for education toward the end of the century. The new financial largess included private bequests for higher education, state and federal governmental assistance to universities, and the provision of local tax money to support high schools. The threats and promises of the capital-intensive studies were then far more real. A Latin scholar in 1888 acknowledged the increasing call for education to emphasize "the scientific habit of mind," but averred that "classical studies" properly taught could do this just as well as "physical studies," and were "even superior, on account of the availability of the apparatus, namely books, and the poverty of the schools in laboratories."[11]

This argument did not greatly impress ambitious reformers such as Harvard president Charles Eliot, however. During the deliberations leading up to the Committee of Ten Report, Oscar Robinson used school poverty to challenge Eliot's "equivalence of studies " position: "In short, with any equipment which high schools are likely to have for a generation at least, I do not believe that geography or history can be made an equivalent of Latin or mathematics." To which the confident Eliot replied, "The lack of means must be admitted; but should we not indicate what is desirable, and ask for the means?"[12] Thus by the time of the Committee of Ten the new studies were often characterized by the material difficulties that they posed for educators, but the signs of progress were already evident:

> Our schools, with few exceptions, need also a better material equipment. Laboratories for the physical sciences are common, but in many cases are poorly equipped for accurate work. The facilities for biological study are still too rare and too meagre. Libraries suited to thorough work in history and literature are too often lacking. Yet all of us can remember when matters were worse, and hence we confidently hope to see them better.[13]

By the early nineteenth century, American instruction in mathematics (and in Greek and Latin) was being defended on the basis of mental discipline and

---

[10]See Rudolph, 135.

[11]William Gardner Hale, *Aims and Methods in Classical Study* (Boston: Ginn & Co., 1888), 12.

[12]These letters, dated respectively October 23 and October 24, 1893, were published in Oscar Robinson, "The Work of the Committee of Ten," *School Review* 2 (1894): 368–369.

[13]Ray Greene Huling, "The New England Association of Colleges and Preparatory Schools," *School Review* 2 (1894): 601–602.

faculty psychology. The roots of these notions can be traced back to John Locke, and even to Plato, but their clearest and most widely read early expression in the United States came in the oft-cited Yale Report of 1828. Significantly, this came in response to rumblings at other colleges (notably, Harvard and the University of Virginia) that the classical curriculum was standing in the way of progress. Yale president Jeremiah Day responded by preparing a manifesto to the Yale corporation on behalf of the faculty, defending the classical college. Since Yale was the largest college in the country, and its graduates and faculty notably vigorous in proselytizing on behalf of its educational ideals, the Yale Report exercised a large influence for many years.[14]

In the report Day emphasized that though other educational institutions might legitimately proceed with other ends in view, a "college" had a special duty "to LAY THE FOUNDATION OF A SUPERIOR EDUCATION," by looking to "the *discipline* and the *furniture* of the mind; expanding its powers, and storing it with knowledge. The former of these is, perhaps, the more important of the two." The caution of the last sentence suggests a less than fanatical attitude, often overlooked by later partisans, but certainly Day did advance the essence of the mental discipline thesis, that certain studies were valuable less for their content than for their capability of preparing the student for a broad range of further accomplishments. To attain such discipline the aim should be "to call into daily and vigorous exercise the faculties of the student." Here we see the idea that the mind was composed of various "faculties" or "powers," which could be improved by "exercise," a clear intimation of the analogy with physical exertion and muscular development which would become much more prominent later in the century. For Day the mental faculties included the powers of attention, thought, analysis, judgment, imagination, discrimination, and memory.[15]

Day's psychological ideas did not imply for him that there should be an absolute monopoly of instruction on the part of Greek, Latin, and mathematics. A college education should seek balanced mental development: "it is necessary that *all* the important mental faculties be brought into exercise." Day readily acknowledged, for example, a place in the college for physical science, through

---

[14]Rudolph, 118-135; Burton J. Bledstein, *The Culture of Professionalism* (New York: Norton, 1976), 239–240.

[15]For the Yale Report I have used the reprint in Theodore Rawson Crane, ed., *The Colleges and the Public, 1787-1862* (New York: Bureau of Publications, Teachers College, Columbia University, 1963). Day's comments on superior education are from p. 85, and his use of the exercise analogy on p. 88. Italics and emphases in original.

16   2. MATHEMATICS EDUCATION IN NINETEENTH-CENTURY AMERICA

which the student "becomes familiar with facts, with the process of induction, and the varieties of probable evidence"; and for English reading, by means of which the student "learns the powers of the language in which he is to speak and write." Again, these evidences of moderation were often lost sight of, then and later. What was most noticed was Day's uncompromising assertion that there were certain crucial faculties which only the classical subjects could fully and properly exercise. In particular, "From the pure mathematics, he [the student] learns the art of demonstrative reasoning."[16]

Day knew there were many who scoffed at the college curriculum for failing to directly support the professional aspirations of the student. Such critics asked, "Will chemistry enable him to plead at the bar, or conic sections qualify him for preaching, or astronomy aid him in the practice of physic?" Day insisted that the curriculum he recommended "taught *how* to learn," and was thus of great worth to the prospective professional man. There was no need for most people to be troubled with such learning; it was quite true that "the carpenter may square his framework, without a knowledge of Euclid's Elements." But the fact remained that such a carpenter "needs the constant superintendence of men of more enlarged and scientific information." Thus Day's views ultimately derived from a firmly held hierarchical conception of society. This is not to say that he believed in a society of fixed class distinctions. This point is made clearly by Jurgen Herbst, who cites Day's words later in the Yale Report that "superior intellectual attainments ought not to be confined to any description of persons." The leaders of society might well come from varied backgrounds. Nevertheless, such leaders would continue to constitute a small minority. Whatever their social origin, there would continue to be many carpenters and few superintendents, and thus college education would continue to be an elitist enterprise.[17]

Theology and mathematics were crucial to Jeremiah Day's career, providing a good representation of the place of both these subjects in the life of early nineteenth century American colleges. In his time, and for some time afterward, the overwhelming majority of American college presidents were clergymen;[18] and given the prominence of mathematics in the curriculum it was not unusual for these presidents to have some degree of specialized knowledge

---

[16]Ibid., 86. Italics in original.

[17]Jurgen Herbst, "The Yale Report of 1828," *International Journal of the Classical Tradition* 11 (Fall 2004): 228. Day's quotes are from Yale Report, 89 (italics in original) and 92.

[18]Rudolph, 14.

of that subject. The importance of religion for the colleges and their presidents declined through the nineteenth century, but it remained a powerful rhetorical resource. Of the late nineteenth century college presidents Burton Bledstein has observed, "Theirs was a religion of convenience."[19] A similar remark might be made regarding mathematics: the college presidents gradually lost their close association with the subject during the nineteenth century, but it retained a vestigial presence in their lives, and was occasionally valuable as a tool of argument. We will see what might be called *the mathematics of convenience* exhibited by both Daniel Coit Gilman, president of Johns Hopkins University, and by Charles Eliot, president of Harvard.

Day, whose initial position on the faculty at Yale was as professor of mathematics and natural philosophy, exemplifies the earlier period when acquaintance with mathematics was often less casual on the part of college presidents. He retained his mathematics professorship for some years after becoming president, and wrote several textbooks on mathematical topics. His views on mathematics complemented and sustained his more general views on education. In his most successful textbook, *An Introduction to Algebra*, first published in 1814, he made many of the arguments he would reiterate in the Yale Report of 1828.

Day was at pains to distinguish the aims of studying mathematics during the course of "a liberal education" from the case of studying it "merely to obtain such a knowledge of the *practical* arts, as is required for transacting business." He did not at all deny "the practical applications of the mathematics, in the common concerns of business, in the useful arts, and in the various branches of physical science." Indeed, he cataloged them in detail, with some evident pride. Nor did Day propose that pure mathematics should stand on its own in the curriculum. Its virtues became even more evident when it was "connected with the *physical* sciences, astronomy, chemistry, and natural philosophy." But here again educational problems were considerably simplified for him by his belief in parallel hierarchies governing knowledge and society. The liberally educated few would be able to apply their knowledge of "principles" as needed to guide the practical efforts of the more superficially educated many: "The mariner calculates his longitude by tables, for which he is indebted to mathematicians and astronomers of no ordinary attainments." The production of "expert mathematicians" ought not, however, be the explicit aim of a college; rather, mathematics was in the curriculum "for the purpose of forming *sound reasoners*." Day was well aware that mathematics was no static and completed

---
[19]Bledstein, 198.

subject, and that research was ongoing, but "original discoveries are not for the benefit of *beginners*, though they may be of great importance to the advancement of science." The methods and subject matter of mathematics education were not fixed for all time, but should lag a respectable distance behind the research frontier; Day reported that his treatment of algebra owed much to earlier but not ancient mathematicians, such as Newton, Maclaurin, and Euler.[20]

Altogether Jeremiah Day's pronouncements in his *Algebra* and in the Yale Report were expressed with many nuances not fairly categorized as intransigent conservatism, but some aspects became distinctly inconvenient for the reforming educators of the late nineteenth century, Charles Eliot in particular.

2.2. Charles W. Eliot: A Harvard Man Confronts Mathematics

Charles Eliot is a titanic figure in the history of American education. Jerome Karabel has remarked, "No college president, before or since, has exerted a greater impact on the shape and character of American higher education."[21] His imprint can be found in the secondary schools also, as the present work will emphasize. Only a very partial portrait of Eliot's entire range of accomplishments will emerge in considering his views and activities with regard to mathematics. Although as a college student Eliot may have been exposed to some of the deeper mathematical mysteries, in later life he displayed little intimacy with the subject or genuine interest in its practice. In his career as a university president mathematics was but one building block in the educational edifice he endeavored to assemble, at Harvard and nationwide.

Eliot's predecessor as Harvard president, Thomas Hill, had been like Jeremiah Day both clergyman and mathematician, writing several textbooks on elementary geometry and arithmetic.[22] Eliot was decidedly of a new generation. As early as 1869 he made known his opinion that the clerical role in education ought to be reduced; his attitude toward mathematics was more complicated. He was sufficiently interested in the subject as a Harvard undergraduate during 1849–1853 to take not only the required classes in algebra,

---

[20] Jeremiah Day, *An Introduction to Algebra* (New Haven: Howe & Deforest, 1814). The quotations used are from pages 3, 4, and 6 of the preface and from pages 6 and 7 of the main text. All italics in original.

[21] Jerome Karabel, *The Chosen: The Hidden History of Admission and Exclusion at Harvard, Yale, and Princeton* (Boston: Houghton Mifflin, 2005), 39.

[22] On Hill see William G. Land, *Thomas Hill, Twentieth President of Harvard College* (Cambridge, Mass.: Harvard University Press, 1933); William G. Land, "Hill, Thomas," *DAB*, 5 (pt. 1):45–46; and Luke Fernandez, "Hill, Thomas," *ANB*, 10:802–803.

geometry, and trigonometry in his first two years, but also advanced electives in pure and applied mathematics as a junior and senior. This brought him into contact with Benjamin Peirce (1809–1880), who occupied the pinnacle of American mathematics for many years. Like most Harvard students during the Peirce era, Eliot agreed that Peirce "was no teacher in the ordinary sense of that word," but he nevertheless found him "inspiring." Eliot did well in Peirce's classes, but never showed much inclination to follow his teacher into research in the abstract realms. As an old man Eliot recalled an encounter in which Peirce had gently chided him for expressing skepticism regarding certain "functions and infinitesimal variables" upon which Peirce had been expounding. Eliot was more comfortable with "facts or realities."[23]

Soon after graduation Eliot was appointed a tutor of mathematics of Harvard College. He and his classmate James Mills Peirce, son of Benjamin, handled much of the college's more elementary mathematics instruction for several years. In itself appointment as a mathematics tutor did not necessarily imply either desire or qualification for a lifelong career in mathematics, a state of affairs whose demise would be applauded by a more professionally conscious mathematician in 1887:

> The old system of college tutorships, now in its decadence, is responsible for a great deal of harm in mathematics as well as in other subjects. The young man just out of Harvard was often employed as a tutor, while pursuing law or divinity as a student: the now most eminent English scholar in the country, being valedictorian of his year, was selected as such a tutor in *mathematics* a year after his graduation. Of course he went through the form perfectly; but his heart cannot but have been elsewhere. Within a few years this has been changed in nearly all good colleges. Men are selected to teach college students with some reference to their own tastes and studies; and the corresponding requirement that those studies shall be kept up is more insisted upon.[24]

---

[23]Eliot's views on the clergy and education are found in Charles W. Eliot, "The New Education," *Atlantic Monthly* 23 (1869): 366. His interaction with Benjamin Peirce is mentioned in R. C. Archibald, ed., "Benjamin Peirce," *AMM* 32 (1925): 1–4 and Henry James, *Charles W. Eliot* (Cambridge: Riverside Press, 1930), 1:49.

[24]T. H. Safford, *Mathematical Teaching and Its Modern Methods* (Boston: Heath, 1887), 28. Italics in original.

The English scholar alluded to was undoubtedly Francis J. Child (1825–1896). Child graduated first in the Harvard class of 1846, served as a tutor in mathematics during 1846–1848, and subsequently enjoyed a long and distinguished Harvard career as a scholar of English philology, best known for his monumental *English and Scottish Popular Ballads*.[25] Eliot, like Child, proved to be more interested in areas other than mathematics. Chemistry in particular, with its laboratory work, was much more to his liking. In 1858 he became an assistant professor of both mathematics and chemistry, and was soon complaining at his "slight disgust" at having to supervise mathematics recitations when he would much prefer to teach chemistry. The resignation of a chemistry professor in the Lawrence Scientific School enabled him to drop his mathematics duties entirely in 1861.[26] Eliot's later comments on mathematics would reflect an admiration for the achievements of investigators such as Benjamin Peirce and Simon Newcomb, together with a resentment of the subject for standing in the way of educational progress.

By the time Eliot began to make public pronouncements about educational issues (in 1869, shortly before he was appointed president of Harvard), he had bolstered his interest in practical education and laboratory methods of instruction by serving as a professor of chemistry at the new Massachusetts Institute of Technology (MIT). He believed that MIT and other seats of the "new education" were filling an important national need not being met by the colleges. Moreover, in making this claim Eliot implicitly suggested an explanation for a problem that was coming to trouble more and more American educators: a perceived decline in the relative numbers of college students. Eliot saw students and their parents turning away from the classical colleges in search of more practical training. This diagnosis would be echoed by others in the next decades, and the underlying issue would be much on the minds of educators associated with the Committee of Ten. One such educator, Ray Greene Huling, a high school principal from New Bedford, Massachusetts who served on the Committee of Ten's Conference on History, Government, and Political Economy, praised the report's push for uniformity in secondary school curricula. Huling saw this as likely to result in "a larger current of young lives from the secondary schools, particularly from the high schools, into the institutions of higher learning," a result he hailed as a "felicitous consummation."[27]

---

[25]See George Harvey Genzmer, "Child, Francis James," *DAB*, 2:66–67.
[26]James, 1:67, 69, 87, 88.
[27]Ray Greene Huling, "The Reports on Secondary School Studies," *School Review* 2 (1894): 273.

2.2. CHARLES W. ELIOT: A HARVARD MAN CONFRONTS MATHEMATICS

Mathematics, in Eliot's 1869 partition of the American educational world into new and traditional, was comfortably ensconced in both camps:

> We wish to review the recent experience of this country in the attempt to organize a system of education based chiefly upon the pure and applied sciences, the living European languages, and mathematics, instead of upon Greek, Latin, and mathematics, as in the established college system.[28]

At that time he had not seen the secondary school mathematics curriculum as requiring much reform; the preparatory schools were accustomed to teach arithmetic, algebra, and geometry to their college bound students, and he deemed this to be adequate for admission to the scientific and technical schools as well. He complained only of the poor quality of much arithmetic instruction.[29] Later on, with the deepening of his involvement with secondary school issues beginning in the 1880s, Eliot joined a growing body of educators who were looking at mathematics more critically. By 1892 he was grumbling that "mathematical reasoning is a peculiar form of logic which has very little application to common life, and no application at all in those great fields of human activity where perfect demonstration is not to be obtained."[30]

Eliot in 1869 was far from advocating the abolishment of the classical curriculum in the colleges. At that time his preference seems to have been to let the colleges proceed in their traditional manner, but to build up the independent technical schools, and most especially to insist that the graduates of these schools were entitled to all the respect given to college graduates. In support of this latter contention he began sounding a theme that would become attached to his name throughout his career. Eliot's critics, especially, would come to classify the following passage as an instance of "the doctrine of equivalence of studies":

> It is not to be imagined that the mental training afforded by a good polytechnic school is necessarily inferior in any respect

---

[28] Eliot, "The New Education," 204.

[29] Ibid., 360.

[30] Eliot, "Shortening and Enriching the Grammar-School Course," in Charles William Eliot, *Educational Reform* (New York: The Century Co., 1898; repr., New York: Arno Press, 1969), 259. This talk was given at an NEA meeting in February 1892.

to that of a good college, whether in breadth, vigor, or wholesomeness.[31]

Mental training, it should be observed, retained a high place for Eliot as a criterion of educational value. He insisted, however, that the claims of the colleges to produce "a rounded man, with all his faculties impartially developed" (recall Day aiming to exercise *all* the mental faculties), was a mirage. The world was too complicated, individual abilities were too varied; "the division of mental labor" should be frankly acknowledged and promoted, and the particular skills cultivated by the technical schools were increasingly valuable.[32]

Eliot extended and qualified his ideas of 1869 over the rest of his busy career. Quite naturally, as president of Harvard, the promotion of the independent technical schools became a lesser priority than that of pushing the colleges to accommodate the requirements of the "new education." In an 1884 address given at Johns Hopkins University he declared that because of "the general growth of knowledge, and the rise of new literatures, arts, and sciences during the past two hundred and fifty years," the definition of a "liberal education" was in need of expansion. Those conservative educators who defended an unchanging curriculum were guilty of ignoring history, since even the traditional subjects had changed markedly over the years, mathematics in particular. In this instance he wished to enlist mathematics as an ally in attaining his most cherished goal of raising the natural sciences to a condition of curricular equality with the classical subjects. If it was rightly understood that even the supposedly static mathematics curriculum had in fact been evolving, then the way would be prepared for the sciences.[33]

Indeed, Eliot became so enthusiastic regarding the progress of "modern mathematics—algebra, analytic geometry, the differential and integral calculi, analytic mechanics, and quaternions," that he made the preposterous assertion that "when we examine closely the matters now taught as mathematics in this country, we find that they are all recent inventions, of a character... distinct from the Greek geometry and conic sections which with arithmetic represented mathematics down to the seventeenth century." But later in the same speech Eliot saw mathematics more as an obstacle than as an ally, since the limited amount of time in the curriculum had to be recognized. In order for more

---

[31]Eliot, "The New Education," 215. For later criticism of Eliot on equivalence see Sizer, 135–140, and Edward A. Krug, *The Shaping of the American High School 1880–1920* (New York: Harper & Row, 1964), 84–85.

[32]Eliot, "The New Education," 218.

[33]Eliot, "What is a Liberal Education?" in Eliot, *Educational Reform,* 89, 110.

students to be able to partake of the valuable mental training provided by the natural sciences there was no alternative but to reduce requirements for mathematics and languages, in both schools and colleges.[34]

This was a feature that distressed one of Eliot's listeners at Johns Hopkins, Simon Newcomb, who hastened to defend mathematics:

> It can hardly be claimed that we pay disproportionate attention to either mathematics or languages in this country. Not only is our mathematics education far behind that of France and Germany, but a much better mathematical training than our average student gets is absolutely necessary to an adequate comprehension of modern physical science.[35]

For Eliot, however, the essence of science instruction was the laboratory, not the heavily mathematical projects that appealed to Newcomb.[36] For his part, the intellectually omnivorous Newcomb seems to have had little appreciation for the dilemma of the finite curriculum that perplexed Eliot.

Reform-minded late-century educators such as Eliot struggled to accommodate demands that new subjects be given greater representation in the curriculum of both the secondary schools and colleges. Differences with earlier educators became more evident, a prime example provided by Jeremiah Day's belief that students of mathematics must "make themselves masters" of the subject. That is, as Day put it more vividly, "In mathematics as in war, it should be made a principle, not to advance, while any thing is left unconquered behind."[37] But Eliot and his ilk held that defending "mastery" or "thoroughness" was often simply an excuse to keep other subjects in an inferior position. Arithmetic was the mathematical topic that most caught the eye of the reformers, since it monopolized so much time in the primary and secondary schools. Eliot complained in 1888 that "Many an educated New Englander remembers to this day the exasperation he felt when he discovered that problems in Colburn's Sequel [an arithmetic text] over which he had struggled for hours, could be solved in as many minutes after he had got half way through Sherwin's Algebra."[38]

---

[34]Ibid., 92, 112.

[35]Simon Newcomb, "President Eliot On a Liberal Education," *Science* 3 (June 13, 1884): 704.

[36]Eliot, "Shortening and Enriching the Grammar-School Course," 259.

[37]Jeremiah Day, *An Introduction to Algebra* (New Haven: Howe & Deforest, 1814), 9.

[38]Eliot, "Can School Programmes Be Shortened and Enriched?" *Proceedings of the Department of Superintendence of the National Educational Association* (1888): 109.

William Torrey Harris seconded Eliot's views on arithmetic, calling for "omission of problems that can be more economically solved by higher methods." In making his argument Harris, whether consciously or not, introduced a military metaphor directly contrary to the one Day had used some 70 years before:

> Let the pupil flank his higher arithmetic by learning the elements of algebra and geometry ... The method of flanking strong intrenchments by attacking the line of communications in the rear—that is to say in this metaphor, by attacking a higher study which contains the theory of the lower study—has obvious advantages in certain cases.[39]

In pronouncements such as those of Eliot and Harris one can readily discern the early stages of the emphasis on efficiency which was to become prominent in early twentieth century education, and which historians have noted as a hallmark of the progressive era generally. Indeed, Eliot has been cited as the founder of the "Economy of Time" movement in education.[40] The evolution of Eliot's choice of slogans is itself an indicator of the trend. In 1869, focusing mainly on the colleges, he had talked blithely of "broadening and deepening the course of study."[41] By 1888 the zero-sum nature of the curricular game had become more evident to him; broadening and deepening could not be pursued indefinitely without threatening to prolong a student's education intolerably. Thus he began to talk of "shortening and enriching."[42] Variants of the latter phrase proved popular among educators.

The preference for algebra over arithmetic displayed by Eliot and Harris was in some ways quite congenial to mathematicians. Simon Newcomb, for one, had no love for arithmetic texts full of miscellaneous rules and problems; he welcomed a greater emphasis on algebra as a unifying tool. The most explicit statement of Newcomb's position seems to be that in an unpublished manuscript written sometime after his service with the Committee of Ten:

---

[39]William Torrey Harris, response to Eliot's "Can School Programmes Be Shortened and Enriched?" Ibid., 117–118.

[40]Hugh Hawkins, *Between Harvard and America: The Educational Leadership of Charles W. Eliot* (New York: Oxford University Press, 1972), 244.

[41]Eliot, "The New Education," 363.

[42]As in the titles of these articles: "Can School Programmes Be Shortened and Enriched?" and "Shortening and Enriching the Grammar School Course," in Eliot, *Educational Reform,* 151–178, 253–272. These essays were originally published in 1888 and 1892, respectively.

2.2. CHARLES W. ELIOT: A HARVARD MAN CONFRONTS MATHEMATICS

> To deny the use of this [the algebraic] method to the beginner in arithmetic is simply to deprive him of an important instrument of progress, not merely of a [sic] instrument which is to lessen the mental work which he has to forward. To require a beginner to think of quantities without the use of algebraic symbols is not so impolitic as to expect a child to think without the use of words but it is a step in that direction.[43]

But this quotation makes clear that mathematicians advocated more algebra in the curriculum not merely for the sake of efficiency; more important, they saw algebra as connecting the student with the central body of their live and growing subject. In the early twentieth century the general educators who looked kindly on algebra as a time saver began to be superseded by more narrowly focused advocates of educational efficiency, struggling with the great flood of students into the secondary schools. These educators began to ask whether the mathematics problems students were solving, whether by algebra or arithmetic, were worth solving at all. Mathematicians and general educators were thus only temporarily in alliance regarding the issue of efficiency.

In his attack on arithmetic Eliot was not only promoting the concept of efficiency but was distinguishing his conception of mental discipline from the strong ascetic strain that had been prominent among many nineteenth-century American educators. Such educators, appealing to traditional Protestant morality, were not concerned to minimize student "exasperation"; quite the reverse.[44] Die-hard defenders of the classical subjects were among the most ardent in this regard. One educator in 1893 was much consoled at the thought that the student was not yet able to escape arithmetic:

> It is refreshing to know that there is one subject which he must master for himself slowly, sometimes painfully, and always with much labor. If arithmetic stands in the way of drawing or music, of German or Latin, of book-keeping or civics, let us be thankful that it is an obstacle which will develop the sound mental muscle that is needed for the work further on.[45]

---

[43]SNP, Box 100. This is from the first page of the section titled "The Algebraic Method" in the folder labeled "Book: Education." Portions of this manuscript seem to have been written as late as 1907.

[44]See Kolesnik, 90, and Laurence R. Veysey, *The Emergence of the American University* (Chicago: University of Chicago Press, 1965), 23–24.

[45]Roland S. Keyser, "The Readjustment of the School Curriculum," *School Review* 1 (Mar. 1893): 135.

An advocate of Greek and Latin wrote in a similar vein in 1895:

> Power is not grasped by toying with the easy, but by contending with the difficult. Muscle physical or intellectual is not gained by manipulating straws, and reading French novels, but by lifting ponderous weights, and delving into the arcana of language.[46]

Eliot's contrary contention was that true mental discipline could only come in a subject in which the student was genuinely interested.[47] No doubt he had reflected his patrician background in declaring in 1869 that "labor is not exercise," and in recommending "catching butterflies" as much better for the health than "digging potatoes,"[48] but he never wished pain and toil on any student, of whatever class. Further, he was always conscious of being a seller in an educational buyer's market: "for it is not the proper business of universities to force subjects of study, or particular kinds of mental discipline, upon unwilling generations," especially since, he added, "it would be so easy for the generations, if repelled, to pass the universities by."[49]

From his own time to the present Eliot has been known as a champion of the elective system, or elective principle, as he preferred calling it.[50] This notion, in Robert Kohler's succinct description,

> was how American educators solved the problem of the fixed classical curriculum, which was no longer able to accommodate the rapid growth of scholarly knowledge nor to respond to the chorus of demand for a more practical kind of training, especially for business careers.[51]

Eliot's last remark in the previous paragraph lends some credence to the claim of Charles and Mary Beard that a major perceived virtue of the elective system was that it "eased the preparatory strain on the plutocracy." But the Beards also concede, using one of Eliot's favorite words, that the elective system "*enriched* the curriculum with the new sciences, natural and humanistic."[52] Kohler

---

[46]Andrew F. West, "The Plan of a Six-Year Latin Course," *School Review* 3 (1895): 343.

[47]Hawkins, *Between Harvard and America*, 90–91; and Krug, *High School*, 21.

[48]Eliot, "The New Education," 217.

[49]Eliot, "What Is a Liberal Education?" 120.

[50]Krug, *High School*, 20.

[51]Robert E. Kohler, "The Ph.D. Machine: Building on the Collegiate Base," *Isis* 81 (1990): 646.

[52]Charles A. Beard and Mary R. Beard, *The Rise of American Civilization* (New York: Macmillan Co., 1930), 2:472. Italics added.

and other later historians have further sketched the diversity of motives influencing the nineteenth-century push for electives.[53] In Eliot's case, as we have seen, at least the following sentiments were surely involved: sincere concern for the personal development of the students; anxiety regarding the viability of educational institutions and the educational system as a whole; and desire that new subjects, especially the natural sciences, take a more prominent place in the curriculum.

In summary, in the passage of time from Jeremiah Day's era to that of Charles Eliot we can observe the emergence of several related factors affecting educational thinking in general and mathematics in particular. First, the pressure for parity with the classical subjects on behalf of "new" subjects had become much more intense and widespread in Eliot's time than it had been in Day's. The economic constraints that had long limited the curricular ambitions of schools and colleges had begun to slacken. The sciences, which benefited especially from this development, were also the beneficiaries of much ardent propaganda, some of which explicitly disdained mathematics. Second, in the course of pushing for expanded curriculum content, many educational polemicists found it more and more desirable to modify the mental discipline thesis. Eliot, we have seen, claimed merely to seek for true or effective discipline as opposed to spurious varieties, but such distinctions prepared the way for later educators who were more explicitly hostile to mental discipline.[54] Grim disciplinary arguments for mathematics could no longer be relied on, and educators increasingly catered to student interest . Third, questions of efficiency and allocation of resources became increasingly central to educational debates. This was in part a natural consequence of the curricular competition, but reflected larger trends as well. And finally, there were two demographic issues, one perhaps more perceived than real, the other more real than initially perceived. The first was the supposed shortage of college students, which had troubled educators since mid-century and continued to engage the Committee of Ten. The other was the major upsurge in the number of students attending secondary schools. But this latter phenomenon only became clear to many educators after the turn of the century, with Charles Eliot and the Committee of Ten of 1893 in particular displaying less than full awareness of its implications.[55]

---

[53]See Kohler, 646–648.

[54]See Veysey, 25.

[55]See Hawkins, *Between Harvard and America,* 253; Martin Trow, "The Second Transformation of American Secondary Education," *International Journal of Comparative Sociology* 2 (Sept. 1961): 145–151; and Sizer, 144, 146.

2.3. Emerging Challenges to Mathematics in the Curriculum

From the time of the Yale Report of 1828 onward Greek and Latin (the "dead languages," as critics enjoyed designating them) were far more perilously exposed than mathematics to the complaints of educational reformers.[56] Nevertheless, by the time of the Committee of Ten mathematics had been the recipient of a number of pointed criticisms. As the tradition of mental discipline became less secure, defenders of mathematics deemphasized the disciplinary value of mathematics and increasingly appealed to other alleged virtues. But even before mathematical educators had reformulated the defense of their subject, critics of the place of mathematics in the curriculum were espying deficiencies in mathematics with regard to newer educational ideals.

Already in the Yale Report the practical usefulness of mathematics was at issue, as Jeremiah Day felt compelled to justify teaching conic sections to the prospective professional man; and in 1852 Henry Philip Tappan, president of the University of Michigan, defended the utility of "the higher mathematics" by noting that "he who is seeking after another planet in the heavens, will tell you that the calculus is the most practical thing he can find," no doubt alluding to the much publicized discovery of Neptune in 1846.[57] But skepticism about smug claims for the practicality of mathematics markedly increased at the end of the century, in conjunction with the growing competition for time within the curriculum. Some of this criticism came from advocates of studies that were themselves under attack by utilitarians. An example is provided by a publication of 1888 by William Gardner Hale, a Latin scholar who was aggrieved by what he saw as fraudulent claims of practical relevance offered by his former partners in the classical curriculum:

> The uses of mathematics which the average untechnical man makes in his daily life are the operations of addition, subtraction, multiplication, and division, applied to calculations of which the reckoning of interest is probably the most complicated; and these matters he learned, not in college, nor in the high or corresponding private school, but in the grammar school. As for geometry, the average man uses very little of it, while algebra commonly drops wholly out of his life, as clean

---

[56]Herbert M. Kliebard, *Forging the Curriculum* (New York: Routledge, 1992), 7; and Francis Wayland, *Report to the Corporation of Brown University*, in Crane, 139, 146.

[57]Henry Philip Tappan, "Inaugural Discourse," in Crane, 165.

2.3. EMERGING CHALLENGES TO MATHEMATICS IN THE CURRICULUM

forgotten as our friends the objectors to the classics tell us Greek and Latin are.[58]

Another educator in 1894 condemned the dominance of arithmetic, which he attributed to a false view of its usefulness:

> Arithmetic absorbs more of the time and energy of the elementary schools than all other subjects. This comes from the prevailing notion that arithmetic is the key to success in business. This is wholly unfounded. The common laborer, the artisan, and the tradesman have slight use for arithmetic.[59]

More rarely mathematics was attacked on cultural grounds. In 1895 Harvard professor of pedagogy Paul Hanus divided all school subjects into seven categories: (1) languages and literatures; (2) history; (3) art; (4) philosophy; (5) mathematics; (6) natural science; and (7) manual training. Since a primary aim of education was to "subject the pupil to social and ethical incentives," the first four categories of subject matter had superior educational value, because of their high social and ethical content. According to Hanus, "mathematics is especially narrow in the range of its possible incentives."[60]

But the most effective and zealous nineteenth-century criticisms of the standing of mathematics in education came from advocates of scientific and technical training. The simplest complaint was that mathematics was equally guilty with Greek and Latin of obstructing the natural sciences from their rightful place within the educational system. David Starr Jordan, ichthyologist and president of Stanford University, looked back nostalgically from 1911 to the days when the grip of the classical curriculum was first loosened:

> The advent of the elective system, thirty years ago, bore a wonderful fruitage. Men, soul-weary of drill, turned to inspiration. Teachers who loved their work were met by students who loved it. The students of science thirty years ago came to it by escape from Latin and calculus with the eagerness of colts brought from the barn to the pasture.[61]

---

[58]Hale, 18.

[59]George I. Aldrich, "Mathematics in the Elementary Schools," *Journal of Education* 40 (1894): 346.

[60]Paul Hanus, "Educational Aims and Educational Values," *Educational Review* 9 (1895): 331.

[61]David Starr Jordan, "The Making of a Darwin," *Nature* 85 (January 12, 1911), 358.

Mathematics and science were here considered not to be allies at all but to belong to different realms. Even physicist Henry Rowland, possessing considerable respect for mathematics, felt compelled to draw distinctions that had the effect of downgrading mathematics in the interest of his own subject. Addressing a gathering at Johns Hopkins in 1886, he declared that natural history, chemistry, and mathematics each had important, but incomplete, educational virtues: natural history trained "the powers of observation and classification"; chemistry taught the lesson of "care in experiment"; and mathematics exercised the "exact and logical powers of reasoning"; but physics and astronomy stood above all these in the educational hierarchy, because they "combine[d] all this training in one."[62]

Much fundamental criticism of mathematics vis-à-vis natural science had issued from mid-nineteenth-century English philosophers and scientists, and these views had made their way across the Atlantic by several different routes. The general theme was the educational inferiority of deduction to induction, or more moderately, the claim that inductive science was now entitled to a more prominent place in the curriculum, with deductive mathematics being correspondingly reduced. Much of the discussion was largely contemporaneous with agitation regarding evolution, and in several cases was conducted by the same people. Evolutionary notions did enter into the educational arguments to some degree, although sometimes only in vague appeals to "progress."[63] The general prestige of experimental and observational science was probably more useful to educational polemicists than the new biological notions in particular.

The Scottish philosopher William Hamilton (1788–1856) had declared that the study of mathematics failed to "educate to any active exercise of the power of observation," and failed to "cultivate the power of generalization." These words were quoted approvingly on the American scene by MIT president Francis Amasa Walker in 1887, as part of an argument on behalf of "the attention given by a class of interested children in the study of natural history," as opposed to that given by "pupils who are driven reluctantly through an arid waste of mathematics."[64]

---

[62]Henry A. Rowland, "The Physical Laboratory in Modern Education," in *Physical Papers*, 617. On Rowland's respect for mathematics see *Physical Papers*, 612, 630.

[63]See Kliebard, *Forging the Curriculum*, 29.

[64]Francis A. Walker, "Arithmetic in the Common Schools," in Walker, *Discussions in Education*, 213–214. Compare Hamilton's words on generalization with the following from a mathematical adept in 1900: "The word 'generalization' really cannot be fully understood without studying modern mathematics; nor can the beauty of generalization be in any other

English naturalist T. H. Huxley (1825–1895) was a strong proponent of more and earlier science study in the schools. In 1869 he charged that mathematics was inferior to the natural sciences because the former was "that study which knows nothing of observation, nothing of experiment, nothing of induction, nothing of causation." This aroused James Joseph Sylvester, future head of the Johns Hopkins University mathematics department. Strikingly, he did not bother to defend deduction, but instead argued from his own experience that the process of mathematical research (as opposed to the form in which the products of such research were traditionally presented) was in fact awash in all the admirable attributes claimed by Huxley to be possessed only by the natural sciences.[65]

Sylvester's improvisational and experimental manner of teaching at Hopkins lent support to his view, although he never made clear what implications it might have for more elementary mathematics instruction. What he emphatically did convey was how insulting it was to be told that one's field did not partake of the attributes of natural science, illustrating the growing power of science as a rhetorical resource. Terms such as "observation," "experiment," "induction," and especially "laboratory," provided instant respectability to numerous endeavors. This last word was a favorite of Daniel Coit Gilman, first president of Johns Hopkins and the man who hired Sylvester.

Gilman and others helped spread the laboratory metaphor far and wide through the American educational scene in the late nineteenth century. A foundation had already been laid by the earlier importation of the inductive and objective teaching of such European educators as Johann Pestalozzi and Friedrich Froebel, and by the rise of industrial and manual education after the Civil War. By the early 1890s the "laboratory method," the "inductive method," the "heuristic method," and the "method of discovery" were often contrasted in pedagogic writings with the "lecture method," the "didactic method," the "text-book method," the "dogmatic method," and the "method of instruction."[66] In 1886 Henry Rowland of Johns Hopkins lambasted "the present state of education in the schools and colleges," in which "words, mere

---

way so well appreciated." Charles Sanders Peirce, *Selected Writings*, Philip P. Weiner, ed. (New York: Dover, 1958), 270.

[65]Parshall and Rowe, 81–82.

[66]For example: Albert L. Arey, "Methods of Teaching Physics," *The Academy* 6 (Feb. 1891): 38; E. C. Hewett, "The Laboratory Method," *The Public School Journal* 13 (Nov. 1893): 162-163; Charles B. Scott, "Laboratory Methods in Elementary Schools," *National Educational Association Proceedings* (1894): 191-197; "The Inductive Method of Language Teaching," *The Educational Exchange* 7 (Jan. 1894): 10–12; J. W. Cook, "Arithmetic in

words, are taught." A new order of education was needed: "If they study the sciences, they must enter the laboratory and stand face to face with nature."[67] Strikingly similar phraseology was used by the Boston Superintendent of Schools in 1887:

> This workshop or laboratory method of instruction brings the learner *face to face with the facts of nature*. His mind increases in knowledge by direct personal experience with forms of matter, and manifestations of force. *No mere words intervene.* Abstract definitions, statements, and rules are put aside.[68]

Harvard president Charles Eliot was also highly supportive:

> The old-fashioned method of teaching science by means of illustrated books and demonstrative lectures has been superseded, from the kindergarten through the university, by the laboratory method, in which each pupil, no matter whether he be three years old or twenty-three, works with his own hands, and is taught to use his own senses.[69]

Advocates of subjects other than science did not fail to realize the value of making allusions to the laboratory. Representatives of more bookish specialties claimed participation by speaking of the "laboratory and library methods of instruction,"[70] or by observing that "the library serves as a sort of laboratory for the humanistic studies."[71] Even bolder claims were made as well, creating some skeptical backlash:

> It was a great and important advance in the method of the study of science when the laboratory was substituted for the text-book. But no sooner has this change been effected in science than we begin to employ it in the learning of other

---

the Higher Grades," *The Public School Journal* 13 (July 1894): 636; Truman Henry Safford, "Text-Books in Arithmetic," *Pedagogical Seminary* 2 (1892): 163.

[67]Rowland, "The Physical Laboratory," 616–617.

[68]Edwin P. Seaver, quoted in Francis A. Walker, "A Plea for Industrial Education in the Public Schools," in Walker, *Discussions in Education*, 156. Italics added.

[69]Eliot, "The Unity of Educational Reform," in Eliot, *Educational Reform*, 318–319. This essay was originally published in 1894.

[70]Charles C. Ramsay, "Some Necessary Reforms in the College," *Educational Review* 9 (1895): 10.

[71]West, *Liberal Education*, 119. West's original talk was delivered in 1900.

branches. History, literature, geography, and even arithmetic are to be taught by the laboratory method.[72]

Another educator complained that "in many quarters a method must be labeled 'inductive' in order to 'go.' "[73] As we shall see, mathematical educators were among those vigorously appropriating the metaphorical trappings of natural science, especially from the late 1880s onward.

Perhaps most influential of all in encouraging the scientific critique of mathematics education was Herbert Spencer (1820–1903). Spencer, whose work became popular in the United States almost as soon as he began publishing in the 1850s, touted science as the "knowledge of most worth," and the key to "complete living." Of particular relevance to mathematics instruction was Spencer's approval of inductive and discovery methods of instruction, with children urged to concentrate on careful observation of material objects before being exposed to abstract concepts. Charles Eliot was an early admirer of Spencer's educational program.[74]

Spencer had an even more enthusiastic American supporter of his views in Edward Livingston Youmans (1821–1887), termed by Richard Hofstadter as "the self-appointed salesman of the scientific world-outlook."[75] Youmans had arranged for the publication of Spencer's work in the United States beginning in 1864, and in 1867 published an anthology that contained writings of Spencer, Huxley, Mill, Whewell and other thinkers who believed that an increased role in education should be found for natural science.[76] Youmans's introduction to this work, entitled "On Mental Discipline in Education," was a rhetorical barrage that is well worth unpacking for the insight it gives into evolving mid-century views on mental discipline generally and on the place of mathematics education in particular.

The central problem of education, Youmans declared immediately, was that knowledge continually grows while institutions strive for stasis. "The friends of educational improvement," of which Youmans clearly considered himself a

---

[72]Hewett, 162.

[73]Lewis B. Avery, "A Misuse of the Inductive Method," *The Public-School Journal* 13 (May 1894): 521.

[74]For Spencer's influence on American education see Cremin, *Transformation of the School*, 91–93. For Eliot's admiration for Spencer see James, 1:349–351.

[75]Richard Hofstadter, *Social Darwinism in American Thought* (Boston: Beacon Press, 1955), 14.

[76]E. L. Youmans, ed., *The Culture Demanded by Modern Life* (Akron, Ohio: The Werner Co., 1867). Subtitled, "A Series of Addresses and Arguments on the Claims of Scientific Education."

member, realized that present educational institutions were woefully out of date, and were standing in the way of "the general progress of society." The proper role of education was to provide "preparation for the duties and work of the age in which we live." Unhappily this laudable goal was regularly misinterpreted by certain unnamed "adherents of the traditional system" as being nothing more than a surrender to "low and sordid utility." Now Youmans clear implication throughout this essay was that utility was in fact a noble aim of education, but rather than arguing directly for this proposition he elected to accept, hypothetically, the claim that *"Mental Discipline* is the true object of a higher culture."[77]

Youmans then proceeded to argue that the assertion that "the ancient classics and mathematics" were of "incomparable fitness to develop all the mental faculties," was quite false; a scientific education could beat a classical education at its own game. This underlines the important point that mental discipline and the classical curriculum were not connected by logical necessity but by historical contingency, and suggests that the status of their relationship at mid-century was at a delicate point of balance. On the one hand Youmans deferred to the continuing power of the mental discipline thesis as the ground on which any educational argument had to be made. But on the other hand the traditional system, or at least the traditional system as seen by Youmans, had "yield[ed] the point of the usefulness of the knowledge it imparts."[78] The implication that mental discipline was in fact subordinate to goals of utility and progress would help undermine mental discipline. Further, by attacking the classical curriculum by means of the mental discipline thesis, Youmans was contributing to the eventual decline of mental discipline in another way, since the relationship was reciprocal: the classical curriculum was part of the support for mental discipline as well as vice versa.

Youmans's examination of the classical curriculum was none too subtle. First he cast some general aspersions to the effect that it was an object of "mystical reverence," "superstition," and "cant," and was contrary to the "spirit of the age."[79] Then he laid into the claim that studying classical grammar was a good way to train the memory and judgment, in the course of which he contrived to refer to Greek and Latin as "dead" seven times in six pages. When it came to memorizing facts he pointed out "the immeasurable superiority"

---

[77]Youmans, 1–3. Italics in original.
[78]Ibid., 3–4.
[79]Ibid.

2.3. EMERGING CHALLENGES TO MATHEMATICS IN THE CURRICULUM

of scientific facts, because of the "natural and necessary connections" among them.[80]

Youmans began his critique of mathematics in a seemingly deferential manner, acknowledging as "unquestionable" its claim "to an important place in a liberal scheme of education." Mathematics was a "key to universal science" and therefore possessed "broad and solid utility" (evidently much preferable to "low and sordid utility"). Alas, the "devotees of tradition" were not content with this "subordinate" status for mathematics, but insisted on making "extravagant claims" of disciplinary virtues. Youmans countered by pointing out that mathematics provided incomplete mental exercise; given assumed premises it could produce "trains of proof," but had nothing useful to say about the truth of the premises. "The primary question is," said Youmans, unafraid to sound like Mr. Gradgrind in *Hard Times*, "What are the facts, the pertinent facts, and all the facts, which bear upon the inquiry?" The limitations of mathematics in "practical emergencies of thought" drove Youmans to the following harsh conclusion:

> The pure mathematician is therefore liable to a one-sided and erratic judgment of affairs. An exclusive mathematical discipline must, therefore, be held as an actual disqualification for the work of life.[81]

Through the rest of his more than 50-page essay Youmans proceeded to heap further insults on the classical curriculum while praising science. He saw the classical curriculum as fundamentally European, hence inappropriate to the American love of freedom.[82] To "reason deductively from foregone assumptions," which he held to be the essence of both linguistic and mathematical study, served to "habituate to the passive acceptance of authority."[83] Even when he conceded some value to training in deductive reasoning he found that the physical sciences provided such training of a "more valuable character,"

---

[80]Ibid., 5–10.

[81]Ibid., 10–11. In *Hard Times*, published in 1854, Charles Dickens had Mr. Gradgrind declare: "Now what I want is facts. Teach these boys and girls nothing but the facts. Facts alone are wanted in life." Dickens was participating in an educational debate in England which corresponded in some ways to that in the United States. See Kliebard, *Forging the Curriculum*, 28.

[82]Youmans, 53–54.

[83]Ibid., 12.

since they came "nearer to the realities of experience."[84] Youmans also anticipated emphases on economy and efficiency in education that would become highly important at the end of the century. Rejecting "the old metaphysical method" of studying the mind purely by introspection, he advocated a "modern psychology" which admitted that mental phenomena were embodied in physical changes in the brain.[85] This led him to conclude that there were strict physical limits to learning, and to therefore insist on a "law of mental economy," and to castigate "mental waste."[86]

It is important to note that some late nineteenth-century mathematical educators essentially accepted the scientific critique of mathematics education. J. Howard Gore, professor of mathematics in Washington, D.C. at Columbian University (which later became George Washington University) agreed that the sciences were essentially inductive studies while mathematics was deductive. Since both processes were vital to human thinking he found the sciences justified in claiming a greater share of the college curriculum than they presently had. In the interest of economy (his article was titled "The Waste of Mathematics"), mathematics was then obligated to shorten its courses. Gore thought that the proper way to do this, except for prospective specialists, was to limit instruction to only those portions of mathematics that were applicable to the sciences. He evinced no fear that this plan held any political difficulties.[87]

Levi L. Conant went even further in accepting a reduced role for his subject. Conant spent most of his career teaching mathematics as a utilitarian subject, first at the Dakota School of Mines and then at the Worcester Polytechnic Institute. From this experience he came to the conclusion that mathematics education faced unavoidable limitations: "About the subject of mathematics there is nothing to inspire the enthusiasm or to awaken the spontaneous interest that may be made to accompany work in natural science." The student of mathematics "finds himself here brought face to face with" not nature, alas, but "something essentially abstract and theoretical." This meant, according to Conant, that inevitably there were only a few students who enjoyed mathematics for its own sake. Such students required little pedagogic attention. A larger group sought to learn mathematics for its practical applications; such students should be subjected to "remorseless rigor," with no pretense that mathematics could be made into a "mere holiday excursion." What then should be done

---

[84]Ibid., 28.
[85]Ibid., 13–14.
[86]Ibid., 19–20.
[87]J. Howard Gore, "The Waste of Mathematics," *School Review* 2 (1894): 27–28.

with the large group of students with neither aptitude nor practical need for mathematics? Conant's answer was blunt:

> Do not compel them to study mathematics at all, beyond the necessary arithmetic of the grammar school and the merest elements of algebra and geometry ... the proposition that higher mathematics, or even the whole of the ordinary mathematics of secondary schools, must constitute an essential part of the education of every man and woman who pursues study above the limits of the grammar grade is utter folly.

This was a radical view for 1893, not at all endorsed by the Committee of Ten, although many non-mathematical educators would be echoing it in the early twentieth century. Some college and university mathematics professors who wished to devote most of their efforts to educating the mathematical elite may have sympathized with Conant. But other university mathematicians, exemplified by Chicago's E. H. Moore, did not feel inclined to accede to the reduced curricular role for mathematics envisioned by Conant. It is suggestive that Conant should have been a mathematics professor in a technical institute, secure in the prospect of an increasing demand for engineers and other appliers of mathematical knowledge. In contrast, a professionally conscious university pure mathematician might well have experienced some anxiety at the thought of retaining responsibility for no more than the small remnant of students Conant described as "actuated only by the purest love of mathematics."[88]

2.4. Mathematics in Nineteenth-Century Educational Institutions

By the time of the Committee of Ten there had grown up a complex of educational institutions, almost all of which taught mathematics in some form. Elementary education was accomplished in institutions usually designated as common schools, or primary schools, or grammar schools. Some of these schools held students well into what would now be considered the province of secondary education, but increasingly this more advanced level of schooling was being dominated by private academies and public high schools. The former had begun in the late eighteenth century as alternatives to the more narrow offerings of the grammar schools, and sometimes had a distinctly practical or scientific orientation. Gradually some of these academies (which also went under names such as seminaries or institutes) had evolved into college preparatory schools, or more vividly, fitting schools; purely terminal institutions were thus referred to as

---

[88] Levi L. Conant, "The Teaching of Mathematics," *School Review* 1 (1893): 211–216.

finishing schools. Public high schools were a more recent innovation, but by the 1880s were enrolling more than twice as many students as the private academies. Some high schools offered two curricula, classical or English, mathematics being a common feature of both. A school, or a program within a school, that aspired to prepare students for college was most reliably identified not by the presence of mathematics, or even Latin, but by the offering of Greek.[89]

Many colleges did not in fact rely on the secondary schools to supply them with students. Although in some regions, principally in the state universities of the middle west, the practice had grown up of automatically admitting graduates of certified secondary schools, it was more common to admit students to college based purely on an examination, with Greek, Latin, and mathematics being the crucial components. Since these admission examinations were devised independently by individual colleges, and were often highly specific, a burden was placed upon those secondary schools preparing students for multiple colleges. This created what was known as the problem of uniformity from the point of view of the secondary schools. From the point of view of an individual student determined to enter a specific college it was often more efficient to prepare for the examination with a private tutor, or by enrolling in the preparatory department of the college itself; such departments were a common appendage of many colleges at the time. There were also institutions that were primarily secondary schools but that nevertheless insisted on occasionally conferring a bachelors degree. Thus was the line blurred between the secondary schools and the colleges. Many educators found this chaotic system distasteful; calls for improving the articulation between the colleges and the schools became frequent and continued for some years.[90]

There were other educational institutions that in the late nineteenth century still held somewhat ambiguous relations with secondary schools and colleges. The normal schools, almost all supported by states or cities, were becoming more prevalent as an avenue to a school-teaching career, especially for women.

---

[89]On fitting schools and finishing schools see Huling, "New England Association," 608, and James H. Baker, "The High School as a Finishing-School," *National Educational Association Addresses and Proceedings* (1890): 633–640. On the high prestige of Greek see Krug, *High School*, 4–6.

[90]On certifying secondary schools see Krug, *High School*, 151–152 and Rudolph, 284. On the issue of uniformity see Krug, *High School*, 7 and Sizer, 57–58. On college preparatory departments see Rudolph, 281–282. One high school giving bachelors degrees was Philadelphia's Central High School. See James Chancellor Leonhart, *One Hundred Years of Baltimore City College* (Baltimore: H. G. Roebuch & Son, 1939), 15–16. On the problem of articulation see Sizer, 176 and Huling, "New England Association," 594–596.

2.4. MATHEMATICS IN NINETEENTH-CENTURY EDUCATIONAL INSTITUTIONS 39

Some of these had clearly established themselves as post-secondary institutions, but many had such minimal entrance requirements as to be little more than alternatives to a high school education; usually they insisted at least on knowledge of arithmetic.[91] A long-enduring conflict over the relative importance of subject matter training and pure pedagogic training had already surfaced, with some normal school educators holding that these schools were "a place for the acquiring of a training strictly peculiar to teaching and independent of academic matter."[92] This latter attitude did not endear the normal schools to college and university mathematics instructors, almost all of whom exhibited strong loyalty to their subject matter. Nevertheless, some among these mathematics instructors gained early teaching experience in normal schools, one example being David Eugene Smith, whose work will be referred to periodically in later chapters.[93]

There were also the scientific and technical schools, either attached to traditional colleges (such as the Lawrence Scientific School at Harvard and the Sheffield Scientific School at Yale), or existing independently (such as the Massachusetts Institute of Technology and the Worcester Polytechnic Institute). These institutions had begun to appear before the Civil War, spurred by concerns for the vitality of American applied science, and more were stimulated by the Morrill Act of 1862. Although there had been some expectations that these schools would be post-collegiate institutions, they soon evolved into collegiate competitors.[94] Admission requirements were less stringent than for the colleges, primarily with respect to the classical languages, for the advocates of these schools often felt that science and engineering were being suppressed in the colleges by the classical curriculum. This enabled Simon Newcomb, for

---

[91]On normal schools see H. G. Good, *A History of American Education* (New York: Macmillan, 1956), 213–214; Florian Cajori, *The Teaching and History of Mathematics in the United States* (Washington: Government Printing Office, 1890), 350–352; William J. Reese, *The Origins of the American High School* (New Haven: Yale University Press, 1995), 250; and Krug, *Salient Dates*, 67–71.

[92]James M. Green, "The Academic Function of the Normal School," *The Educational Exchange* 7 (Nov. 1894): 15.

[93]Eileen F. Donoghue, "The Emergence of a Profession: Mathematics Education in the United States, 1890–1920," in George M. A. Stanic and Jeremy Kilpatrick, eds., *A History of School Mathematics* (Reston, VA: NCTM, 2003), 160–161.

[94]See Rudolph, 231–232, 244–245. See also Francis A. Walker, "Immediate Problems in Technological Education," in Walker, *Discussions in Education* (New York: Henry Holt and Co., 1899), 7–8; and Eliot, "The New Education," 206–207.

example, to enter the Lawrence Scientific School in the 1850s, although his lack of training in Latin and Greek would have kept him out of Harvard College.[95]

That the classical curriculum continued to dominate, however, was reflected by the distinctly lesser prestige attached to the degrees conferred by the scientific schools, usually the Bachelor of Science or the Bachelor of Philosophy. The disparity between admission requirements for the colleges and those for the scientific and technical schools was still a source of contention at the time of the Committee of Ten. Yale's Sheffield Scientific School, for example, in 1892–1893 required English, United States History, geography, Latin, and mathematics; Yale College required Latin, Greek, ancient history, mathematics, and a modern language.[96] Charles Eliot urged the scientific and technical schools to raise their requirements so as to

> ...give a real support to secondary education. At present they are eating into the secondary schools. They receive secondary school pupils one year, even two years before they have finished a fair secondary school course.[97]

The technical schools were reluctant to follow such advice in practice, but did feel compelled to show some rhetorical deference to the continued prestige of the classical curriculum. For example, the official requirement for admission to the Worcester Polytechnic Institute in 1892 was

> ...evidence of proficiency in the common English branches, viz: History of the United States, Geography, Grammar and Arithmetic,—in French, Plane Geometry, and Algebra to quadratic equations.

After a description of these requirements in detail, the following apologetic statement was appended, evidently an attempt to ward off criticism from the educational elite:

> While the conditions of admission comprise only the requirements mentioned above, yet it is advisable that the candidates before applying take a full High School course, including, if possible, such knowledge of Latin, Greek, and Ancient and

---

[95] Simon Newcomb, "Formative Influences," *Forum* 11 (1891):187; Hawkins, *Between Harvard and America*, 23–24; and Eliot, "The New Education," 210.

[96] Rudolph, 232, Sizer, 57–58.

[97] C. W. Eliot, remarks reported in Huling, "New England Association," 618.

Medieval History as is generally required for admission to college.[98]

This prompts two observations. First, in this passage, as in most educational discourse in the late nineteenth century, a high school "course" referred to the entire sequence of class work encountered by a student while enrolled in high school. Such a course would have consisted of multiple "courses" as the word is usually used today. "Course" seems to have begun taking on its modern meaning at the turn of the twentieth century, in connection with the adoption of "units."[99] Second, mathematics did not need to be cited in this list of recommended subjects, since it had readily made the transition from the classical to the technical course. For a time mathematics enjoyed an ambiguous position: for the traditionalists it was a comfortable part of the classical curriculum, while for at least some of those seeking educational reform it could be portrayed as a tool of modern science. There can be no doubt that this helped mathematics to immediately attain a central position within the most recent institutional innovation on the American educational scene prior to the Committee of Ten: the research university. The pioneering institution of this type was the Johns Hopkins University in Baltimore.

2.5. Establishment of Mathematics at the Johns Hopkins University

In 1874 mathematical astronomer Simon Newcomb published an article in the *North American Review* in which he bemoaned the sad state of "Exact Science in America" compared with Europe, finding mathematics in an especially dire condition. He found journals, libraries, learned societies, and educational institutions all inadequate to support distinguished mathematical research. Moreover, he was pessimistic about the future: "The prospect of mathematics here is about as discouraging as the retrospect."[100] This situation would change remarkably rapidly, however, and by 1903 Newcomb was able to write of mathematics that "to-day, we are scarcely behind any nation in our contributions to the subject."[101] Newcomb saw the change as all originating from the founding of the Johns Hopkins University in Baltimore in 1876, an

---

[98]*Twenty-Second Annual Catalogue of the Worcester Polytechnic Institute* (1892): 20, 23.

[99]Kliebard, "Constructing a History," 160.

[100]Simon Newcomb, "Exact Science in America," *North American Review* 119 (1874): 288.

[101]Simon Newcomb, *The Reminiscences of an Astronomer* (Boston and New York: Houghton Mifflin & Co., 1903), 404.

event a later approving commentator has declared to be "perhaps the single, most decisive event in the history of learning in the Western hemisphere."[102] In the case of mathematical learning in particular, development was far from smooth after 1876, and the direct influence of Hopkins would be considerably attenuated by the turn of the century, but the initial example and impetus provided by Hopkins are undeniable.

Newcomb himself could claim some credit for the early development of the Hopkins approach to higher learning. In 1876, shortly before the university opened, he had again published his thoughts on the backwardness of American science, prompting President Daniel Coit Gilman to consult him with regard to the promotion of research at the new university. Subsequently Newcomb became one of the first non-resident lecturers appointed by Gilman, giving 20 lectures on the history of astronomy during 1876–1877. Gilman even considered offering Newcomb the chairmanship of the Hopkins mathematics department, keeping him as a reserve candidate while negotiating with the man he eventually hired, James Joseph Sylvester.[103]

Daniel Coit Gilman was trained in the classical tradition at Yale College, subsequently serving as librarian and professor of geography at Yale's Sheffield Scientific School, and then as president of the University of California. He gradually evolved from a relatively conservative proponent of bringing scientific study into the colleges to become a more venturesome advocate of the centrality of research to higher education, but his rhetoric never settled into a comfortable groove. Laurence Veysey has remarked the lack of connection in general between Gilman's pronouncements and the development of the university over which he presided, with Gilman appealing variously to abstract investigation, utility, and liberal culture as justifications for the proper form of the university.[104] He displayed no special affinity nor deep acquaintance with

---

[102]Edward Shils, "The Order of Learning in the United States," in Alexandra Oleson and John Voss, eds., *The Organization of Knowledge in Modern America, 1860–1920* (Baltimore: Johns Hopkins University Press, 1979), 28.

[103]See Simon Newcomb "Abstract Science in America," *North American Review* 122 (1876): 88-123 and Hugh Hawkins, *Pioneer: A History of the Johns Hopkins University, 1874–1899* (Ithaca: Cornell University Press, 1960), 42, 73. Gilman acknowledged the influence of Newcomb's 1876 essay in the *Johns Hopkins University Thirteenth Annual Report* (1888): 8.

[104]Veysey, 161. On Gilman's career see also Francesco Cordasco, *Daniel Coit Gilman and the Protean Ph.D.: The Shaping of American Graduate Education* (Leiden: Brill, 1960), 2-4.

mathematics. His occasional rhetorical nods toward the subject are best understood in the context of the ardent generalism with regard to scholarly endeavors which he developed at an early age. Thus Gilman wrote to his father during his sophomore year at Yale that "I am more interested in my studies than ever before, particularly in mathematics." But by his senior year he was touting his interest in surveying, astronomy, optics, logic, and ancient history, and avowing to "have never before taken so much interest in entering enthusiastically into all the college requirements."[105] Similarly, Gilman's proclamation of fifty years later that "Before the queen of the sciences—abstract mathematics—I bow my head and kneel uncovered," was but one entry in a list of disclaimers to specialized knowledge. He was, he truly remarked, no expert in any branch, but rather an observer, counselor, defender, and admirer.[106]

Probably Gilman's most noted reference to mathematics is the concise hiring procedure he claimed, some years after the fact, to have recommended to the original Hopkins trustees:

> Enlist a great mathematician and a distinguished Grecian; your problem will be solved. Such men can teach in a dwelling-house as well as in a palace. Part of the apparatus they will bring; part we will furnish. Other teachers will follow.

Mathematicians from 1888 right down to the present day have found this a flattering sentiment, apparently able to ignore its ominous implications for the capital improvements budget. It is true that by some reckonings Gilman's first two faculty appointments were Gildersleeve in Greek and Sylvester in mathematics, but whether Gilman's words were anything more than a politic bow to the verities of classical education is doubtful. The main purpose of the anecdote as Gilman used it in 1888 was to introduce the achievements of Basil Gildersleeve and his philological colleagues and to assure the trustees that "language and literature hold their time-honored place."[107] Nevertheless, whether

---

[105] Fabian Franklin, *The Life of Daniel Coit Gilman* (New York: Dodd, Mead & Co., 1910), 10.

[106] Daniel Coit Gilman, *The Launching of a University* (New York: Dodd, Mead & Co., 1906), 238.

[107] Gilman's remarks are found in *Johns Hopkins University Thirteenth Annual Report* (1888): 29–30. The portion on mathematicians and Grecians is quoted in Cajori, *Teaching and History*, 261; and Robert Edouard Moritz, *On Mathematics and Mathematicians* (New York: Dover, 1958), 122; is partly quoted in William L. Duren, Jr., "Mathematics in American Society 1888–1988, A Historical Commentary," in Peter Duren, ed., *A Century of Mathematics in America* (Providence: American Mathematical Society, 1989), 2:400; and is paraphrased in Roger Cooke and V. Frederick Rickey, "W. E. Story of Hopkins and Clark,"

44 2. MATHEMATICS EDUCATION IN NINETEENTH-CENTURY AMERICA

by accident or by design Gilman at one stroke changed the face of American mathematics by his appointment of Sylvester.

Sylvester had had a checkered career in his native England, with a brief unhappy sojourn at the University of Virginia, but his research production had given him a substantial reputation. In 1875 he was living in retirement, uninvolved with mathematics, but the news of the founding of the Johns Hopkins University reawakened his academic interests. He aggressively thrust his candidacy upon Gilman, drawing on the support of a circle of friends on both sides of the Atlantic eager to see him attain a position worthy of his research accomplishments.[108]

One friend of Sylvester who was especially eloquent on his behalf was Harvard's Benjamin Peirce:

> If you enquire about him, you will hear his genius universally recognized but his power of teaching will probably be said to be quite deficient...But as the barn yard fowl cannot understand the flight of the eagle, so it is the eaglet only who will be nourished by his instruction.

Peirce went on to recommend that Sylvester should be hired for the sake of

> the one pupil who will derive from his master, knowledge and enthusiasm—and that one pupil will give more reputation to your institution than the ten thousand, who will complain of the obscurity of Sylvester, and for whom you will provide another class of teachers.[109]

Peirce, whose teaching had been the subject of similar comments ("few indeed were the pupils who did not consider Peirce about the worst teacher ever"),[110] here candidly expressed the desire of the research elite to withdraw from responsibility for teaching the mass of pupils. The consequences of such attitudes

---

in Peter Duren, 3:33; all with evident approval. E. T. Bell is also approving in *Men of Mathematics* (New York: Simon and Schuster, 1937), 394, but has the story backwards, stating that it was Gilman who received the advice, the advisors being unnamed. According to Cordasco, 73–76, Rowland in physics was in fact the first faculty member hired, although his official appointment was not ratified by the trustees until several weeks after the appointments of Gildersleeve and Sylvester.

[108]Hawkins, *Pioneer*, 34. On Sylvester see Karen Hunger Parshall, "America's First School of Mathematical Research: James Joseph Sylvester at the Johns Hopkins University 1876–1883," *Archive for History of Exact Sciences* 38 (1988): 153–196.

[109]Peirce to Gilman, September 18, 1875. Quoted in Parshall, "Sylvester," 167.

[110]Henry Aaron Yeomans, *Abbott Lawrence Lowell* (Cambridge, MA: Harvard University Press, 1948), 40. See also Parshall and Rowe, 18–19.

would have to be faced by all the research universities in the years ahead, and would naturally create friction with those seeking more democratic or utilitarian education.

Sylvester's presence helped Gilman attract a promising young associate for Sylvester, William E. Story, then serving as a tutor at Harvard, and three enthusiastic graduate fellows for the first year: Thomas Craig, George Bruce Halsted, and Joshua W. Gore. Five other graduate students and seven undergraduates rounded out the department.[111] From this modest beginning emerged a flowering of mathematical research during the Sylvester years, 1876–1883.

Karen Parshall and David Rowe have described in detail the workings and achievements of the Hopkins mathematics department under Sylvester. Notable pure mathematical work was accomplished by Sylvester and his students in invariant theory, partitions, and matrix algebra. Story contributed in algebraic and non-Euclidean geometry, and Benjamin Peirce's son Charles initiated investigations in mathematical logic. Later testimony counted Sylvester's personal enthusiasm as a crucial ingredient. Far from well organized, often deficient in his knowledge of relevant contemporary mathematical literature, Sylvester nevertheless inspired the fervor in a select group of students which Benjamin Peirce had predicted. Sylvester's lecturing, which was almost entirely devoted to the research frontier as understood by him, featured frequent digressions as new problems caught his interest, and overall gave the students the impression of being immediately involved with the process of mathematical investigation.[112]

Sylvester's lectures were supplemented, from 1878 on, by the "Mathematical Seminary," a German inspired feature which Gilman was encouraging throughout Hopkins. In mathematics the seminar (to use the modern locution) generally met monthly, and consisted of one or more short presentations by faculty or students on original research. Parshall and Rowe acclaim Sylvester's seminar as a "sort of laboratory for the creation of new mathematics," citing Sylvester's insistence on induction rather than deduction as the motive power in mathematical research.[113] They do not present any evidence that the laboratory metaphor was applied to Sylvester's seminar at the time, although it would be surprising if it was not. Gilman saw the laboratory and all it stood

---

[111]Ibid., 76–79.

[112]Ibid., 80–81, 99–146. Also see Fabian Franklin, "James Joseph Sylvester," in *People and Problems* (New York: Henry Holt and Company, 1908), 20–21.

[113]On the Hopkins mathematics seminary see Parshall and Rowe, 86–87; and Cajori *Teaching and History*, 276. On Hopkins seminaries more generally see Cordasco, 90–92; and Veysey 153–158.

for as the central feature of the new research university he was trying to establish. He liked to equate the seminars at Hopkins on linguistic and historical topics with the laboratories of the natural sciences, and enthused often on the educational virtues of laboratories.[114] The word was becoming a staple of educational rhetoric and would remain so into the early twentieth century. One of his students even referred to Sylvester's ideas as having "sprung from the laboratory of his mind."[115]

Laboratory or not, it is doubtful that the seminar played such a central role within the mathematics department as the often cited seminar of Herbert Baxter Adams did within history (where the laboratory metaphor was explicitly applied), or Gildersleeve's seminar in Greek. These latter two seminars seem to have met more frequently, as much as twice a week.[116] Almost all the testaments to Sylvester's inspiring leadership come from his lectures, not the seminar; in the seminar his inability to pay attention to any but his current line of research often proved irksome.[117] Nevertheless, Parshall and Rowe's essential point stands: Sylvester, by eschewing polished lecture presentations and slavish following of textbooks, exposed a small but ardent group of students to the reality of mathematical research.

Another initiative urged by Gilman was the establishment of a mathematics journal under the auspices of the university. In pressing for this Gilman was strongly influenced by Newcomb, who had cited the lack of journals for original research as a major factor in the backwardness of American science, especially mathematics. Newcomb, together with Sylvester, (who was initially unenthusiastic), Rowland, and Story solicited interest in the proposed journal by circulating a letter to the mathematical community which restated some of the major points from Newcomb's essays. Newcomb was named cooperating editor for astronomy, and was himself a frequent contributor to the early issues. The journal served especially to highlight the contributions of pure and applied mathematicians associated with Hopkins, but attracted enough high quality submissions from other American and European researchers to establish a solid

---

[114]Larry Owens, "Pure and Sound Government: Laboratories, Playing Fields, and Gymnasia in the Nineteenth-Century Search for Order," *Isis* 76 (1985): 183–185. See also Robert H. Kargon, "Henry Rowland and the Physics Discipline in America," *Vistas in Astronomy* 29 (1986): 132–134.

[115]Cajori, *Teaching and History*, 266.

[116]See Richard Hofstadter, *The Progressive Historians* (New York: Vintage, 1970), 38; Peter Novick, *That Noble Dream* (Cambridge: Cambridge University Press, 1988), 33; and Cordasco, 91.

[117]Cajori, *Teaching and History*, 276.

reputation. Altogether the *American Journal of Mathematics*, which exists to the present day, greatly transformed the American scene Newcomb had decried in 1874: "Of mathematical journals designed for original investigations, such as we find in nearly every country in Europe, we have none and never had any."[118]

2.6. An 1890 Survey of the Educational Scene in Mathematics

On the eve of the Committee of Ten, in 1890, one of the future members of the mathematics subcommittee of that body published a landmark book titled *The Teaching and History of Mathematics in the United States*. The author was Florian Cajori (1859–1930), who had been born in Switzerland and had come to the United States at age 16. He received both a bachelor of science degree and a master of science degree from the University of Wisconsin, but also spent 18 months studying mathematics and physics at Johns Hopkins during 1884–1885, where Simon Newcomb was one of his professors. Cajori's book has been rightly pronounced the "first comprehensive history of mathematics in the United States," and it remains more ambitious than most successors in attempting to encompass developments in both elementary and advanced mathematical education within the same book.[119] Cajori's research was done largely during 1888 while he was temporarily employed in Washington, D.C. by the United States Bureau of Education. The book as a whole is an eclectic mixture; Cajori gathered his material from library research, formal opinion surveys, informal correspondence, and personal observations. He placed educational institutions at the center of historical development in mathematics, but in accord with his own special interests he treated several miscellaneous topics, such as the history of logarithms, more purely intellectually. Despite these quirks the book is, as Karen Parshall has observed, a useful "snapshot" of American mathematics education at one point in time, and as such has value for assessing the Committee of Ten.[120]

Cajori used the resources of the Bureau of Education to conduct an extensive survey of mathematics teaching at contemporary educational institutions.

---

[118] Newcomb, "Exact Science," 287–288. See John C. French, *A History of the University Founded by Johns Hopkins* (Baltimore: Johns Hopkins University Press, 1946), 50–51 and Parshall and Rowe, 89, 93–94.

[119] Uta C. Merzbach, "The Study of the History of Mathematics in America: A Centennial Sketch," in Peter Duren, 3:641. Cajori took at least two courses under Newcomb: "Analytical and Celestial Mechanics," and the "Seminary." *Johns Hopkins University Circulars* (Mar. 1885): 69.

[120] Karen V. H. Parshall, "A Century-Old Snapshot of American Mathematics," *The Mathematical Intelligencer* 12 (1990): 7–11.

A substantial section of his book is devoted to reporting the results of this survey: he recorded replies from 168 institutions that he classified as "universities and colleges," 44 that he classified as "normal schools," and 180 that he classified as "academies, institutes, and high schools."[121] The classification problem was not entirely trivial, but Cajori did not explain his criteria. Designations such as college, academy, institute, and high school were still in flux in 1890; indeed, one of the major themes of the agitation building up to the Committee of Ten was the need to bring order to the chaos by better drawing the line between secondary and higher education. Cajori displayed some awareness of the current state of terminological ambiguity; for instance, he placed the Franklin Female College of Topsham, Maine in the category of academies, institutes and high schools, very likely a correct judgment. But he was not infallible. Even though he had lived in Baltimore he put Baltimore City College in the same class as Johns Hopkins University, although the former was purely a secondary school.[122] Other classifications seem dubious as well.

Some prestigious institutions apparently did not deign to reply to the survey (e.g. Harvard, Yale, Princeton, Phillips Andover, Phillips Exeter, and the Lawrenceville School), but the roster of those that did reply is nevertheless impressive. It is also evident that Cajori and the Bureau of Education were diligent in requesting information from institutions besides those serving white males. Among such institutions responding were Howard University, Smith College, Mount Holyoke Female Seminary, Bryn Mawr College, Fisk University, the State Colored Normal School of North Carolina, and the New Hope Female Academy of the Choctaw Nation.[123]

Cajori carefully recorded the names and titles of all respondents to his questionnaire, making it possible to note the interesting fact that five individuals who would later serve with him on the Mathematics Conference were respondents for their respective institutions: Simon Newcomb of Johns Hopkins University; Andrew Ingraham of the Swain Free School of New Bedford, Massachusetts; Truman Safford of Williams College; George Olds, then of the University of Rochester; and W. A. Greeson, of the High School in Grand Rapids, Michigan. This is evidence that the members of the Mathematics Conference were indeed taking an active interest in mathematical pedagogy just prior to being named to the conference. For some of his questions Cajori recorded a brief response from each individual respondent, so that it is occasionally even

---

[121]Cajori, *Teaching and History*, 297–360.
[122]Ibid., 298, 354. On Baltimore City College see Leonhart, 15.
[123]Cajori, *Teaching and History*, 297–301, 350, 353.

possible to compare the views of the conference members. No great deviations are discernible. For instance, Newcomb, Ingraham, Safford, and Olds were all in agreement that the proper foundation for teaching the calculus was the method of limits, and that infinitesimals should only be introduced when based on limits.[124] When asked whether they favored the memorizing of rules in algebra both Safford and Olds answered no; and when further asked what reforms they favored for teaching algebra Safford mentioned "more thoroughness," and Olds mentioned "greater attention to detail."[125]

The overall value of Cajori's survey is diminished by several factors. As already noted, Cajori's classification of institutions is subject to doubt, and this then undermines the significance of the responses, since the questions asked were partly dependent on the classification. When this is coupled with the vagueness of some of the questions, the concision of many of the answers, and the fact that not all respondents answered all the questions, one must declare many of the responses essentially meaningless, absent much supporting information. For instance, Cajori inquired, "Are topics assigned to students for special investigation?" Simon Newcomb replied, in full, "They are in the seminaries," while Andrew Ingraham replied with an unqualified "Yes." To put it mildly, this hardly conveys the nature of the relative mathematical activity at the Johns Hopkins University and the Swain Free School of New Bedford, circa 1888. As Cajori himself admitted in a footnote to the entire set of responses to this question: "Widely different views seem to be implied in the above answers as to what constitutes a 'topic for special investigation.'"[126]

But although the survey cannot support definitive conclusions, it can serve to highlight and confirm some of the key pedagogic concerns of the time, along with revealing much about Cajori's own interests and biases. We have seen that he had tried to determine something about the level of mathematical research activity at colleges and universities; he also asked of these institutions, "What mathematical journals are taken? Are there any mathematical seminaries or clubs, and how are they conducted?" Other questions concerned teaching methods. Of the colleges and universities he asked: "Is the mathematical teaching by textbook or lecture?" Of all but the normal schools he asked: "To what extent are models used in geometry?" Of only the normal and secondary schools he inquired about the order of instruction: "Which is

---

[124]Ibid., 313–314. Greeson, because of Cajori's classification of his school, was not asked this question.
[125]Ibid., 332–333. Neither Newcomb nor Ingraham responded to this question.
[126]Ibid., 305–307.

taught first, algebra or geometry?" He asked about mathematical requirements for admission (curiously, only of secondary and normal schools) and whether "students entering your institution are thorough in the mathematics required for admission." (The thoroughness question was asked of all institutions, and garnered a great quantity of negative responses, from all levels. This agreement on the inadequate preparation of students in mathematics may well be the only statistically significant result in the entire survey). With the colleges and universities he raised the much discussed issues of the classical curriculum and the intellectual ability of women: "Do scientific or classical students show greater aptitude for mathematics? Which sex?" He indulged in some special pleading on behalf of pet ideas, notably the desirability of teaching the metric system and the history of mathematics.[127]

To further illustrate the range of issues involved in trying to interpret Cajori's survey, and the kind of insights that can be teased out of it even in the absence of statistical rigor, it is worthwhile to analyze the responses to one particular question at some length: "How does analytical mathematics compare in disciplinary value with synthetical?" This provoked a striking segmentation of responses. Slightly more than half of the reported respondents opined that analytical mathematics was superior to synthetical, many of these opinions being very unambiguous: "In my judgment the analytical is so far superior to the synthetical that there is left little room for comparison. Permit me to say that reason wants *light*, not *darkness*." The advocates of synthesis, though much smaller in number, also included unequivocal partisans: "Analytical mathematics is very far inferior to synthetic in disciplinary value." Finally, there was a substantial minority who saw the two methods as essentially of equal value, and several of these were, if one may so express it, very forceful in their equivocation: "Each has its special function; as well ask whether braces or tie-rods are of most service in a bridge truss." Two varieties of hedging were especially popular, namely that synthesis was better for discipline, while analysis was better for "investigation" or "use," or that the former was superior for beginning students, while the latter was preferable for advanced students. Thus Simon Newcomb: "The latter is probably the more valuable discipline in early stages of a mathematical education; but after the elements of geometry are mastered, probably the reverse."[128]

---

[127]Ibid., 301, 302, 316, 335, 352, 359, 360.

[128]The quote favoring analysis over synthesis was from J. M. Bandy of Trinity College, North Carolina. Ibid., 310. Italics in original. By my count, 67 of 129 reported respondents agreed. The quote favoring synthesis was Sylvester's Johns Hopkins student G. B. Halsted,

2.6. AN 1890 SURVEY OF THE EDUCATIONAL SCENE IN MATHEMATICS

In attempting to interpret these responses one must inquire into the meaning of Cajori's question. To the ancient Greeks analysis and synthesis distinguished two complementary ways of doing mathematics: proceeding from a sought for result backward to an assertion already known to be true; or starting from known results or accepted axioms and finally demonstrating the sought for result. The latter process appeared to many to be well exemplified by the coordinate-free study of lines, circles, triangles and other geometric objects in the manner first codified by Euclid. Thus it had become conventional by the nineteenth century to refer to this as "synthetic" geometry, while geometry employing coordinates and algebra, pioneered in the seventeenth century, was referred to as "analytic" geometry. It is thus likely that many of Cajori's respondents interpreted him to be asking simply about geometry instruction. Some mathematical educators also distinguished treatments of other parts of mathematics, algebra for instance, as being "synthetical" or "analytical," depending on whether or not an attempt was made to develop the subject in an axiomatic fashion in the manner of Euclid's geometry. Analysis often came to be linked with encouraging students to discover mathematical truths on their own, while synthesis became associated with presenting students with a perfected system of knowledge. In the case of arithmetic, the traditional approach via rules came to be identified with the synthetic style, and the movement away from rules came to be designated as analytic:

> Analysis is the separation of things into their elements or parts. In *Arithmetic*, it is the method of solution by reasoning according to the *nature* of the question, without reference to *special rules*.[129]

More generally, by the second half of the nineteenth century synthetic and analytic as applied to mathematics had come to be nearly synonymous with ancient and modern. Thus Charles Eliot in 1884 distinguished between "the mathematics of Euclid and Archimedes," and "the analytical mathematics now almost exclusively in use."[130]

---

then at the University of Texas. Ibid. I count 20 respondents endorsing synthetic mathematics. The equivocator was S. H. Peabody of the University of Illinois. Ibid., 308. I count 42 respondents who attributed no clear superiority to either method. Newcomb's quote was from p. 309.

[129]Joseph Ray, *Practical Arithmetic by Induction and Analysis*, one thousandth edition-improved (Cincinnati: Van Antwerp Bragg & Co., 1857), 261. Italics in original.

[130]Eliot, "What Is a Liberal Education?" 93.

Hence the three basic positions expressed by Cajori's survey respondents correspond very plausibly to three general attitudes regarding education: advocates of analytical mathematics become supporters of progress and reform; advocates of synthetical mathematics become defenders of tradition; and the equivocators become cautious moderates trying to meld old and new, but usually reluctant to challenge the commonly existing order of instruction, in which synthetic geometry was taught prior to analytic geometry. Whether or not this classification is accurate for each individual respondent, the overall picture that emerges is consistent with other information from the period: a large but not overwhelming contingent supporting reform initiatives, a small but dedicated core of conservatives, and a substantial group of moderates able to see virtue in ideas from both sides.

The meaning Cajori attached to analytical and synthetical mathematics was not, however, clear to all respondents. One such was Calvin Woodward of Washington University in St. Louis, a noted advocate of manual training: "Do you mean graphical (or geometrical) by synthetical? I think descriptive geometry has the finest disciplinary value." Some probably reached back to the original Greek concepts mentioned above. Others saw the question as probing a profound dichotomy within all knowledge: "Analyzing the whole into its elements is valuable, but building the whole from elements is *very* valuable." References were made to the explications of philosophers. Truman Safford's candid admission of confusion foreshadowed the direction in which mathematical opinion would eventually move: "Both methods are essential and I am not aware of any difference. Perhaps I do not understand the question." Later mathematicians would become progressively less able to understand the question; not only would they be unable to distinguish the disciplinary value of synthetical and analytical mathematics, they would be unable to see any difference between synthetical and analytical mathematics at all.[131]

As opposed to the confusion over analytical and synthetical there seems to have been none regarding the nature and desirability of discipline as an educational goal. Nevertheless, a sign of the coming breakup of the consensus on discipline can perhaps be seen in those who expressed the view that discipline was not the only goal of mathematics education: "I think synthetical has much

---

[131]Woodward's quote is from Cajori, *Teaching and History*, 309. On Woodward see Cremin, *Transformation of the School*, 26–29. The educator who mused on wholes and elements was H. Boring of Eminence College and Normal School, Kentucky. Ibid. Italics in original. J. F. Downey of the Minnesota State University quoted philosopher Sir William Hamilton. Ibid.

2.6. AN 1890 SURVEY OF THE EDUCATIONAL SCENE IN MATHEMATICS   53

the greater disciplinary value; analytical has much the greater value for practical application."[132]

It may have been noted from the foregoing that Cajori asked somewhat different questions of institutions depending on their place in his three-level educational hierarchy. There is a clear pattern. The notable inclusiveness of Cajori's approach, earlier remarked, cannot hide a formidable and fundamental bias in favor of those institutions he classified as colleges and universities. He asked far more questions of these institutions than of those he classified as secondary and normal schools, and it was the colleges and universities to which he directed most of the more controversial policy questions as to how mathematics ought to be taught as opposed to how it was being taught. Thus he asked both colleges and secondary schools whether they used geometric models, but only the colleges were asked to supply ideas on the reform of geometry teaching. Similarly, only colleges were asked to comment on reform of algebra; the secondary schools were asked merely whether they taught it prior to geometry. The only reform question asked of the secondary schools was on arithmetic, further evidence of Cajori's hierarchical bias. The normal schools were treated with even more condescension, being asked purely for facts about their programs, with no questions whatsoever alluding to current educational controversies or proposed reforms.[133]

Thus Cajori's survey provides a distinct portrait of the educational system of 1890, as seen through the eyes of an educator who identified strongly with the colleges and universities. While he acknowledged that the multiple levels of the system should not be considered in isolation from each other, since the problems within the system often overlapped levels, he expected proposed solutions to come almost exclusively from the top. The normal and secondary schools were expected to provide data, not ideas. Moreover, Cajori's total lack of any attempt to justify or explain the structure and details of his survey suggests how indisputable and natural his view of the educational system appeared to him. To be sure, it must be acknowledged that Cajori was far from assuming any dictatorial role; he was hesitant to offer any policy prescriptions himself, and made hardly any attempt even to assess his survey results. Late nineteenth-century secondary school educators often complained of college domination of educational policy-making, and Cajori's attitude lends some support to these

---

[132]O. D. Smith of the Alabama Polytechnic Institute, Agricultural and Mechanical College (Auburn). Cajori, *Teaching and History*, 307.

[133]Ibid., 335, 357.

complaints, but in a more subtle form than sometimes noted. He was no educational bully, but the passive and almost unconscious assumption of college superiority he exemplified may have been harder for the school educators to combat in the long run.

CHAPTER 3

Simon Newcomb: One Mathematician's Educational Theory and Practice

To provide greater specificity to our account of American nineteenth-century mathematics education we now focus on Simon Newcomb (1835–1909), whose encounters with mathematics are well documented from early in his childhood. Newcomb was a major contributor to nineteenth century mathematical astronomy, who in addition made part-time forays into economics, academic administration, and mathematics education. Newcomb's career path brought him into contact with institutions and individuals of central importance to the evolution of American mathematics education and education generally. At Harvard in the 1850s he met the young Charles Eliot and studied with Benjamin Peirce, as that venerable institution was in the early stages of attempting to accommodate more technical education. Later, in the 1870s, Newcomb played an advisory role as Daniel Coit Gilman assembled the initial faculty at the research-oriented Johns Hopkins University. In the 1880s Newcomb himself took the helm of the Hopkins mathematics department, at the same time becoming involved with high school mathematics through textbook writing and polemical essays. Thus the Eliot of the 1890s, by then a senior statesman on the educational scene, found it appropriate to appoint Newcomb as the chairman of the mathematics conference of the Committee of Ten.

But Newcomb's utilitarian approach to mathematics appeared old fashioned to later generations of more specialized mathematicians, more intrigued than he by the internal patterns of mathematics irrespective of their usefulness in other fields. This was reflected in an event contemporaneous with the Committee of Ten, the Mathematics Congress associated with the Chicago World's Fair of 1893. Newcomb had scoffed at this congress, while E. H. Moore, representing a younger generation of mathematicians, worked vigorously to make it a resounding success. Looking more closely at Newcomb therefore provides valuable insight into the evolving position of mathematics within the larger nineteenth-century educational environment, as the internal development of the field at the professional level created repercussions from the schools to the universities.

3.1. Newcomb's Educational Roots

Simon Newcomb began life in remote northeastern Canada rather than in the United States, but his situation was more representative of education in English speaking North America during the first half of the nineteenth century than the Boston Brahmin upbringing of his close contemporary Charles Eliot. School enrollments had begun to increase, but many children continued to encounter little in the way of formal education, especially in the rural areas. Newcomb, born in 1835, was one such rural child who rarely attended school, even though his own father was a school master. That Newcomb's first concentrated experience of education was as a teacher rather than as a student would not have greatly distinguished him either. Geraldine Jonçich Clifford has found, from extensive perusal of personal histories, that throughout much of nineteenth century America "the possession of a modicum of formal learning equipped a young man or woman to take up teaching—an occupation requiring almost nothing in the way of a capital investment." Teachers no older than fourteen or fifteen were not uncommon.[1]

As Newcomb noted in his autobiography, the trade of schoolmaster in Nova Scotia and Prince Edward Island at the time his father practiced it was an insecure one, requiring frequent changes of venue in search of pupils in this thinly populated region. The tradition of the itinerant teacher in Nova Scotia was a holdover from the middle of the eighteenth century, when many of these individuals were more valued for their "congenial companionship" than for their often "meager store of knowledge."[2] John Newcomb seems to have sought a higher standard, with not entirely happy results; the son attributed part of the family's wandering existence to the strictness of the father's pedagogic views, which seem to have frequently conflicted with those of the clientele. Newcomb noted especially his father's dislike for "the learning of anything, especially of

---

[1]Geraldine Jonçich Clifford, "Home and School in 19th Century America: Some Personal-History Reports from the United States," *History of Education Quarterly* 18 (Summer 1978): 13.

[2]Patrick Wilfred Thibeau, "Education in Nova Scotia before 1811" (Ph.D. diss., Catholic University of America, 1922), 67.

arithmetic and grammar, by the glib repetition of rules."[3] This suggests that by the 1840s such notions were not confined to the republican citizenry of the United States, among whom Patricia Cline Cohen has located the emergence.

Newcomb noted two reasons for his very intermittent school attendance: the necessity of his working to keep his family financially afloat, and his father's concern that Simon might damage his health by too much mental exertion.[4] Newcomb had displayed what his parents considered to be a charming but overly intense interest in mathematical knowledge, as recorded in the following anecdote told by his father:

> When a little over four and a half, one evening, as I came home from school, you ran to me, and asked, "Father, is not 4 and 4 and 4 and 4, 16?" "Yes, how did you find it out?" You showed me the counterpane which was napped. The spot of four rows each way was the one you had counted up. After this, for a week or two, you spent considerable number of hours every day, making calculations in addition and multiplication. The rows of naps being crossed and complexed in various ways, your greatest delight was to clear them out, find how many small ones were equal to one large one, and such like. After a space of two or three weeks we became afraid you would calculate yourself "out of your head," and laid away the counterpane.[5]

It would thus appear that Newcomb's later advocacy of making arithmetic reasoning visible (as in the Committee of Ten Report of 1893), had deep roots indeed.

Building upon this admirably concrete introduction to arithmetic, Newcomb continued, according to his father, to work his way through an arithmetic textbook: "When my school ended here [Bedeque, Prince Edward Island], you were six and a half years of age, and pretty well through the arithmetic. You had studied, I think, all the rules preceding including the cube root."[6] In standard nineteenth-century textbooks the extraction of square and cube roots were always considered more advanced topics than proportion and the Rule of Three, and generally occupied a late chapter under the title "evolution." To raise a

---

[3]Simon Newcomb, *The Reminiscences of an Astronomer* (Boston and New York: Houghton Mifflin & Co., 1903), 6–7.

[4]Simon Newcomb, "Formative Influences," *Forum* 11 (1891): 184.

[5]Newcomb, *Reminiscences*, 10. Newcomb was quoting from a letter written to him by his father in 1858. The original letter is in SNP, Box 8.

[6]Newcomb, *Reminiscences*, 11.

positive whole number to a power was to "infold" it or "involve" it in a larger number; to reverse the process by extracting the root was then to "unfold" or "evolve." Square and cube root algorithms of ancient lineage (and already remote from specialized mathematical practice by 1800), made use of all the elementary operations (addition, subtraction, multiplication, and division) and so were considered to be an appropriate culmination of the arithmetic course. Simon Newcomb's Mathematics Conference of the Committee of Ten would seek to place root extraction in a much more subordinate position in the curriculum.

As a teenager, Newcomb found interest in a small collection of mathematical books inherited from his grandfather, of which an English translation of Euclid most caught his attention. He recounted how his attempt to grasp the forty-seventh proposition of Euclid's First Book (commonly known as the Pythagorean theorem) gave him "the first conception of mathematical proof that I had ever met with."[7] Newcomb made a point of acknowledging that at the same time he was enthralled by Euclid he was also captivated by phrenology. He explained this as follows:

> It may appear strange to the reader if a system so completely exploded as that of phrenology should have any value as a mental discipline. Its real value consisted, not in what it taught about the position of the "organs," but in presenting a study of human nature which, if not scientific in form, was truly so in spirit.[8]

Thus did Newcomb, writing in 1903, continue to defend the doctrine of mental discipline, a doctrine which by that date was itself considered in some quarters to have been exploded.[9]

In 1853, at the age of 18, Newcomb left Canada to seek brighter prospects in the United States. He made his way to Maryland, where he taught school for two years. His teaching experiences prompted him to put pen to paper at age 19 in an attempt to articulate his pedagogic views. The resulting unpublished manuscript contains several striking anticipations of his later public pronouncements, especially in the importance he attached to illustrating abstract principles with concrete examples, and provides clear evidence that Newcomb was already becoming acquainted with the issues and terminology that would

---

[7]Newcomb, *Reminiscences*, 17–18.
[8]Ibid., 14.
[9]See Laurence R. Veysey, *The Emergence of the American University* (Chicago: University of Chicago Press, 1965), 54 and Edward A. Krug, *The Shaping of the American High School 1880–1920* (New York: Harper & Row, 1964), 206–207.

be paramount in educational controversies through the rest of the nineteenth century: "Education has two great objects, first the acquisition of knowledge, and secondly training and disciplining the mind." The young Newcomb did not see any necessary conflict between these two objectives, but did note the difficulty, quite likely writing from personal experience, of motivating the student to partake of the educational fare being offered. In arithmetic, for example, students who were induced to work solely for the sake of matching the answers in the book would have little reason to retain the knowledge thus gained; but tell them

> that the very object of that science is to enable them to transact their business with certainty, and to guard against the designs of those who would impose on them, and he will immediately see something to interest him in the study.[10]

Thus we encounter again the notion of student interest, which would occupy many educators for the rest of the century and beyond.

In his spare time during his residence in Maryland Newcomb tried to advance his knowledge of science and mathematics on his own. He tackled textbooks on mathematics by Charles Davies, finding this author's book on calculus rather unsatisfactory; he was more complimentary regarding books by Étienne Bezout (1739–1783). Then, having concluded "that mathematics was the study I was best fitted to follow," he took the ambitious step of tackling Newton's *Principia*, in English translation. He claimed to have found the experience salutary, for much the same reason as with phrenology:

> The path through it was rather thorny, but I at least caught the spirit here and there. No teacher at the present time would think of using it as a text-book, yet as a mental discipline, and for the purpose of enabling one to form a mental image of the subject, its methods at least are excellent.[11]

Once again, Newcomb here suggested the possibility of conveying the "spirit" of science quite apart from its technical details. Similarly, Newcomb's first contact with the *Méchanique Céleste* of Laplace fired him with enthusiasm without providing him anything approaching mastery of the text.[12]

---

[10]SNP, Box 114. Mixing of plural and singular pronouns in original.

[11]Newcomb, *Reminiscences*, 53–54. The Davies book was *Elements of the Differential and Integral Calculus* (Philadelphia: A. S. Barnes & Co., 1840).

[12]Newcomb, "Formative Influences," 187–188.

It was during his sojourn in Maryland that Newcomb made his initial acquaintance with Joseph Henry of the Smithsonian Institution. Henry was the recipient of Newcomb's first attempt at publication, "A New Demonstration of the Binomial Theorem." Henry and an anonymous mathematical referee both demurred, but in such a kind and constructive way as to encourage the 20 year old Newcomb to pursue a mathematical career.[13]

In 1856 Henry and his colleague J. E. Hilgard of the Coast Survey office helped Newcomb obtain a position in Cambridge, Massachusetts as a calculator for the office of the *American Ephemeris and Nautical Almanac*. This office had been established in 1849 to produce comprehensive positional information on astronomical phenomena for the use of navigators and astronomers. Newcomb asserted that the location of this office was chosen so that experts from Harvard University would be readily available for consultation, most especially "Professor Benjamin Peirce, who was recognized as the leading mathematician of America."[14] The definite article used by Newcomb to refer to Peirce little exaggerates the narrow distribution of advanced mathematical knowledge in the United States in this period.[15] Newcomb would play a significant role in widening this distribution, but the efforts of his generation would remain relatively modest compared to those of the following generation, who would truly make it absurd to speak of "the" leading mathematician.

The young Newcomb was surprised and disappointed to find that he could perform the duties of a computer for the *Nautical Almanac* without any real knowledge of celestial mechanics.[16] Newcomb was here clearly confronted with one of the most fundamental problems of mathematics education, a problem haunting all debates to the present day between utility and discipline as justifications for studying mathematics. If useful mathematical work can be performed by individuals who do not understand the basis behind the work, why expend resources to impart this understanding? However, his *Nautical Almanac* experience does not seem to have inspired much reflection on the part of the mature Newcomb, pedagogue and economist though he was.

---

[13] Newcomb, *Reminiscences*, 54–55. It has been suggested that this referee was probably Benjamin Peirce. See Arthur L. Norberg, "Simon Newcomb's Early Astronomical Career," *Isis* 69 (1978): 211.

[14] Newcomb, *Reminiscences*, 63.

[15] See Karen Hunger Parshall and David E. Rowe, *The Emergence of the American Mathematical Research Community, 1876–1900: J. J. Sylvester, Felix Klein, and E. H. Moore* (American Mathematical Society, 1994), 1–52.

[16] Newcomb, "Formative Influences," 188.

Newcomb's duties for the *Nautical Almanac* in Cambridge were so undemanding that working in his spare time he was able to obtain a Bachelor of Science degree in 1858 at Harvard's Lawrence Scientific School, working under Peirce's direction. It was during this period that Newcomb made the acquaintance of Charles Eliot, who was then serving as a mathematics tutor at Harvard.[17]

Newcomb returned to Washington in 1861 to take up a position at the Naval Observatory. After declining a call by President Eliot in 1875 to head the Harvard Observatory, Newcomb in 1877 achieved a long held ambition of becoming the Superintendent of the *Nautical Almanac* Office, which by this time had moved to Washington.[18] Newcomb considered the *Nautical Almanac* more appropriate than the Naval Observatory as a venue in which to ply his favored trade of mathematical astronomer.[19] Here he assembled the resources to carry out a long held ambition of reanalyzing the motions of all the planets to created a reformed astronomical ephemeris, a massive endeavor involving less in the way of new observations than it did a wide survey and reexamination of older observations made all over the world.[20] It was this type of project that inspired tributes by later commentators regarding Newcomb's organizational prowess in managing the staff of eight or ten computers required to carry out the work.[21]

Newcomb had his own apprentice experience in Cambridge as a guide, as well as George Biddell Airy's "manufacturing" approach to running the Royal Observatory at Greenwich, which had impressed Newcomb during a visit in 1870. Under Airy's hierarchical system the Astronomer Royal "could himself work out all the mathematical formulae and write all the instructions required to keep a small army of observers and computers employed." This army included in its lowest ranks those who "needed to understand only the four rules of arithmetic; indeed, so far as possible Airy arranged his calculations in such a

---

[17]Newcomb, *Reminiscences*, 74. James, 1:67.

[18]W. W. Campbell, "Simon Newcomb," *Memoirs of the National Academy of Sciences* 17, 1st mem. (1924):8.

[19]Newcomb, *Reminiscences*, 115.

[20]Newcomb had begun working on parts of this project as early as 1866. Arthur L. Norberg, "Simon Newcomb and Nineteenth Century Positional Astronomy" (Ph.D. diss., University of Wisconsin, 1974), is devoted to describing the full details of Newcomb's program.

[21]Albert E. Moyer, *A Scientist's Voice in American Culture: Simon Newcomb and the Rhetoric of Scientific Method* (Berkeley: University of California Press, 1992) 74–75; G. W. Hill, "Simon Newcomb as an Astronomer," *Science* n.s. 30 (1909): 355; E. W. Brown, "Simon Newcomb," *BAMS* 16 (1910): 347.

way that subtraction and division were rarely required."[22] Newcomb adopted the Airy system with gusto, delighting in finding a man who could act as a "foreman in supervising the work," and extolling another assistant as "the most perfect example of a mathematical machine that I have ever had at command," although unhappily this latter individual eventually "broke down."[23]

As with his experiences in the 1850s, these later experiences with the organized application of mathematics appear to have had a singularly small impact on Newcomb's pedagogic thinking. He noted Airy's view that for most astronomical work "plodding industry, properly directed, was more important than scientific training," but drew no conclusions.[24] He seems to have had no special anxiety about reforming the educational system so as to produce more mathematical foremen or machines. Newcomb, like most of his nineteenth century colleagues, did not conceive of the era of mass education, where educational experiences would as a matter of course be thought of as training large numbers of students for later employment. This left a vacuum into which later proponents of "vocational" and "socially efficient" education would rush at the expense of thinkers like Newcomb.[25]

In the 1880s Newcomb managed to find time amidst his responsibilities at the *Nautical Almanac* to involve himself in mathematics education at two different levels: he became chairman of the mathematics department at Johns Hopkins University, and he became a writer of mathematics textbooks for the secondary school market. In part these activities seem to have been motivated by financial concerns, but they also allowed him to explicitly express his ideas on teaching procedures and curriculum.

3.2. Newcomb at Johns Hopkins

As noted in the previous chapter, Daniel Coit Gilman made a dramatic stroke by appointing J. J. Sylvester as the first head of the Johns Hopkins University mathematics department. The Sylvester era at Hopkins turned out to be potent but brief. Betting on the prospect (shortly to become reality) of at last attaining a prestigious and secure professorship in England (the Savilian Chair at Oxford), Sylvester resigned from Hopkins in 1883. He strongly recommended

---

[22]Newcomb, *Reminiscences*, 286–288.
[23]Newcomb, *Reminiscences*, 223–225.
[24]Ibid., 287.
[25]See Theodore R. Sizer, *Secondary Schools at the Turn of the Century* (New Haven: Yale University Press, 1964), 130, and Herbert M. Kliebard, *Forging the American Curriculum* (New York: Routledge, 1992), 175.

to Gilman that Felix Klein of Germany be named to succeed him. The hiring of Klein came tantalizingly close to reality, but in the end Gilman refused to meet Klein's salary requirements. After also failing to attract England's Arthur Cayley, Gilman settled on Newcomb.[26]

More recently there has been much musing and muttering about the wisdom of Gilman's conduct by some members of the American mathematical community and fellow-traveling historians, privileged to look back at these events across many intervening years. Parshall and Rowe, unable to conceal their regrets, begin the paragraph wherein they reveal Gilman's parsimonious final offer to Klein with "Unfortunately for the mathematics program at the Johns Hopkins..."[27] Constance Reid's brief account of the incident is wistfully titled, "The Road Not Taken."[28] But for the most dyspeptic view of all one should consult Clifford Truesdell: "The mathematicians at Hopkins after the departure of Sylvester and before the arrival of Bateman—nearly thirty years—seem to have been research zombies."[29] Such attitudes are perfectly natural, especially in view of the subsequent career of Klein, now generally considered to be a central figure in the organization of modern German and indeed world mathematics. As Hugh Hawkins has noted, when it still appeared Klein would come to Hopkins "Sylvester described his expected successor like a John the Baptist describing the coming Messiah."[30] Most historically aware mathematicians and mathematically aware historians would today declare that Klein resoundingly earned deification, and that Newcomb was a mere mortal.

Nevertheless, it is worth looking at the issue more dispassionately in the context of the time, especially for what it reveals about the transformation of American mathematics since the 1880s. In fact, from this perspective the more recent commentators who have bewailed the fate of Johns Hopkins University in 1884 become prime exemplars of this transformation.

Gilman's choice of Newcomb in 1884 to head the Johns Hopkins mathematics department was an eminently reasonable one. Newcomb was without question a major figure by that date, well known in Europe as well as America. He

---

[26] Parshall and Rowe, 138–143.

[27] Parshall and Rowe, 143.

[28] Constance Reid, "The Road Not Taken: A Footnote in the History of Mathematics," *The Mathematical Intelligencer* 1 (1978): 21–23.

[29] Clifford Truesdell, "Genius and the Establishment at a Polite Standstill in the Modern University: Bateman," in *An Idiot's Fugitive Essays on Science: Methods, Criticisms, Training, Circumstances* (New York: Springer-Verlag, 1984), 416.

[30] Hugh Hawkins, *Pioneer: A History of the Johns Hopkins University, 1874–1899* (Ithaca: Cornell University Press, 1960), 136.

was clearly recognized as the direct inheritor of the mathematical-astronomical tradition of Benjamin Peirce, which was still the strongest line of development of mathematical investigation in the United States. Sylvester's rather chaotic introduction of pure mathematical topics at Hopkins could easily be seen as a brief uncharacteristic interlude. This proved not to be the case. Pure mathematics did triumph, and one consequence of this triumph is that a great many modern observers see Sylvester's Hopkins sojourn not as eccentric, but as a precursory episode whose fulfillment was thwarted by careless handling.

To be sure, there is ample room to question Gilman's judgment, especially his willingness to let Newcomb divide his time between Hopkins and the *Nautical Almanac*. This presented obstacles to the cultivation of graduate students at Hopkins, and surely contributed to the very mild impress that Newcomb seems to have made on his colleagues in the mathematics department. These ramifications will become especially clear in comparison with the University of Chicago under E. H. Moore, which possessed a basic unity of purpose unseen at Hopkins under Newcomb. Moore's department produced a sequence of doctoral students who identified strongly with his professional vision for mathematics, and many of these students proceeded to inspire yet further generations of like minded mathematicians. Newcomb failed to initiate any such self perpetuating school of mathematicians, or mathematical astronomers for that matter.

Yet it is unfair to attribute all this to Newcomb personally. Moore benefited greatly from the concomitant growth aspirations of other departments of mathematics just as he was sending his brood of students out into the world. Newcomb's students, like those of Sylvester, faced a much less welcoming milieu.[31] Thus in comparing Newcomb and Moore we are not purely comparing individual leadership abilities, but taking a snapshot of the entire supporting mathematical community, and demonstrating just how remarkably that community changed between say 1885 and 1905.

There are certainly reasons to doubt Newcomb's commitment to his responsibilities at Johns Hopkins. Writing from Washington soon after he succeeded Sylvester, Newcomb seemed to regard the position as a momentary diversion: "Did you know that I had taken upon me the department of mathematics in Johns Hopkins University as an experiment?"[32] A few months later he was writing to his old friend Alexander Agassiz, candidly discussing the question of whether any of the Hopkins mathematicians might fill a faculty opening at Harvard. In this letter Newcomb treated Craig and Story as quite dispensable,

---

[31] Parshall and Rowe, 145.
[32] Newcomb to O. C. Marsh, Oct. 27, 1884, letterpress copy in SNP, Box 6.

and in general conveyed little proprietary attitude toward the Hopkins department.[33] This contrasts with the attitude E. H. Moore would take at Chicago, with his aggressive plans for departmental growth and his excitement at the prospect of colonizing other institutions with departmental graduates.

But here again the fact that later generations have seen Moore as exemplary should not be allowed to obscure the achievement of Newcomb at Hopkins in the face of very real difficulties. Most fundamentally, the financial predicament of the university during Newcomb's tenure made planning for departmental expansion a vain exercise. The faculty desire for promotion and comfortable salaries naturally came into conflict with the strained resources of the university, which were even more precarious than during Sylvester's time. Moreover, the university faced pressure to expand the undergraduate program, as a direct aid for the financial situation, as a means of sustaining support in the Baltimore area, and as a means of supplying qualified students for its graduate departments. Newcomb had then to meet the needs of "the average American boy who goes to a college to burn his Euclid when he gets through his course," while at the same time placating the aspirations of faculty and graduate students who had been attracted to Hopkins to pursue advanced research.[34]

Finally, Newcomb had to dispatch Ph.D. recipients into a world still not quite ready for them. While a discerning eye at that time might have seen favorable trends developing nationally for young mathematicians, a major expansion of opportunities was still in the future. Hugh Hawkins, writing more generally than about mathematics, has cast the situation in a romantic light: "Part of the glory of Hopkins in the 1880s was this band of young men waiting for American higher education to catch up with their aspirations."[35] Newcomb at the time could not afford to dwell on the glory, only on the reality, as in the cautious counsel he offered a German correspondent who queried him about the American job market in 1885:

> I really do not know where I should advise you to turn for a position in teaching the higher mathematics. At the Johns Hopkins University I already have three assistants. No more are needed. At Harvard University I think they are also fully

---

[33]Newcomb to A. Agassiz, Apr. 16, 1885, letterpress copy in SNP, Box 6.
[34]Newcomb to G. Stanley Hall, Apr. 1889, letterpress copy in SNP, Box 6.
[35]Hawkins, *Pioneer*, 210.

supplied with their own graduates. Nearly all our colleges prefer either their own graduates or members of some religious sect.[36]

A letter Newcomb wrote to his junior colleague Fabian Franklin in 1892 further illustrates Newcomb's departmental problems and procedures. Franklin was on sabbatical in Germany, prompting Newcomb to surmise, no doubt correctly, that "you will come back full of Klein's mathematics, and anxious to teach it to graduate students." Newcomb expressed sympathy for such a desire, but found it necessary to alert Franklin to disagreeable realities. Because of rising undergraduate enrollments in introductory courses, and no prospects of additional teaching help, "you must therefore do your full share of undergraduate work," although Franklin was of course free to teach graduate courses on top of his undergraduate load. Newcomb then proceeded to discuss the problem of the current class of mathematics majors, found to be "woefully deficient in the elements of the calculus," and explained that in order to combat this state of affairs "we have introduced a plan of pretty much abolishing long lectures and explanations, and making the men spend most of the hour in actual practice at the blackboard."[37]

Newcomb's usual course load during his Hopkins tenure was one graduate level lecture course related to astronomy and one seminar. The seminar was likely oriented toward astronomy also; it was explicitly so recorded from 1887. Newcomb never taught any undergraduate mathematics or astronomy courses at Hopkins.[38] Of those mathematics graduate students at Hopkins who entered during the Sylvester era from 1876 to 1883, nine achieved a doctorate, all nine dissertations being in pure mathematics.[39] During the next ten years, with Newcomb in charge, doctorates were awarded to ten more students, two of whom (W. S. Eichelberger and C. L. Poor) were recorded as being in "mathematics and astronomy," and thus clearly directed by Newcomb.[40] Poor and Eichelberger went on to have moderately prominent careers, and it is evident that Newcomb

---

[36]Newcomb to F. Montesor, Jan. 13, 1886, letterpress copy in SNP, Box 6.

[37]Newcomb to Franklin, Feb. 10, 1892, letterpress copy in SNP, Box 6.

[38]*Johns Hopkins University Circulars* (1885–1894). In 1889 Newcomb was among the 15 (out of 55) faculty members who had no contact with undergraduates. See Hawkins, *Pioneer*, 256.

[39]Parshall and Rowe, 97.

[40]*Johns Hopkins University Register* (1894–1895): 51–60.

played a role in both, but no influence of Newcomb's pedagogic interests is readily discernible.[41]

In 1893, just about the time of his service with the Committee of Ten, Newcomb became embroiled in a personnel controversy at the *Nautical Almanac*, resulting in his being asked to step down from his Hopkins post by the Secretary of the Navy. Newcomb's pedagogic activities played a role. He had attempted to fire one of his *Nautical Almanac* assistants, Joseph R. Morrison, for incompetence, but Morrison fought back by accusing Newcomb of having neglected the *Nautical Almanac* in favor of Hopkins, and of having "employed [Morrison] to write several chapters of a mathematics text-book during office hours."[42] The resulting investigation cleared Newcomb of all charges, but Secretary of the Navy H. A. Herbert, long critical of the administrative deficiencies of government scientists, did not miss the opportunity to rein in Newcomb's outside activities.[43]

After his retirement from the government in 1898 Newcomb was reappointed at Hopkins, but he did not resume the level of involvement he had had during 1884–1893; he never again taught a regular lecture course at Hopkins, although he did once again become the chief editor of the *American Journal of Mathematics*. In 1900 Newcomb retired from Hopkins as an emeritus professor.[44]

3.3. Newcomb as Textbook Writer and Educational Theorist

Newcomb first entered the book writing business in 1878 with *Popular Astronomy*, based on lectures he had delivered at Hopkins. In the same year he published *Astronomy*, his revision for the American market of a book by Irish astronomer Robert S. Ball. This latter book was his first association with the ambitious publisher Henry Holt (1840–1926). Yet another astronomy book with Holt followed in 1879, this one written in cooperation with E. S. Holden, a former assistant at the Naval Observatory. This was Newcomb's first explicit venture into the textbook market, and both he and Holt were keen to move quickly to exploit this market further. A letter from Holt to Newcomb

---

[41]On Eichelberger see *NCAB*, C:75. On Poor see Richard Berendzen and Richard Hart, "Poor, Charles Lane," *DSB*, 11:83–84.

[42]*Astronomy and Astro-Physics* 12 (Oct. 1893): 664–665.

[43]See SNP, Box 104, Folder "Autobiography: Rejected Chapter."

[44]On Newcomb's late Hopkins period see Campbell, 13; and Raymond Clare Archibald, *A Semicentennial History of the American Mathematical Society* (New York: American Mathematical Society, 1938; repr., New York: Arno Press, 1980), 124.

from November 1879 shows the two men already immersed in planning future textbook ventures:

> I feel an interest in the proposed ~~elementary~~ mathematical course, and think there would be a decided gain in bringing out the elementary Algebra and elementary Geometry together. I also think the idea of the circular asking the opinion of teachers, a good one. Possibly it would be worthwhile to send on the MS. of the Algebra before you come, & it can be got ready very soon.[45]

The manuscript referred to by Holt was one originally written by Newcomb for the purpose of introducing his daughter Anita to algebra.[46] It is clear that Newcomb soon saw publishing possibilities for this book, and that he quickly conceived of putting out not just one but a whole series of mathematics textbooks. A good part of Newcomb's motivation for writing the mathematical series was financial. In 1881 he remarked to a friend that "An increase in salary from the Government being as good as hopeless, I have been for the last two years engaged in an enterprise for eking it out in the future by preparing a course of text books in mathematics."[47]

Algebra for Schools and Colleges appeared in June 1881, followed by *Elements of Geometry* in August. By 1887 "Newcomb's Mathematical Series" from Henry Holt and Company also contained a more elementary algebra book, two textbooks on trigonometry, an analytic geometry book, and a book on differential and integral calculus. In addition, the logarithmic tables included with the trigonometry were sold separately, a book combining chapters from the geometry and the first trigonometry book was marketed, and "keys" were sold for the two algebra books, containing problem solutions and more detailed explanations of fine points for the aid of teachers. All but one of the series went through multiple editions, with the original algebra going through twelve editions.[48]

---

[45] Henry Holt to Newcomb, Nov. 8, 1879, SNP, Box 27. Strikeouts and orthography as in original. The astronomy books mentioned are Simon Newcomb, *Popular Astronomy* (New York: Harper and Brothers, 1878). R. S. Ball, *Astronomy*, specially revised for America by S. Newcomb (New York, H. Holt & Co., 1878); and Simon Newcomb and Edward S. Holden, *Astronomy for Schools and Colleges* (New York, Henry Holt & Co., 1879).

[46] Raymond Clare Archibald, "Simon Newcomb 1835–1909, Bibliography of his Life and Work," *National Academy of Sciences Memoirs* 17 (1924): 55.

[47] Newcomb to Alexander Agassiz, May 15, 1881, SNP, Box 14.

[48] Archibald, "Newcomb Bibliography," 55–56.

3.3. NEWCOMB AS TEXTBOOK WRITER AND EDUCATIONAL THEORIST 69

Newcomb used these books to work out in more detail some pedagogic ideas that had appealed to him for some time, in particular the notion that abstract concepts should be introduced to students gradually by means of more concrete examples. This is especially clear in the first book in the series, *Algebra for Schools and Colleges*, in the preface of which Newcomb explicitly expounded this position.[49] The simplest instance in this book is Newcomb's introduction of positive and negative whole numbers by depicting them as arrayed on a line: zero is in the middle, with the positive numbers increasing indefinitely to the right, and the negative numbers decreasing indefinitely to the left. This approach has become so accepted as to seem inevitable, but it was not standard in the middle of the nineteenth century. Indeed, algebra textbooks were especially likely to avoid any admixture of geometric reasoning, however simple. For example, a survey of nineteenth century textbooks shows that the standard method of root extraction was commonly justified by geometry in arithmetic textbooks, but almost never so justified in algebra textbooks.[50]

By the 1880s Newcomb was not alone in using the number line in an algebra textbook; his close contemporary G. A. Wentworth was also an advocate. But Newcomb's use was more emphatic and self-conscious. After his introduction of the number line he advised in a footnote: "The student should copy this scale of numbers, and have it before him in studying the present chapter." He went on to explain the elementary operations of addition, subtraction, multiplication and division in terms of the number line, and continued to make occasional appeals to it throughout the text. Furthermore, to those who might object that to use geometry to assist algebraic understanding was to mark oneself as mathematically naive, Newcomb noted in the preface that precisely such geometrical aids had been a prerequisite to the great advances made by mathematicians in understanding "the complicated relations of imaginary quantities." He was referring here to the study of quantities involving the square root of -1, often referred to as imaginary numbers. The representation of such quantities along an axis perpendicular to the ordinary number line, (the whole system eventually referred to as the complex plane) had facilitated the enormous expansion

---

[49]Newcomb, *Algebra for Schools and Colleges*, 2d ed. (New York: Holt, 1881), iii.

[50]See, for example, one of the bestsellers of the period, Joseph Ray, *Primary elements of Algebra, For Schools and Academies* (Cincinnati: Wilson, Hinkle & Co., 1866): 7–12. Another example is D. H. Hill, *Elements of Algebra* (Philadelphia: J. B. Lippincott & Co., 1861), 13–17. In these books no appeal whatsoever is made to geometry in introducing numbers. See also Peggy A. Kidwell, Amy Ackerberg-Hastings, and David Lindsay Roberts, *Tools of American Mathematics Teaching, 1800–2000* (Baltimore: Johns Hopkins University Press, 2008), 134.

of research in this area in the nineteenth century, toward what would become known as the theory of complex variables, or complex function theory.[51]

In the preface to the algebra book Newcomb also took the opportunity to reiterate another favorite theme: "the backward state of mathematical instruction in this country, as compared with the continent of Europe." The problem, Newcomb claimed, was that too few Americans appreciated that "all mathematical conceptions require time to become engrafted upon the mind." That is, American mathematical instruction often tried to cover too much material too rapidly, not allowing time to prepare the ground for the most difficult concepts. This argument led to the happy conclusion that the situation could be readily assuaged by individual teachers. Newcomb offered to help these teachers by providing a book that followed "the French and German plan of teaching algebra in a broader way, and of introducing the more advanced conceptions at the earliest practicable period in the course." For example, Newcomb was keen to introduce algebra students to the concept of limit, although it would not be much utilized until these students took a calculus course. In the preface Newcomb went on to explain his advocacy of "subdividing each subject as minutely as possible," "thoroughness ... rather than multiplicity," and avoidance of excessive drill: "it is better that the student should go on rather than expend time in doing what it is certain that he can do."[52] This last would prove to be a sticking point with teachers, as Newcomb would soon find out.

The many editions, revisions, and rearrangements of the texts in Newcomb's Mathematical Series bespeaks an aggressive effort by Holt to market the series to secondary schools and colleges; nevertheless, the sales were not very great. The bestseller was the separate volume of logarithmic tables, with 12,258 copies printed as of 1921; the twelve editions of *Algebra for Schools and Colleges* yielded only 8,800 copies; *Elements of Geometry* sold only 3,902 copies. In comparison, more than 1,000,000 copies of G. A. Wentworth's *Elements of Algebra* were issued between 1881 and 1900, and the same author's *Elements of Geometry* sold over 500,000 copies.[53]

---

[51] See Morris Kline, *Mathematical Thought from Ancient to Modern Times* (New York: Oxford University Press, 1972), 628–632. Newcomb's comments on the number line are found in *Algebra for Schools and Colleges*, 2d ed., 5, 42–43, iii. Wentworth's use of the number line can be found in G. A. Wentworth, *Elements of Algebra* (Boston: Ginn, Heath, & Co., 1881), 5–7.

[52] Newcomb, *Algebra for Schools and Colleges*, 2d ed., iv–vi, 358–367.

[53] The Wentworth figures are from *NCAB*, 10:106. The Newcomb sales figures are from Archibald, "Newcomb Bibliography," 55–56.

3.3. NEWCOMB AS TEXTBOOK WRITER AND EDUCATIONAL THEORIST

Despite all the effort by Holt, it was extremely difficult for a newcomer to achieve great success in the textbook market at that time. The burgeoning school population had created a superheated competitive environment, with many losers and only a few clear winners. It has been estimated, for example, that at least 68 algebra textbooks were being offered for sale in 1892. New firms were entering and exiting the textbook business at a rapid rate. Complaints of corruption in book selection decisions were widespread. The chaos of this industry brought forth strategies to bring it under control familiar to students of other aspects of American economic history in the late nineteenth century: formation of trade associations, mergers, government intervention.[54] Holt, not being exclusively devoted to textbooks, was only a minor player in this game. The firm's general aim seems to have been to establish a niche at the high end of the market by commissioning prestigious scholars to write books appropriate for the elite secondary schools and the colleges. One of Holt's most successful ventures was *Psychology: Briefer Course* by William James, which sold 47,531 copies in the ten years following its publication in 1891.[55]

In Newcomb's case, his books were competing with some strongly entrenched earlier entries in the field, and with new books by authors having more experience with the realities of secondary school education. Holt, unsurprisingly, was anxious to provide what the market wanted, which meant, in the case of texts for "schools and colleges," being sensitive to prevailing teaching methods and to the content of the entrance examinations used by the colleges of the northeast, the region where Holt did most of its marketing at this time. A Holt representative wrote to Newcomb in 1883 with recommendations. The tone of the letter, seemingly deferential at first, soon became peremptory, reflecting the fact that Newcomb was a newcomer to the textbook field:

> We present for your consideration the following suggestions in regard to your "Algebra for Colleges." They come to us from teachers who, by long experience in fitting boys for college, have learned what kind of instruction and how much drill are necessary.
>
> Prominence, out of proportion to their importance in the science, must be given to such subjects as Factoring, Simultaneous Equations, and Quadratics, while there can scarcely be too much drill on these or any processes involved in preparatory

---

[54] John Tebbel, *A History of Book Publishing in the United States*, vol. II, *The Expansion of an Industry 1865–1919* (New York & London, R. R. Bowker, 1975), 561, 565, 571.

[55] Tebbel, 313.

work. Practical problems must be numerous, special attention being given to such classes of them as appear most often on college examination papers, e.g., the watch, and the hare and hound problems.[56]

There followed a comparison of Newcomb's book with that of Wentworth, a book "quite popular with teachers of all classes," showing the inadequacies of Newcomb's book with regard to the numbers of exercises on the topics mentioned. The letter concluded on the following note:

> The exercises in Part I can be made by some experienced teacher engaged in preparatory work after a study of sets of papers used at the entrance examinations of Yale, Harvard, Princeton and Amherst. Can you arrange to have it done at once, or shall we try to, charging the expense against royalty?[57]

Newcomb did not find these suggestions entirely pleasing, reflecting both a lack of enthusiasm for making "importance in the science" subservient to the tradition of drill, and dismay at having to take time to come up with appropriate problems. The preserved correspondence from Holt and Company indicates some testy negotiation on the details of the revisions, with Mr. Holt himself occasionally stepping in to calm the waters stirred up by his underlings, some of whom seemed to greatly enjoy maneuvering for advantage by emphasizing Newcomb's lack of teaching experience. Note the apparently casual but cutting use of alleged third-party testimony in the following:

> Touching the Geometry, [a teacher at St. Paul's School] said with a little remodeling here and there it could be made a most excellent class-book; but he questioned "whether Prof. Newcomb had the patience, or experience in teaching, which would enable him to do this properly. It is a pity the whole

---

[56]Signed "Bristol" on behalf of Henry Holt & Co. to Newcomb, Mar. 24, 1883, SNP, Box 27. An example of a watch problem: "At what time are the hands of a watch together between 3 and 4?" G. A. Wentworth, *Elements of Algebra* (Boston: Ginn, Heath, & Co., 1881), 145. An example of a hare and hound problem: "A hare takes 4 leaps to a greyhound's 3; but 2 of the greyhound's leaps are equivalent to 3 of the hare's. The hare has a start of 50 leaps. How many leaps must the greyhound take to catch the hare?" Ibid., 146. Such problems were "practical" only in a very specialized sense.

[57]"Bristol" on behalf of Holt & Co. to Newcomb, Mar. 24, 1883, SNP, Box 27. G. A. Wentworth (1835–1906), a student at Harvard College about the time that Newcomb was studying at the Lawrence Scientific School, spent more than 30 years as head of the mathematics department at Phillips Exeter Academy. He wrote numerous mathematics textbooks. See David Eugene Smith, "Wentworth, George Albert," *DAB*, 10:655–66.

3.3. NEWCOMB AS TEXTBOOK WRITER AND EDUCATIONAL THEORIST   73

series could not have gone through the hands of an experienced teacher." A propos: Has the shorter Trigonometry been through such hands?[58]

Newcomb did ultimately make many of the recommended changes. With regard to *Algebra for Schools and Colleges* in particular, a comparison of the second edition of 1881 with the sixth edition of 1888 strongly suggests that all changes that Newcomb made were confined to the very beginning and the very end of the book, reflecting the financial incentives for preserving unchanged as many as possible of the original plates. In the preface to the second edition Newcomb had offered an unapologetic defense of the lack of certain topics (greatest common divisor of polynomials, square roots of binomial surds, and Sturm's theorem): these topics did not "advance the student's conception of algebra," and "in studying them, power is expended which can be devoted to more profitable objects." These words are lacking from the sixth edition. In their place, Newcomb explained that one of the previously missing topics (greatest common divisor of polynomials) was in fact worthy of inclusion, but had been "postponed to what the author considers its proper place," (conveniently in the last chapter). "It has, however, been presented in such a form that it can be taught to pupils preparing for colleges where it is still required for admission." Newcomb was clearly responding to the college admission issue raised by the Holt critique, but the phrase "still required," intimates that he saw the college policies in question as out of date and thankfully in decline.[59]

The most substantial revision in the later edition of Newcomb's algebra was a new 85 page appendix of "Supplementary Exercises," "partly original and partly selected from the best recent German collections of problems," including five pages of factoring problems and thirteen pages of quadratics. Whether by inadvertence or by design, none of the problems appears to involve either watches or hares and hounds.[60]

---

[58]Henry Holt & Co. to Newcomb, May 10, 1884.

[59]I have compared Newcomb, *Algebra for Schools and Colleges*, 2d ed., v. with Newcomb, *Algebra for Schools and Colleges*, 6th ed. (New York: Holt, 1888), v. I have not been able to find a copy of the first edition of this text. Since the second edition followed the first by only three months (June versus September 1881), it is assumed here that changes between the two were not major. See Archibald, "Newcomb Bibliography," 55. The significant point is that the second edition appeared before the Holt critique cited above, and the sixth edition appeared after this critique.

[60]Newcomb, *Algebra for Schools and Colleges*, 6th ed., 459–545. The quotation is from page 460.

Other suggestions for improving the "Mathematical Series" continued to come to Newcomb from his publisher during the 1880s. It was hinted, for example, that Newcomb might do well to add arithmetic textbooks, especially to meet the needs of "the elementary schools of the barbarous regions," which seem to have included any part of the United States outside of the northeast.[61] But Newcomb had little interest in arithmetic texts of any variety. By the middle of the nineteenth century textbook writers often offered arithmetic in several levels of sophistication. Charles Davies, for instance, in 1847 had three different arithmetics in print: *First Lessons in Arithmetic*, *Arithmetic*, and *University Arithmetic*. The first was "designed for beginners," and the second "for the use of schools and academies." The last, also for "our advanced schools and academies ," claimed the more exalted subtitle of "the science of numbers," by virtue of its more detailed treatment of topics, and moreover had been prepared "to adapt it to the business wants of the country," by including "articles on Weights and Measures, foreign and domestic; on Banking, Bank Discount; Interest, Coins and Currency, Exchanges, Book-keeping, &c."[62]

It was precisely such volumes as this last which Newcomb found especially aggravating, as he expressed to Henry Holt in 1884:

> I have not the nerve to even attempt the infliction upon the youthful mind of the so-called University Arithmetic filled with all kinds of business problems, mostly incomprehensible to a man not in business, the production of which ought to be prohibited on the penalty of a fine and imprisonment.[63]

To Newcomb such arithmetic books seemed to be perversely designed to obscure the power of abstraction in mathematics; instead of emphasizing general principles, these books wallowed in a multiplicity of special techniques and terminology.[64] As we have seen, other educators shared Newcomb's disdain for common practices in arithmetic instruction, and his preference for the superior unifying character of algebra. Such ideas would be emphasized in the Report of the Committee of Ten.

---

[61]Henry Holt & Co. to Newcomb, Nov. 3, 1883, SNP, Box 27.

[62]Charles Davies, *The University Arithmetic Embracing the Science of Numbers and Their Numerous Applications* (New York: A. S. Barnes & Co., 1847), v–vii.

[63]Newcomb to Henry Holt, Feb. 12, 1884, letterpress copy in SNP, Box 6.

[64]Newcomb expounded at greater length on his dislike of "university arithmetics" on pages 13–15 of an unpublished, undated manuscript entitled "Mathematical Education," SNP, Box 111.

3.3. NEWCOMB AS TEXTBOOK WRITER AND EDUCATIONAL THEORIST

Given these views of Newcomb, it is thus somewhat surprising that he should also have turned down a more innovative proposal from Henry Holt in 1885. In September 1884 the president of the Massachusetts Institute of Technology, Holt's great friend Francis Amasa Walker, read a paper before the American Social Science Association on industrial education. The paper was published in the *Journal of Social Science* soon after.[65] Holt wrote to Newcomb of his interest in Walker's "ideas about introducing young people to some simple algebraic and geometrical ideas before they have got through an elaborate arithmetic."[66] As Holt was aware, Walker had already "tried to induce Newcomb to make a series of progressive books in mathematics, progressing according to the difficulty of the subject, without regard to the distinctions between arithmetic, algebra, geometry, etc. But I couldn't make him budge."[67] Holt nevertheless insisted on pressing Newcomb further, proposing that if a series was not practical yet

> it might be worth while to try a flyer on a single book to be called perhaps an "introduction to mathematics," which would be supposed to follow the little arithmetic in which children learn their multiplication table etc. It should give them the rationale of notation, addition, subtraction, multiplication and division, and then at once produce "x" and base the farther solutions of problems on that symbol with the simple equations. It should also, I think, introduce a few elementary geometrical conceptions in the way, not only as is usual in arithmetics, of the extraction of the roots, but of the mensuration of surfaces and solids; and perhaps some simple notion of triangulation applied to the measurement of distances.[68]

These ideas of Holt and Walker strongly anticipated the calls for unifying mathematics, and for the breaking down of arbitrary boundaries within mathematics, which would be made by later generations of mathematical educators, notably E. H. Moore. All evidence suggests that these ideas were likewise attractive to Newcomb, and yet he turned down the Holt proposal: "The book which you propose as an introduction to mathematics would undoubtedly be

---

[65] Francis Amasa Walker, "Industrial Education," *Journal of Social Science* no. 19 (1884): 117–131.

[66] Henry Holt to Newcomb, Feb. 6, 1885, SNP, Box 27.

[67] Walker to Holt, Dec. 15, 1884. Quoted in James Phinney Munroe, *A Life of Francis Amasa Walker* (New York: Holt, 1923), 282.

[68] Henry Holt to Newcomb, Feb. 6, 1885, SNP, Box 27. Underlining in original.

a good thing from my point of view, but I doubt if it would be from yours." Newcomb went on to explain two features of the national educational scene that would in his view tell against the favorable reception of the book proposed by Holt: "the ironclad character of the curriculum in all our schools," and the baneful influence of leading educational theorist W. T. Harris, whose Hegelian notions Newcomb suspected would be "opposed to any of our modern ideas of teaching."[69]

There is no reason to believe that Newcomb was not being entirely sincere in these remarks. He certainly had been dismayed to discover how much conservatism was present in the nation's educational system; indeed, it was Holt's representatives who had most forcefully conveyed this to him. But some additional factors were likely influencing Newcomb here, and further interpretation is in order.

It must be recalled that by 1884–1885, when Walker and Holt were pressing Newcomb, he had taken on the leadership of the Johns Hopkins mathematics department. This likely affected Newcomb's attitude toward textbook writing in at least three ways. First, his Hopkins salary would have diminished his financial motivation to write more textbooks, a venture which was proving less than spectacularly remunerative. Second, his Hopkins duties would have diminished the time available for him to work on textbooks. Third, his Hopkins duties would have pushed Newcomb's attention toward more advanced topics rather than elementary textbooks, however innovative. In conformance with these points one notes that all the volumes of "Newcomb's Mathematical Course" with one exception had been written prior to Newcomb taking over at Hopkins in the fall of 1884, and that this one exception was the most advanced book in the series, *Essentials of the Differential and Integral Calculus*.[70]

Moreover, all of Newcomb's textbooks, while they contained innovative treatments of some topics (and innovative pushing forward of some topics at the expense of other excised topics and problems) were not remarkable in their basic structure. The books proposed by Walker and Holt would in contrast have been more truly new, requiring considerably more concentrated effort to prepare. Perhaps the Newcomb of the 1870s, more innocent about the educational world and less encumbered with other obligations, would have leapt at the chance to write the innovative books advocated by Walker and Holt. But

---

[69] Newcomb to Holt, Feb. 7, 1885, letterpress copy in SNP, Box 6. Harris (1835–1906), superintendent of schools in St. Louis at the time Newcomb was writing this letter, became United States Commissioner for Education in 1889.

[70] Archibald, "Newcomb Bibliography," 55–56.

for the Newcomb of the 1880s there were clearly limits to his commitment to mathematical pedagogy.

It may also be that Newcomb had come to the conclusion that a man of his elite stature should be writing not textbooks but more theoretical works, which the textbook writers (preparatory school teachers like Wentworth) could then take as guidance. It is certainly a fact that in the late 1880s Newcomb's efforts on behalf of mathematical education began to take a more theoretical turn. In 1887 T. H. Safford, an old friend of Newcomb from the Cambridge *Nautical Almanac* of the 1850s, published a short monograph entitled *Mathematical Teaching and Its Modern Methods*.[71] Newcomb, who had seen this work in proof, wrote to the publisher, D. C. Heath and Company, that he had been encouraged "by a party interested in the subject" to write a similar book. At first he had found the Safford book quite satisfactory, but after further consideration Newcomb had concluded that there was "much to be added in the way of discussing the methods of teaching elementary mathematics," and he therefore inquired whether Heath would be interested in publishing a second book on the subject:

> My exposition would differ from that of Professor Safford's in giving some results of experiment and observation and going more fully into the practical details of different methods of teaching; especially the comparative merits of the lecture system, the textbook, and recitation systems.[72]

Although this project did not materialize, and in particular nothing ever seems to have come of Newcomb's proposal to abundantly record "results of experiment and observation" on teaching methods, Newcomb did putter about with expressing his ideas on the subject over the next few years. The unpublished manuscript entitled "Mathematical Education" appears to be from the period, containing as it does discussions of issues raised in the correspondence with Heath and Holt, and passages very similar to those that would appear in the articles Newcomb published in the *Educational Review* in 1892–1893.

This manuscript began with an attack on "the popular demand for a so called practical education," forcefully proclaiming the ultimate utility of general mental training, one example of which was the amplification of knowledge to be gained by understanding mathematical abstractions. After decrying the

---

[71]T. H. Safford, *Mathematical Teaching and Its Modern Methods* (Boston: D. C. Heath & Co., 1887).

[72]Newcomb to [D. C. Heath & Co.], Mar. [?], 1887, letterpress copy in SNP, Box 6.

popularity of the "University Arithmetics" as one example of the common misapprehension of what constituted true practicality, Newcomb noted that the old epistemological debate on the relative importance of intuition versus experience in gaining new knowledge, most recently exemplified by William Whewell and John Stuart Mill, had taken a decisive turn in favor of experience with the discovery of non-euclidean geometries earlier in the century. The mathematicians had shown that Euclid's parallel postulate was indeed an assumption, not a truth about the world which could be divined by pure reason. To discover whether the parallel postulate holds true in the actually existing world therefore required experimental test: for example, measurement of triangles to see if their interior angles add up to exactly 180 degrees, a property which derives from the parallel postulate. From this Newcomb drew the conclusion that mathematical education must begin with experimental manipulation, especially with measurement of concrete objects.[73]

Furthermore, in complete accordance with Walker's suggestion, Newcomb advocated ignoring the usual distinctions among subjects found in the mathematical curriculum, especially for the youngest students: "The first step in mathematics should not be confined to arithmetic, algebra or geometry but should include combination of all, designed simply to train the young pupil in mathematical thought and language." Newcomb then demonstrated how elementary arithmetic could be made more concrete by using straight line segments to represent quantities, much as he had done in his algebra book, and as he would again encourage in his 1892 *Educational Review* article.[74]

With regard to geometry instruction Newcomb suggested that more attention be given to such informal but practical exercises as dividing triangles into two equal parts by eye and estimating the number of degrees in an angle or the ratio of two line segments. This was an approach he had advocated in his geometry textbook. In both writings he anticipated objections by Euclidean purists, but in "Mathematical Education" he offered more substantial philosophical support by appealing again to the non-euclidean revolution which had vanquished the notion that "geometrical ideas are innate in the mind." In Newcomb's view the earliest geometric teaching should concentrate not on reasoning, nor on simply recounting facts about geometric objects, but rather should rely heavily on encouraging students to actively imagine how geometric figures would look after various transformations. Just as young children acquire a language not by formal instruction in grammar or vocabulary, so too could a

---

[73] Newcomb, "Mathematical Education," 9, 13, 17–23.
[74] Ibid., 27–32.

novice geometry student acquire knowledge simply by the activity of working with geometric objects.[75]

Newcomb also touched on more advanced mathematical topics in this manuscript. Expanding upon the thesis he had proposed in his algebra textbook, he attacked the "English speaking" practice of postponing the introduction of crucial concepts. Students of calculus were utterly mystified when encountering concepts they should have been prepared for much earlier, with the result that "the average student a month after he has finished the calculus finds absolutely nothing left of it in his mind." Newcomb found very doubtful the common belief that such a student "has nevertheless undergone a valuable process of mental discipline."[76]

Newcomb concluded his manuscript with a discussion of "the relative advantages of the oral to the text book system of instruction," possibly a summary of points he had intended to include in his unwritten monograph. He found the German system of oral instruction in elementary mathematics preferable to the American tendency to rely on textbooks. He also made the more novel suggestion that in advanced mathematics Americans relied too much on oral presentations (i.e., lectures), and "have a tendency to make a less use of text books than they really deserve."[77] It is hard to resist a suspicion of special pleading here by the busy Newcomb, eager to reduce his lecturing burden. Hugh Hawkins reports that some of Newcomb's Hopkins students were not greatly impressed by him as a teacher.[78]

In 1892 Newcomb published the first part of a two-part essay on "The Teaching of Mathematics," in the *Educational Review*. Once again Newcomb proclaimed the importance of "embodying mathematical ideas in a concrete form." In contrast to his unpublished manuscript he did not support this with any excursions into philosophy or references to non-euclidean geometries, apparently judging his audience to be more susceptible to more elementary social and psychological observations. Indeed, his initial argument was a blatant appeal to race prejudice, being the claim that "the lower orders will reason about concrete things, even when they could not comprehend an abstract statement," as illustrated by an anecdote about African natives unable to count past six while at the same time being sharp and accurate participants in financial transactions. Newcomb then picked up the point he had made in his algebra textbook, that

---

[75]Ibid., 33–37, 39–42.
[76]Ibid., 47.
[77]Ibid., 49–53.
[78]Hawkins, *Pioneer*, 137.

use of visualization was not a sign of mathematical ineptitude, but had been employed by luminaries such as Cauchy and Gauss: "If the greatest mathematical minds feel such an aid to be necessary to their work, why should not a corresponding aid be offered to the farmer's boy ... ?" Newcomb proceeded to explain his proposal for a "sensible arithmetic" using lines to represent quantities, just as in "Mathematical Education" but with more detail. Newcomb again expressed his distaste for the complicated business problems "so prominent a feature of our advanced arithmetics," and lamented "the prejudices of the so-called practical man against everything scientific in education." As he had in "Mathematical Education" Newcomb further proposed a "sensible geometry," emphasizing informal estimation and drawing exercises. Algebra he promoted as "a kind of language."[79]

The second part of Newcomb's essay, on the teaching of more advanced mathematics, was published in 1893. This he began with his familiar theme of the backwardness of the United States in mathematics: the first 70 years of the nineteenth century had produced only one important American mathematician, his old teacher Benjamin Peirce; the next 20 years had seen "a few American students ... now happily coming to the front," but still Europe continued to dominate. The diagnosis offered in this essay was his most specific yet: calculus, the first "advanced" mathematics course encountered by students, was not being properly taught in America. In particular, there was inordinate confusion surrounding the basis of the subject. There seemed to be three competing methods for laying the foundation of the subject: the method of limits, the method of infinitesimals, and the method of rates.[80] Newcomb, however, was firm in the belief that there was only one true foundation, namely the method of limits, the fundamental notion being as he had defined it in his calculus textbook:

> The **limit** of a variable quantity X is a quantity L, which we conceive X to approach in such a way that the difference $L-X$ becomes less than any quantity we can name, but which we do not conceive X to *reach*.[81]

The other two alleged foundations were only valid inasmuch as they were understood to be based on limits: "an infinitesimal quantity is simply one which

---

[79]Simon Newcomb, "The Teaching of Mathematics (I): Elementary Subjects," *Educational Review* 4 (Oct. 1892): 278, 280, 283–285.

[80]Simon Newcomb, "Mathematical Teaching (II)," *Educational Review* 6 (Nov. 1893): 332–335.

[81]Newcomb, *Calculus*, 17. Emphases in original.

3.3. NEWCOMB AS TEXTBOOK WRITER AND EDUCATIONAL THEORIST

approaches zero as a limit"; and "a rate is nothing but a disguised differential coefficient of a quantity with respect to the time," that is to say, the limit of the ratio of the change in the quantity to the change in the time.[82]

The approach to limits and infinitesimals that Newcomb was advocating here was essentially that of the French mathematician Augustin Cauchy (1789–1857), although he claimed a much more ancient provenance, asserting that it was "as old as Euclid."[83] In the Cauchy approach, an infinitesimal was no longer a tiny fixed number with certain convenient properties (considered paradoxical by some observers), as it had been for many mathematicians of the eighteenth century, but only a shorthand to designate a dependent variable that could be made arbitrarily small by appropriate choice of the independent variable. The Newcomb-Cauchy approach may be compared with that of the calculus text by Charles Davies that had annoyed Newcomb as a young man. Davies, while not using the word "infinitesimal" in the eighteenth century sense, reasoned very much in that mode. In regard to an increment h of a variable x Davies wrote:

> We have represented by dx the *last* value of h, and this value forms no appreciable part of h or x. For if it did it might be diminished without becoming 0, and therefore would not be the *last* value of h.[84]

Such reasoning was unsatisfactory to Newcomb and many of his generation.

In his 1893 essay Newcomb went on to explain that the pedagogic confusion among limits, infinitesimals, and rates had been tolerated in America because,

> Our mathematical teaching has not yet recovered from the stage at which the study of the calculus was looked upon simply as a mental discipline, and the student himself never expected to be able to apply it intelligently.

He insisted, on the contrary, that the calculus was preeminently a practical subject, a point he emphasized by reminding the reader of the triumphs of the method in mechanics, culminating in the prediction of planetary orbits.[85] At almost precisely the same time, Newcomb would carry such views into his chairmanship of the Mathematics Conference of the Committee of Ten.

---

[82] Newcomb, "Mathematical Teaching (II)," 335–336.
[83] Ibid., 335.
[84] Davies, 16. Italics in original.
[85] Newcomb, "Mathematical Teaching (II)," 338–340.

CHAPTER 4

The Committee of Ten and Its Mathematics Conference

The Report of the Committee on Secondary School Studies of 1893, popularly known as the Committee of Ten Report, is a well trod piece of historiographical ground for commentators on education in the United States, and indeed continues to be a useful foil for polemical writing on education down to the present day, but the specific implications for mathematics have received much less attention. The larger significance of the report for mathematics is best understood when placed in context with other contemporaneous events, notably the rapid strengthening of the American mathematical research community.

The Mathematics Conference of the Committee of Ten was dominated by college and university mathematicians whose views were often in accord with leading general educators of the time. This accord was, however, unstable; both the mathematicians and the general educators would soon evolve in ways that would prove less compatible. The proposals of the Mathematics Conference were largely synonymous with those of its chairman, Simon Newcomb, but Newcomb's generation of mathematicians was in the process of passing away, to be supplanted by a new generation committed to forming a vibrant research community devoted especially to pure mathematics. The general educators meanwhile would soon be fixated on the problems of coping with the massive increase in the high school population.

We approach the Mathematics Conference of the Committee of Ten by first summarizing how the Committee of Ten came to be and its initial intent. Then we closely examine the Mathematics Conference membership and its report. Finally, we survey the response to the report, both in its immediate aftermath and in subsequent years. We will find that although the Committee of Ten has sometimes been described as representing one or another pole of educational debate, it in fact exemplified a sincere, though confused, moderation. Much of the confusion centered around the much discussed issue of student differentiation. The Ten sought to simultaneously provide for students preparing for college, for technical schools, and for those using high school as a finishing school, but wished to avoid distinguishing these groups of students before the fact. The Mathematics Conference, without drawing much attention, deviated somewhat on this issue; in some limited circumstances the conference was willing to allow some students to follow differing paths in high school depending on their career goals.

The Mathematics Conference also sought to maintain the well established divisions among arithmetic, algebra, and geometry, while placing greater emphasis on inductive and concrete methods, thus laying the groundwork for more abstract concepts in later school years. The value of mathematics as a mental discipline was not denied, but it was also noted that such value could be exaggerated, and that it needed to be balanced with the cultivation of student interest. Such balancing would not have been so much emphasized by earlier generations of educators.

4.1. Immediate Origins and Basic Facts

In July 1892 the National Educational Association (NEA)[1] appointed ten educators as a Committee on Secondary School Studies to inquire into and make recommendations regarding the relationship between secondary school curricula and college entrance requirements. It is important to note that the NEA at this time had not yet acquired the characteristics of a union representing rank and file school teachers, but was instead primarily a forum for elite educational executives (college presidents, public school superintendents and principals, headmasters of private schools), joined by some younger members ambitious to acquire such responsibilities.

An especially pressing problem exercising these educational leaders in the late 1880s, and the problem most immediately responsible for the formation of a special committee, was the issue of uniformity, or the lack thereof. Discussion of this issue ranged widely, but usually centered on the two complementary problems of lack of uniformity in college entrance requirements and lack of uniformity in high school curricula. Charles Eliot, who had often called attention to these problems, was named chairman of the Committee of Ten. Among the other members were William T. Harris, United States Commissioner of

---

[1]The National Educational Association became the National Education Association in 1908. See Edward Krug, *Salient Dates in American Education, 1635–1964* (New York: Harper & Row, 1966), 80.

Education; Oscar D. Robinson, Principal of the Albany High School; James C. Mackenzie, Head Master of the Lawrenceville School; and James H. Baker, President of the University of Colorado.[2]

The Ten met as a group for three days in November 1892, in New York City. The original issue of uniformity had by this time become more specifically focused on the aim of prescribing desirable secondary school curricula, one of several features of the report that caused many to interpret it as an effort by college educators to dictate to the secondary schools. The Committee agreed, guided by statistics gathered by Eliot, that the secondary school curriculum should consist of nine principal subjects: 1. Latin; 2. Greek; 3. English; 4. Other Modern Languages; 5. Mathematics; 6. Physics, Astronomy, and Chemistry; 7. Natural History (Biology, including Botany, Zoology, and Physiology); 8. History, Civil Government, and Political Economy; 9. Geography (Physical Geography, Geology, and Meteorology). The Ten then appointed a separate conference of ten members for each of these subject areas, and drew up a set of questions to guide the deliberations of these conferences. Each conference was urged to consider when in a student's career its subject should be studied and with how much concentration; what topics within the subject should be introduced, and when; and what were the best methods for teaching these topics. The conferences were also invited to make recommendations regarding college admission requirements for the subject. Probably the most provocative question posed, both then and in subsequent years, concerned the issue which came to be referred to as differentiation:

> Should the subject be treated differently for pupils who are going to college, for those going to a scientific school, and for those who, presumably, are going to neither?

In using the word "presumably" the Committee should not be taken as doubting the existence of students not bound past secondary school. Rather, the Committee was aware of students who entered secondary school presuming they would not seek higher education, but who later discovered a desire or need for additional learning. If such students had embarked on a non-preparatory school program their college aspirations would be greatly impeded. The solution to this problem envisioned by the Committee, especially by Eliot, permeated the entire report: schools and colleges should be so brought into coordination that

---

[2]*Report of the Committee of Ten on Secondary School Studies*, reprinted in Theodore R. Sizer, *Secondary Schools at the Turn of the Century* (New Haven: Yale University Press, 1964), 211. All references to the report will be to the reprint on pages 209 to 271 of the Sizer book.

college admission could be attained by "any youth who has passed creditably through a good secondary school course, no matter to what group of subjects he may have mainly devoted himself in the secondary school."[3]

The nine conferences met separately but simultaneously in December of 1892. Each conference had been instructed to prepare a report to be submitted to the Ten by April of 1893. The Ten were then to produce a summary, the whole to be unveiled in Chicago in July at the International Congress of Education held in conjunction with the World's Columbian Exposition. This plan had to be abandoned when two conferences, Natural History and Geography, failed to produce their reports until July. The Ten conferred by mail, met once more in person in November, and agreed on a final document in early December; all signed, but Baker insisted on appending a minority report as well. The Bureau of Education issued the full report in January 1894, with a title page bearing the date 1893. This volume contained the summary conclusions of the Ten, a minority report by Baker, and the nine conference reports. The report was widely distributed, with 30,000 copies sent out initially by the Bureau of Education, and 10,000 more eventually sold by the American Book Company.[4]

4.2. The Mathematics Conference and Its Report

Simon Newcomb, whose long acquaintance with Charles Eliot has been noted, was among those chosen for the Mathematics Conference. Although expressing some misgivings regarding his lack of contact with elementary teaching, Newcomb accepted the appointment on November 25, 1892.[5] The entire committee is listed in Table 4.1.[6]

It is clear that personal connections with the Ten played a significant role in selecting the Conference members. In addition to Newcomb's connection to Eliot, Byerly was a Harvard graduate and a Harvard professor (both under Eliot); Safford was a Harvard graduate of nearly the same vintage as Eliot, as well as a former colleague of Newcomb at the *Nautical Almanac* Office; Cajori was a one-time student of Newcomb at Johns Hopkins; and Patterson was on

---

[3]Edward A. Krug, *The Shaping of the American High School 1880–1920* (New York: Harper & Row, 1964), 46–52, 237. *Report of the Committee of Ten,* 212–213, 261.

[4]Krug, *High School,* 55–57, 67. Sizer, 148.

[5]Newcomb to Eliot, Nov. 25, 1892, letterpress copy in SNP, Box 6. Eliot immediately expressed his pleasure in Newcomb's acceptance. Eliot to Newcomb, Nov. 26, 1892, SNP, Box 21.

[6]*Report of Committee of Ten,* 217–218.

Prof. Simon Newcomb	Johns Hopkins University, Baltimore, Maryland
Prof. William E. Byerly	Harvard University, Cambridge, Massachusetts
Prof. Florian Cajori	Colorado College, Colorado Springs, Colorado
Arthur H. Cutler	Principal of a Private School for Boys, New York City
Prof. Henry B. Fine	College of New Jersey, Princeton, New Jersey
W. A. Greeson	Principal of the High School, Grand Rapids, Michigan
Andrew Ingraham	Swain Free School, New Bedford, Massachusetts
Prof. George D. Olds	Amherst College, Amherst, Massachusetts
James L. Patterson	Lawrenceville School, Lawrenceville, New Jersey
Prof. T. H. Safford	Williams College, Williamstown, Massachusetts

TABLE 4.1. The Mathematics Conference of the Committee of Ten

the faculty of the Lawrenceville School under headmaster James Mackenzie, one of the Ten.

None of this is to say that these gentlemen were lacking in appropriate credentials to serve on the Conference. We have seen that Newcomb had already been involved with mathematical pedagogy prior to the formation of the Committee of Ten, and in fact the Committee tapped him just as he was exhibiting particular interest in the subject. Safford, Cajori, and Fine had all written on educational topics in the five years immediately preceding the Conference.[7] The members of the Mathematics Conference met at Harvard University on December 28–30, 1892. The ever-busy Newcomb, who had originally agreed to

---

[7]T. H. Safford, *Mathematical Teaching and Its Modern Methods* (Boston: Heath, 1887); H. B. Fine, *The Number System of Algebra Treated Theoretically and Historically* (Boston: Leach, Shewell and Sanborn, 1891); Florian Cajori, *The Teaching and History of Mathematics in the United States* (Washington: Government Printing Office, 1890).

serve only on condition that the place of meeting be no further from Washington than New York City, attended only the first day's sessions. He was elected as chairman, Harvard host William Byerly was elected as vice chairman, and Arthur Cutler was chosen as secretary. Before Newcomb departed, the Conference agreed to consider the mathematics curriculum as divided into four major subtopics: arithmetic, algebra, inventional geometry, and formal geometry. Small subcommittees were chosen to prepare reports for each subtopic, with a final report to be constructed by Newcomb.[8]

It appears from Newcomb's correspondence that he was no mere figurehead, but exercised his prerogative to edit the final report. He wrote to both Safford and Fine explaining that he had altered the sections of the report on which they had worked. He apologized for this action, but expressed no doubt that what he had done was right and proper.[9] Such evidence as this does not indicate any genuinely troublesome dissent within the Conference. Indeed, from the available information it does not seem likely that the members of the Conference represented a very wide range of views on educational issues.

Nevertheless, there can be discerned within the backgrounds of the Conference members some intimations of fractures ahead within the growing population of mathematical educators. Newcomb and Safford were educated in an era when mathematical activity in America was in close alliance with astronomy, an era that was beginning to pass away as these two approached the end of their careers. These two and Cajori had wide-ranging concern with mathematical pedagogy at all levels, although with strongly entrenched assumptions regarding educational hierarchy. The eldest of the three, Newcomb, was able to dabble in pedagogy while pursuing an active research career, while the youngest, Cajori, was much more of a historical/pedagogic specialist; Cajori's approach would more and more be the only feasible option, but it would entail a discernible loss in prestige. Byerly, Fine, and Olds belonged to a younger generation of teacher-administrators, not ignorant of astronomy and the physical sciences, but in their efforts to routinize collegiate mathematics instruction they tended to promote the abstract core of pure mathematics much more than had many of their predecessors. In Fine especially can be seen the growing appeal and primacy of the pure mathematical research ethos. Although he himself was

---

[8]Newcomb to Eliot, Nov. 25, 1892, letterpress copy in SNP, Box 6. Cutler to Newcomb Jan. 2, 1893, in SNP, Box 20. All the other subject-matter conferences met during the same three days, in various locations around the northeast and middle west. See Sizer, 109, 214.

[9]See Safford to Newcomb, Mar. 16, 1893, in SNP, Box 38; and Newcomb to Fine, Apr. 26, 1893, letterpress copy in SNP, Box 6.

4.2. THE MATHEMATICS CONFERENCE AND ITS REPORT

not an important researcher he created conditions where research could thrive. Fine at Princeton, Moore at Chicago, and others would construct a mathematical community in which Safford's confident assertion from 1887 of the social advantages of applied mathematics would be found less and less persuasive:

> If he be one of those rare men who can solve the great problems of abstract mathematics, he will soon find himself obliged to live alone, with hardly any sympathy; and the few disciples he can get will, like himself, be in danger of becoming hopelessly unpractical. This will happen unless he takes a lively interest in some branch of natural science proper.[10]

All of this would have implications for the relationship between mathematicians and educational reform.

The report of the Mathematics Conference was thirteen pages in length, and was divided into five sections: a "General Statement of Conclusions," followed by reports on the four subtopic areas. Overall, the report well illustrates Edward Krug's later general assessment of the Committee of Ten era as a whole: "moderate revisionism."[11] The Conference made respectful allusions to traditional justifications for studying mathematics, while cautiously pushing for new emphases.

In the case of arithmetic teaching, the Conference declared that they sought "radical change," but in fact proposed little not already publicly endorsed by individual members of the Conference or of the Ten. In particular, the Conference demonstrated likely awareness of Charles Eliot's earlier critique of arithmetic by declaring that the subject ought to be "abridged and enriched."[12] This was to be accomplished by promoting proper over improper uses of abstraction, in a manner which Newcomb and Safford had been advocating individually, and by distinguishing true from spurious mental discipline, in line with Eliot.

The Conference's approach to "commercial arithmetic " is especially rich with connections to earlier pronouncements, and revealing of the evolution of attitudes among nineteenth-century educators. Newcomb and Safford had both expressed disapproval of arithmetic courses swollen with special topics from business and commerce, a distaste also evident in the Conference report. Here

---

[10]Safford, 26.

[11]Krug, *High School*, 190. All references to the report are from the reprint "Report of Mathematics Conference," in James K. Bidwell and Robert G. Clason, eds., *Readings in the History of Mathematics Education* (Washington, D.C.: NCTM, 1970), 129–141. I estimate the report contains approximately 4,000 words.

[12]"Report of Mathematics Conference," 131.

indeed was a region of arithmetic ripe for abridgement. In part, the Conference position reflected the recent concern with efficiency in education; commercial arithmetic was held to "waste valuable mental energy." But it is also not hard to detect the heritage of the 1828 Yale Report of Jeremiah Day, one of whose arguments against "mercantile" subjects in the colleges had been that "these can never be effectually learned except in the very circumstances in which they are to be practiced." Writing more than 60 years later, the Conference asserted that "until [the student] is brought into actual contact with the business itself, he can form no clear conception of what it all means."[13]

The Conference did feel obliged to acknowledge "the popular demand for a system of education which should be more practical and better suited to the demands of modern commercial and business life." It conceded that "it may well be" that students in "business colleges" would benefit from a steady diet of commercial arithmetic, and even that for high school students intent on business careers instead of college it "might well be" that they should branch off into "bookkeeping, and the technical parts of commercial arithmetic," while their fellow students delved further into algebra.[14] Such formulations were more diplomatic than the disdain for the whole notion of commercial arithmetic expressed individually by Newcomb and Safford. Safford had declared himself as follows:

> The "higher arithmetic" is quite useless; it is a disguised algebra, and one which is tied down to the American dollar as its main material. The moral effect of this, so far as it has any, is bad; the results, for those who are to go into business, either bad or none at all.[15]

Such assessments were not far beneath the surface rhetoric of the Conference report as well. It indulged in ironic praise of the "very excellent definitions" found in business arithmetic textbooks, and suggested that any student with a sound knowledge of fundamental mathematics, without special training, would readily be able to solve business problems when required. This was a case where abstraction should be sought and valued. Nor was the Conference at all taken with the idea that even if business mathematics was essentially useless its study

---

[13] "Report of Mathematics Conference," 133–134. The quote from the Yale Report is from the reprint in Theodore Rawson Crane, ed., *The Colleges and the Public, 1787–1862* (New York: Bureau of Publications, Teachers College, Columbia University, 1963), 91.

[14] "Report of Mathematics Conference," 133.

[15] T. H. Safford, "Instruction in Mathematics in the United States," *Bulletin of the New York Mathematical Society* 3 (1893): 5.

might still provide "valuable mental discipline." The Conference argument was very much in keeping with Eliot's position:

> While the Conference admits that, considered in itself, this discipline has a certain value, it feels that such a discipline is greatly inferior to that which may be gained by a different class of exercises, and bears the same relation to a really improving discipline that lifting exercises in an ill-ventilated room bear to games in the open air. The movements of a race horse afford a better model of improving exercise than those of the ox in a tread-mill.

Rather than wearying the student with difficult business problems, the Conference recommended "comparatively easy problems, involving interesting combinations of ideas." Although the Conference might here appear to have descended dangerously far down the slippery slope to the mathematical equivalent of reading French novels, they countered such an impression by also urging "more attention to facility and correctness in work," by maintaining that "quickness and accuracy in both oral and written work should be rigidly enforced," and by declaiming against "the vicious habit of slovenly expression."[16]

It is important to note that the Conference and its members were not expressing any distaste for business and commercial activities themselves. Newcomb was a staunch proponent of laissez-faire capitalism; in one of his economics books he offered the comforting assertion that "No man can accumulate a fortune except by benefiting his fellow men."[17] In contrast, Paul Hanus in 1894 sneered at the "imaginary financiering" cultivated in many arithmetic courses, which featured "a perpetual dealing with dollars and cents, as if they were the most important things in life." It would be far better, according to Hanus, to teach more geometry, which "contributes to the development of aesthetic and sympathetic interests in the works of man, and in man himself."[18] Such sentiments were foreign to the Conference. When Safford complained, as cited above, that arithmetic instruction was too "tied down to the American dollar," he was making a point about insufficient abstraction, not about insufficient aesthetic and sympathetic interest.

---

[16]"Report of Mathematics Conference," 133–134, 137, 139.

[17]Simon Newcomb, *A Plain Man's Talk on the Labor Question* (New York: Harper & Brothers, 1886), 188.

[18]Paul H. Hanus, "The Educational Value of Geometry for Grammar School Pupils," *The Public School Journal* 13 (Mar. 1894): 407.

With commercial arithmetic disposed of, the Conference proceeded to recommend some specific methods for improving arithmetic instruction. The attempt to strike a balance between the abstract and the concrete is notable here, with Simon Newcomb's hand clearly evident. The elementary operations were to be taught with "visible figures"; addition of numbers, for example, was to be conveyed by "joining lines together." Further, students should be brought to understand important "elementary theorems of arithmetic" by visual means:

> Thus, when the pupil comprehends clearly, by means of dots arranged in a rectangle, that three fives contain the same number of units as five threes, that is when he sees that the commutative law is true, then it may be expressed to him in the general form, $a \times b = b \times a$.[19]

The concept at issue here was very old, but distinguishing it by the special term "commutative" was a nineteenth-century innovation. Those unacquainted or unimpressed with the mid-century discoveries of new number systems, such as William Rowan Hamilton's quaternions, where $a \times b = b \times a$ is not necessarily true, found it hard to appreciate the attention given to commutativity, but it was increasingly emphasized by late-century mathematicians, along with the companion notions of associativity and distributivity. Still, the Conference likely felt itself to be on safely traditional ground, since as long ago as 1814 Jeremiah Day, while not using any variant of "commutative," had demonstrated that "the product of ba is the same as ab" by arrangement of 15 dots in a rectangle precisely as recommended by the Conference. Newcomb's algebra text of 1881 similarly appealed to this geometric proof, but elevated its status by calling it the "law of commutation." For Newcomb, as for Day, the primary point of distinguishing the concept was as a tool for simplification of arithmetic and algebraic expressions, certainly a respectably traditional part of school mathematics instruction. It is unlikely that the Conference report, with Newcomb as its final editor, was reaching toward anything more radical than this in its reference to the commutative law.[20]

But pure mathematicians were beginning to think of commutativity in a much more abstract fashion. One such mathematician was Conference member Henry Fine, who placed the commutative, associative and distributive laws at

---

[19] "Report of Mathematics Conference," 134–135. Compare Simon Newcomb, "The Teaching of Mathematics (I): Elementary Subjects," *Educational Review* 4 (Oct. 1892): 280–281.

[20] See Jeremiah Day, *An Introduction to Algebra* (New Haven: Howe & Deforest, 1814), 42; and Newcomb, *Algebra for School and Colleges*, 2nd ed. (New York: Holt, 1881), 38–39.

the forefront of his 1891 book *The Number System of Algebra*. Here the symbol manipulation familiar to school children was quite secondary to the overriding insight "that algebra is completely defined formally by the laws of combination to which its fundamental operations are subject."[21] Fine was thus aligning himself with the then recent movement among European mathematicians to attempt to develop all of mathematics on an axiomatic basis. From this point of view it was far more important to know that $3 \times 5 = 5 \times 3$ than to know that the product is 15; for pure mathematicians the charm (i.e. the generality) of the geometric proof of commutativity was precisely that one need not know what the product is. This held the potential for a radical departure from the old-time arithmetic.

The Conference placed emphasis on "giving the teaching a more concrete form." Further, "the method of teaching should be throughout objective ... the metric system should be taught in applications to actual measurements to be executed by the pupil himself; the measures and weights being actually shown to, and handled by, the pupils ... rules should be derived inductively." Geometric propositions were to be subjected to "frequent experimental tests." This set of ideas was most explicit in the Conference's distinction of "concrete geometry" as an area of the curriculum requiring special attention. "As early as possible" the student should "gain familiarity through the senses" with basic geometric shapes, and should engage in activities such as measuring, modeling, and drawing. This should then be extended to "experimental geometry," wherein the student would learn fundamental properties of geometric figures without full logical rigor. The Conference expressed confidence that this encouragement of geometric intuition would simultaneously satisfy three goals: it would "be of great value to all children," it would prepare selected students for later more rigorous mathematical work, and it would "help the elementary instruction in physics, if such is to be given."[22]

That is, the Conference recognized three roles for their field: as a disciplinary subject, as a research specialty, and as a tool for other fields. The Conference claimed that, thanks to the unifying power of concrete methods, there was no necessary conflict among these roles; all could be served, none need be neglected. In a few years E. H. Moore would extend this line of thought, making more explicit the professional anxieties at its root.

The Conference balanced conservative and reforming sentiment with regard to algebra. Consistent with the views of Charles Eliot, it recommended

[21] Fine, 6, 126–129.
[22] "Report of Mathematics Conference," 131, 134–136.

early exposure of arithmetic students to algebraic language and the shifting of some topics from arithmetic to algebra, where they could be more efficiently taught. But the formal study of algebra need not begin until the student was about 14 years old and should concentrate on manipulative skill; rigorous proofs of propositions such as the binomial theorem could be safely postponed until very late in the school course or even reserved for college. The Conference also recommended frequent oral exercises in algebra, in the manner of "mental arithmetic."[23] The original purpose of the latter, pioneered in the 1820s by Warren Colburn and others as a challenge to the dominance of rule-based instruction, had been to develop facility at solving relatively simple problems purely in the mind. It was soon found, however, that mental arithmetic could coexist perfectly well as the oral adjunct of an essentially rule-based system. Further, it had obvious attractions for proponents of mental discipline, who were naturally tempted to tax student capabilities with more and longer problems.[24] By the 1890s advocacy of mental arithmetic usually connoted a distinctly conservative approach to mathematics. One commentator commended the Conference for "demanding a return to the old-time rigid training in ready reckoning,—mental arithmetic."[25] Finally, the Conference made the recommendation that "Especial emphasis should be laid upon the fundamental nature of the equation."[26] Within a few years mathematicians such as E. H. Moore would recommend emphasis on a more abstract notion, the function.

Demonstrative geometry, according to the Conference, "should begin at the end of the first year's study of algebra," and would also build upon the foundation established in concrete geometry. Demonstrative geometry was the proper subject in which to stress exact definitions and logical reasoning. It was also here that the Conference was inclined to insist that mathematical instruction was not merely for training the minds of average students, but would especially benefit those few students of more unusual capabilities. Geometry, it was declared, "possesses remarkable qualifications for quickening and developing creative talent." Further, the attentive student of geometry would learn important lessons about scientific investigation in general. The Conference completed this

---

[23]Ibid., 133, 136.

[24]See Daniel Calhoun, *The Intelligence of a People* (Princeton: Princeton University Press, 1973), 102–104; Bidwell and Clason, 18, and Patricia Cline Cohen, *A Calculating People: The Spread of Numeracy in Early America* (Chicago: University of Chicago Press, 1982), 134–138.

[25]H. H. Seerley, "The Report of the Committee of Ten," *Education* 15 (1894): 240.

[26]"Report of Mathematics Conference," 137.

tilt toward the specialists by concluding the demonstrative geometry section, and the report as a whole, with a call for the introduction into the curriculum of "at least the elements of modern synthetic or projective geometry," which it lauded for illustrating "the esthetic quality which is the charm of the higher mathematics."[27]

4.3. Near Term Response to the Report

The Committee of Ten Report as a whole, as Edward Krug has noted, has been often credited with "that elusive something known as great influence." Ever since its publication it has been cited as the beginning or ending of one or another educational development. Certainly the Report's convenience to educators and historians as a periodization marker is unquestionable, the appearance of "1893" in titles alone being a trustworthy sign. Nor have mathematical educators shied from such usage:

> The development of the mathematical curriculum of the secondary schools is divided into two periods, one from 1893 to 1923 and the other from 1923 to 1940.[28]

In more recent years historians of education have become more wary of attributing great causal efficacy to artifacts such as committee reports.[29] The present writer is mindful of these cautions, but nevertheless finds that the report of the Mathematics Conference provides useful insight into the evolution of American mathematics education for some years after the time of its writing.

It is important to recall the distinction between the Committee of Ten itself and the nine subject conferences reporting to it. Since the Ten had asked each conference to independently recommend allotments of school time for its subject specialty, the issue of reconciling these allotments naturally arose. The Committee eventually produced a series of tables of subject offerings; beginning with a display of the raw numbers provided by the conferences, the Ten pushed and pulled to produce what they considered to be four realistic four-year secondary school "programmes," designated as Classical, Latin-Scientific, Modern Languages, and English. A graduate of any one of these programs, which varied primarily in foreign language requirements, was deemed worthy of admission to

---

[27]Ibid., 132, 140–141.

[28]E. R. Breslich, "Presenting the Report of the Joint Commission to the National Council of Teachers of Mathematics," *Mathematics Teacher* 33 (Apr. 1940): 147.

[29]See Herbert M. Kliebard, "Constructing a History of the American Curriculum," in *Handbook on Research on Curriculum* (American Educational Research Association, 1992), 162.

college. Eliot would have preferred a more unqualified embrace of the elective principle, but was constrained by the rest of the Committee. He considered the four suggested courses of study to be tentative only and not at all the ultimate recommendations of the Report, but they became the focus of criticism of many readers of the Report nonetheless.[30]

This criticism did not, for the most part, center on mathematics, which was treated respectfully throughout the Report, and in particular in the four suggested programs of study. Most contemporary observers apparently agreed with Theodore Sizer's later assessment that "mathematics suffered no retreat."[31] A traditional algebra-geometry sequence was required in the first three years of all four programs. Trigonometry and higher algebra were required in the senior year of the English program (this being the standard program for entry into a technical or scientific school such as the Massachusetts Institute of Technology), and were options in the other three programs. Advocates of the ancient languages, on the contrary, did feel they had suffered a retreat, and made their complaints known immediately. The Report also angered proponents of manual training, clamoring to gain acceptance in the academic world; the Committee had not even deemed them worthy of a conference. A more general complaint was the theme which became known as "college domination." The Report had called for allowing college admission to any graduate of "a good secondary school course." Eliot clearly considered this an eminently democratic proposal, but others saw it as a scheme to force secondary school students into a rigid classical curriculum. Thus psychologist and Clark University president G. Stanley Hall complained in 1901 that the Committee of Ten was responsible for a lamentable increase in the number of students taking Latin in the high schools. Finally, there was unhappiness regarding the Report's alleged tilt toward equivalence of studies. This was the issue over which Committee member Baker had dissented, blaming Eliot as the culprit.[32]

The first significant reaction to the details of the Mathematics Conference report was provided by the Committee of Ten's own summary of that report. Chairman Eliot, with the approval of the other nine, summarized in approximately 900 words what he evidently considered to be the most significant features of the mathematics report. Eliot did not fail to note where the Conference ratified his own ideas. He raised high the Conference's call for "radical change in the teaching of arithmetic," especially the notion that the arithmetic course

---

[30]Krug, *High School*, 58–62. *Report of Committee of Ten*, 255–56.
[31]Sizer, 142.
[32]*Report of Committee of Ten*, 261, 269. Krug, *High School*, 74–75, 81–84.

ought to be "abridged and enriched." Not content with the Conference's own declaration of unanimity on this subject, Eliot endeavored to bolster the point by exclaiming on how "various" were the backgrounds of the Conference members, a claim that must be judged exaggerated. Eliot further buttressed his own line of thinking by giving prominent mention to the Conference's recommendation that pupils of arithmetic should be enabled to use simple algebra. This was such a congenial proposal that he mentioned it twice. The Conference's disdain for "subjects which perplex and exhaust the pupil without affording any really valuable mental discipline " was also approvingly quoted. In the report on demonstrative geometry Eliot seized eagerly on the Conference's commendation of the disciplinary value of insisting that students be able to present geometrical demonstrations orally in complete and exact form. Here again Eliot seems to have felt that the Conference's point needed a bit of a boost. In summarizing a Conference paragraph of 141 words, in which "discipline" appeared once, unmodified, he employed 82 words, with "discipline" appearing three times, once prefaced by "admirable."[33]

It is striking that Eliot's summary made no mention at all of the Conference's discussion of commercial arithmetic. In part he may have considered this a mere detail concerning the abridgement of arithmetic, but he also had a more fundamental reason for not wishing to draw attention to this part of the report. The Conference, let it be recalled, though clearly of a mind to call for the excision of commercial arithmetic root and branch, had bowed to what it saw as "popular demand" by suggesting that those intending to go directly into business after high school might as well take commercial arithmetic rather than algebra in their last years of school. This concession to differentiation, grudging though it was, directly conflicted with the Committee's loud trumpeting of the alleged unanimity of its members, and the members of all the conferences, that there should be no distinctions in teaching those bound for college and those not. The Committee was pleased to observe that avoidance of differentiation would entail a "great simplification in secondary school programmes."[34]

---

[33]Compare "Report of Mathematics Conference," 139 (paragraph "4") with *Report of Committee of Ten*, 234–235 ("It insists...formally perfect."). See also Oscar Robinson, "The Work of the Committee of Ten," *School Review* 2 (1894): 367.

[34]"Report of Mathematics Conference," 133. *Report of Committee of Ten*, 226–227. The Mathematics Conference was the only one to propose differentiation. See Sizer, 118.

It appears likely that the Conference escaped being labeled traitorous on differentiation by the Committee in part because of the fundamental incoherence of the Committee's views. The problem may be observed in the following passage from the Report:

> Thus, for all pupils who study Latin, or history, or algebra, for example, the allotment of time and the method of instruction in a given school should be the same year by year. Not that all the pupils should pursue every subject for the same number of years; but so long as they do pursue it, they should all be treated alike.[35]

The first sentence describes a no-differentiation position which may be termed *weak:* every student taking, say, second year algebra should cover the same material and be instructed in the same way. This was apparently not very controversial; it likely accounts for Committee member Robinson's remarkable assertion that "no teacher in the United States" had ever expressed approval of differentiation, and that the whole issue was not worthy of much attention. Eliot and other Committee members, however, clearly wished to endorse not merely weak but *strong* no-differentiation: there should be no distinct secondary school programs for students preparing to go to college. Since differentiation in this sense was widely practiced across the country, and would grow in the next decades despite the Committee of Ten, it is difficult to believe that Robinson could possibly have espied a unanimous consensus against it throughout the land.[36] Eliot was almost certainly guilty of announcing unanimity on strong no-differentiation when he was only entitled to claim it for the weak version. Moreover, since Eliot did not believe in a fixed curricula for all, as the above passage from the Report reminds us, the meaning of strong no-differentiation was unavoidably confused. In this confusion the Mathematics Conference recommendation that students not bound for college be given the option of taking bookkeeping and commercial arithmetic instead of second year algebra could be seen as an admirable application of the elective principle, without being a violation of no-differentiation.

The dismissal of differentiation proved to be a point of much controversy for the Committee of Ten Report as a whole, culminating in the heated charge of G. Stanley Hall that the Committee was guilty of disregarding "the great army of incapables" who could not cope with the tradition-bound curriculum

---

[35] *Report of Committee of Ten*, 226–227.
[36] Robinson, 367 and Krug, *High School*, 89.

recommended by the Committee.[37] But the slight dissension of the Mathematics Conference on the subject of differentiation does not seem to have drawn much attention. Nor have I found any evidence that the Conference felt aggrieved at being misrepresented by Eliot. It is plausible, although I have no direct evidence, that the Conference considered commercial arithmetic not to be mathematics at all, and that therefore it did not see itself as proposing any differentiation within mathematics.

Other educators were not as willing as Eliot, however, to let the Conference's animadversions on commercial arithmetic pass unnoticed. Both James M. Greenwood, superintendent of the Kansas City schools, and Edwin P. Seaver, superintendent of the Boston schools, objected strenuously, and were especially incensed at the Conference assertion that young children did not understand business. On the contrary, they avowed, the students of their acquaintance were almost all thoroughly familiar with business dealings and terminology, being broadly exposed to such from a young age. "A boy of twelve that could not understand about stock in a corporation would hardly be worth raising from an intellectual standpoint," declared Seaver.[38]

It is clear that Seaver, and even more Greenwood, were sensitive to issues of professional jurisdiction. The Conference, very much under the sway of the professional mathematicians among its members, argued that commercial arithmetic was distinctly inferior and subordinate to the more abstract parts of mathematics. Greenwood saw this as part of a malign tendency which he detected throughout the conference reports: "The conferences generally want everything taught by specialists. Soon a university graduate from Heidelberg will be necessary to teach the alphabet." But the Mathematics Conference alarmed him especially in this regard. The members were "recognized as mathematicians of no mean ability," but they knew nothing of what grade school pupils should learn. Seaver too was exasperated at what he considered the smug ignorance of school procedures exhibited by the Conference, against which he brandished his own experience: "For forty years I have had something to do with the teaching of arithmetic." Another piece by Greenwood during 1894 enlarged on the same theme, and included a contemptuous swipe at an unnamed individual who must surely be Simon Newcomb:

---

[37]Krug, 85.

[38]J. M. Greenwood, "The Report of the Conferences," *Western School Journal* 10 (1894): 78; J. M. Greenwood, "Discussion of Report of Committee of Ten," *NEA Proceedings* (1894): 453–454; Edwin P. Seaver, "Mathematics," *Journal of Education* 39 (Mar. 1894): 164.

I have in mind a noted mathematician and astronomer, now holding a high position in this country, who, if put in charge of forty pupils in any graded school in this country, would be discharged at the first meeting of the board of education, on account of incompetency.

It is thus no surprise to find Greenwood criticizing the Conference for replacing instruction in comfortably concrete topics such as "banking, insurance, and discount," in favor of dubious abstractions: "Had the conference given more attention to obvious things and less to the 'commutative law,' more good of permanent value would have been accomplished." Other commentators, seeing no great threat from the university specialists, were much more agreeable to the diminishment of commercial arithmetic, and found nothing to remark regarding the commutative law.[39]

It is fascinating to see how preconceptions with respect to the basic trustworthiness of the Committee of Ten endeavor allowed observers to see entirely opposite notions in the Report. Friendly commentator H. H. Seerley specifically applauded the Mathematics Conference for promoting a return to mental arithmetic, while Greenwood, ever suspicious of the Committee and its conferences, was aghast that the Mathematics Conference had "omitted entirely" this indispensable subject. Neither educator had read with sufficient care. B. F. McClelland, a more neutral reader, correctly stated that the Conference had not recommended mental arithmetic per se, but had recommended oral algebra problems in the manner of mental arithmetic.[40]

The topic of cube root as treated by the Conference is especially revealing of the stratification of views on mathematics education at the end of the nineteenth century. Methods of ancient provenance for extracting square and cube roots of numbers, based on the expansions of $(a + b)^2$ and $(a + b)^3$ respectively, had been taught as advanced topics in arithmetic in America for many years. The expansion of $(a + b)^2$ into $a^2 + 2ab + b^2$ is readily depicted geometrically as a dissected square of side $a + b$, and the corresponding expansion of $(a + b)^3$

---

[39]Greenwood, "Report of the Conferences," 78. Seaver, 164. J. M. Greenwood, "They Know, You Know," *Public-School Journal* 13 (1894): 391. B. F. McClelland, "A Discussion of the Committee of Ten on Arithmetic," *Public-School Journal* 13 (May 1894): 518–521; Seerley, "Report of the Committee of Ten," *Education* 15 (1894): 239–241; David E. Smith, "Report of the 'Committee of Ten' on Mathematics," *School Review* 3 (1895): 520–524; Frank Fitzpatrick, "Report of the Committee of Ten," *The North-Western Journal of Education* 5 (July 1894): 12.

[40]Seerley, 240; Greenwood, "Report of the Conferences," 98; McClelland, 520.

can be similarly depicted as a dissected cube. This geometry was widely used as an aid for teaching the root extraction methods in the nineteenth century. Indeed, wooden "cube root blocks" were extensively marketed by educational supply companies. Writing in 1884 MIT president Francis Amasa Walker saw the use of geometry to teach root extraction as one of the few sensible ideas in the otherwise dreary standard arithmetic course.[41]

The Mathematics Conference, which embraced objective teaching, the use of concrete models in geometry, and the intermingling of geometry and arithmetic, might seem likely to welcome cube root extraction and its blocks. Not so. Instead the Conference emphatically declared that cube root was among those arithmetic topics that should be "omit[ed] entirely" because they "perplex and exhaust the pupil without affording any really valuable mental discipline." They conceded only that cube root might be lightly touched upon in an algebra course.[42] The Conference did not make its reasons explicit, but it is clear that the traditional root extraction methods had become unappealing from the point of view both of abstract mathematics and of practical calculation. The cube root blocks contained no information not already present in the algebraic formula for $(a+b)^3$, and although this formula could be generalized to higher dimensions for the purpose of extracting higher roots, the blocks themselves were a dead end. Moreover, other methods of root extraction, such as using logarithms or the technique known as Newton's method, were far more closely linked to the ongoing development of mathematics, and were more practical as well. As early as 1851 an engineering handbook had offered the following advice: "The rules for the cube and higher roots are very tedious in practice: on which account it is advisable to work by means of logarithms."[43]

Despite the disdain of both the mathematicians and the engineers, arithmetic teachers clung to the traditional root extraction methods for a long time. It was considered a good capstone topic in arithmetic because it exercised several more elementary skills. The geometric appurtenances served not to connect the topic with the rest of mathematics but rather to emphasize that a student

---

[41] Francis A. Walker, "A Plea for Industrial Education in the Public Schools," in Walker, *Discussions in Education* (New York: Henry Holt and Co., 1899), 133.

[42] "Report of Mathematics Conference," 131, 133.

[43] Oliver Byrne, *The Pocket Companion for Machinists, Mechanics, and Engineers* (New York: Dewitt & Davenport, 1851), 30. The Newton, or Newton-Raphson method, is a powerful and computationally efficient iterative procedure with many applications. See Herman Goldstine, *A History of Numerical Analysis From the 16th Through the 19th Century* (New York: Springer-Verlag, 1977), 64–68, 278–280.

could learn the method without engaging with the more abstract realms. This is clear from an 1857 advertisement for cube root blocks:

> The extraction of the cube root can be explained most easily by the use of the Cube Root Block. In fact, no person who is unacquainted with Algebra or Geometry can know the reason for this rule without the aid of some such illustration.[44]

This declaration of independence for a piece of mathematical knowledge was alien to the Mathematics Conference, which declared cube root to be a minor topic in algebra. It is then not surprising to find the Conference treatment of cube root mentioned as a special point of vexation by Greenwood and Seaver, critics who suspected the Conference of plotting professional aggrandizement. Greenwood was especially exasperated:

> The cube is taught to the little children in the primary grades along with the sphere, the cylinder, and the cone, and yet the conference would pass the extraction of cube root over to the realm of algebra, forgetting apparently that the subject has many practical applications in measurement of various kinds, and besides it is just as easily and elegantly taught to boys and girls with "the blocks" as it is by the algebraic method.[45]

Greenwood's view was not unanimous. Once again, those who were untroubled by the basic intent of the Conference found little to quibble with on this particular issue.[46]

The Committee of Ten set the fashion for convening national bodies of experts to do educational studies. The first offshoot committee began work even before the Committee of Ten Report was released: the Committee of Fifteen on Elementary Education. Sponsored by the NEA, this committee included United States Commissioner of Education William Torrey Harris, one of the Committee of Ten, as well as E. P. Seaver and J. M. Greenwood, whom we have encountered as critics of the Committee of Ten Report. The report of the Committee of Fifteen was released in February of 1895, and consisted of three subcommittee reports: one on the training of teachers, one on organization of

---

[44] *The Teacher's Guide to Illustration: A Manual to Accompany Holbrook's School Apparatus* (Hartford, 1857), 34.

[45] Greenwood, "Report of the Conferences," 78. See also Seaver, 164.

[46] See McClelland, 519 and Seerley, 241.

city school systems, and one on correlation of studies. Only this last report, written by Harris, made any reference to mathematics.[47]

Harris's report agreed in some respects with that of the Mathematics Conference of the Committee of Ten, but likely this was coincidental; the disagreements were more fundamental. Harris criticized the overemphasis on arithmetic, and called for earlier introduction of algebra, but these notions were already part of Harris's thought before the Committee of Ten. The whole report was imbued with Harris's distinctive philosophical views, notably his basing of education on what he called the "five windows of the soul," which led him to divide the curriculum into five basic categories: mathematics, geography, history, language, and literature. Although he thought that arithmetic was frequently overdone, and that it should be supplemented by some algebra, Harris still felt that it was the mathematics most appropriate to the elementary grades; specifically, arithmetic was far preferable to geometry. In sharp distinction from the Mathematics Conference, and showing clearly that the Conference report was not a source for his own report, Harris explicitly declared concrete geometry to be undesirable in the elementary school. "Practice with blocks in the shape of geometric solids," was all very well in the kindergarten, but he denied the Conference contention that such practice could provide a useful intuitive foundation for later rigorous work in geometry. Here Harris, a major exponent of Hegel in the United States, was clearly in the grip of his allegiance to German idealistic philosophy. Geometry for him was intrinsically a study appropriate to secondary rather than elementary education because "its demonstrations reach universal and necessary conclusions," producing "knowledge transcending experience."[48] Playing with blocks could not possibly yield significant insight into such an exalted subject.

The mathematicians of the Mathematics Conference did not agree, and Simon Newcomb in particular believed that the discovery of non-euclidean geometry earlier in the century invalidated philosophical positions such as that

---

[47]National Educational Association, *Report of the Committee of Fifteen on Elementary Education* (New York: American Book Co., 1895; repr., New York: Arno Press, 1969). See also Krug, *High School*, 97–105 and Sizer, 191–193.

[48]*Committee of Fifteen*, 20–24, 35, 44. William Torrey Harris, response to Eliot's "Can School Programmes Be Shortened and Enriched?" *Proceedings of the Department of Superintendence of the National Educational Association* (1888): 117–118. William T. Harris, "What Shall the Public Schools Teach," *Forum* 4 (1888): 575–576, and Sizer, 85. Krug, *High School*, 22.

of Harris. For Newcomb, geometrical demonstrations produced necessary conclusions only relative to the axioms; that is, relative to basic geometrical conceptions. But "geometrical conceptions are the result of experience." Hence a rich sensory exposure of the student to geometric objects was eminently valuable as a basis for studying standard demonstrative geometry. Here we see how general educators such as Harris could appear to professional mathematicians as egregiously ignorant.[49]

It is considerably more difficult to discuss the effect of the Committee of Ten Report on educational practice as opposed to educational discussion. In 1906 Edwin Dexter, Director of the School of Education at the University of Illinois, made one attempt in this direction. He collected data so as to compare school curricula from 1894, presumably unaffected by the Report, with curricula from 1904. Wisely observing that it would not be possible to disentangle influence from prophecy, Dexter concluded that in general there was a low correlation between the Report's recommendations and the subsequent changes in school programs.[50]

It must be noted, however, that Dexter's view was a narrow one. He measured what he considered measurable, with no apology for ignoring other aspects. What he measured almost exclusively, in accord with the fixation on economy of time which had become pervasive, was the percentage of schools offering specific subjects, for specific amounts of time, during specific years of the high school course. In taking this approach Dexter unconsciously demonstrated that one of the most significant influences of the Committee of Ten Report was its adoption of what one later historian has called the "time basis": the measurement of instruction in a given subject by the number of recitation sections per week. The Committee itself had expressed some reservations about this tack.[51] Dexter appeared to have no such reservations, and his testimony shows how rapidly this measure had become central to educational discussions.

In mathematics in particular Dexter determined that most of the Conference report was merely "a discussion of the general pedagogy of the subject," thus unworthy of comment except for the following "specific recommendations":

---

[49]For Newcomb's views on geometry see Simon Newcomb, "Mathematical Education," SNP, Box 111, page 24. This was likely written about the time of the Committee of Ten. Newcomb's low opinion of Harris can be found in Newcomb to Henry Holt, Feb. 7, 1885, letterpress copy in SNP, Box 6.

[50]E. G. Dexter, "Ten Years' Influence of the Report of the Committee of Ten," *School Review* 14 (1906): 255, 269.

[51]On the "time basis" see *Report of Committee of Ten*, 247 and I. L. Kandel, *History of Secondary Education* (Boston: Houghton Mifflin, 1930), 474.

1. Algebra should be taught throughout the first year and one-half time the second and third.
2. Geometry should be given one-half time for the second and third years.
3. Trigonometry and advanced algebra, one-half time each the fourth year.
4. Bookkeeping and commercial arithmetic, if taught at all, should be optional with algebra the second and third years.

Measuring against these specific aims, it is not surprising that Dexter found the Conference influence minimal. But it is doubtful that the Conference was as concerned with specific time allocations as Dexter clearly was. Thus although the marked increase in algebra instruction revealed by Dexter's statistics might have been counted by a more lenient judge as ratifying the Conference's general position, it failed to meet Dexter's stricter standard because "its arrangement in the course was not that of the committee."[52]

Dexter's final comment on the mathematics report is the most interesting of all:

> The fourth recommendation of the committee had to do with bookkeeping and commercial arithmetic. With the advent of commercial courses these subjects have lost any close affiliation with mathematics which they may once have had, and a discussion of them in connection with the committee report seems out of place.[53]

Two points should be made. First, it has earlier been suggested that the divorce of commercial subjects from mathematics described here by Dexter may have been precisely the disposition desired by the mathematicians of the Mathematics Conference. Second, the "commercial courses" referred to by Dexter, that is separate curricula designed for students going directly into business after high school, were precisely what Eliot and the Committee of Ten had campaigned so hard against when they declared against "differentiation." Thus this rather offhand remark by Dexter is in its way more conclusive than his carefully marshaled statistics of an essential failure of the Committee of Ten. And Dexter's lack of comment on this fact, his conscious or unconscious ignoring of how central the differentiation question had been to the Committee of Ten, is striking evidence of how rapidly the educational environment had changed.

---

[52]Dexter, 261.
[53]Ibid., 262.

Other studies of the Committee of Ten, early and late, support the impression that the Mathematics Conference made few swift converts. Solberg Sigurdson discerns some slow drift in the general direction of Conference recommendations, especially on arithmetic. It is certainly true that some topics disparaged by the Conference, cube root for example, did eventually leave the curriculum, though whether the Conference can claim any credit for this is very doubtful. John Elbert Stout, writing in 1921, confirmed Dexter's observation of an increase in algebra instruction in the late nineteenth century, but showed that this phenomenon had commenced before the 1890s. In this respect the Conference mirrored the Committee of Ten as a whole, which had ratified an already established trend toward enhanced status for the new academic subjects (i.e., science, history, English, and the modern foreign languages).[54]

Looking beyond specific topics and methods, there can be little doubt that the Mathematics Conference report increased awareness of pedagogical issues among many mathematical educators. By virtue of its national distribution it was a convenient point of reference, allowing an educator to demonstrate cognizance of the latest thinking on instructional questions. The prefaces of mathematics textbooks frequently cited the report.[55]

4.4. Long Term Response to the Report

The reputation of the Committee of Ten has continued to evolve to the present day. As Herbert Kliebard has noted, interpretation was long dominated by those who have seen the Report as primarily an obstacle to educational progress: an elitist exercise proceeding from an absolute faith in old-fashioned mental discipline. Analysis of the Mathematics Conference supports Kliebard's critique of this *progressive* view. The Conference report did not exalt mental discipline; rather, it sought to distinguish "a really improving discipline" from mere drill, and was willing to cater to student interest whenever practical. In many respects the Mathematics Conference epitomizes the "moderate revisionism" discerned by Krug, who sees the Committee of Ten Report making a sincere effort to propose high school curricula that would meet the needs of all students, not merely those bound for college.[56]

The progressive interpretation of the Committee of Ten, although long dominant as Kliebard asserts, has never entirely monopolized discussion. Indeed,

---

[54] John Elbert Stout, *The Development of High-School Curricula in the North Central States from 1860 to 1918* (Chicago: University of Chicago Press, 1921), 92–99.

[55] Jones, 168.

[56] Kliebard, "Constructing a History," 162–163; Krug, *High School*, 86–87.

there is a class of commentators who look back to the Report as a fount of wisdom from which later generations, to their cost, have increasingly failed to imbibe. Such commentators frequently view the Committee of Ten and associated events in precisely the opposite manner to that adopted by the progressives, and thus may be fitly called *anti-progressives;* rather than scenting nefarious college domination and elitism, they applaud the Ten's emphasis on rigorous academic training for all secondary school students.[57]

Progressive and anti-progressive interpreters have both displayed lack of historical understanding of the evolution of the mental discipline thesis during the nineteenth century. Thus we find anti-progressive commentator John Francis Latimer and progressive commentator Ellwood P. Cubberley both claiming that the Ten were advocates of "a new psychological theory" that had originated between 1860 and 1890.[58] In Chapter 2 we saw that the mental discipline thesis originated well before 1860, and that by 1893 it was a much less stable notion than it had been earlier in the century. The Committee of Ten does not represent the apex of unquestioned belief in mental discipline, be this considered good or bad. In particular, the desire to cater to student interest (anathema to many anti-progressives) was already present, as can be seen in both Charles Eliot and in the Mathematics Conference. The progressive tale wherein mental discipline was suddenly rendered untenable by psychological experiments around the turn of the century is also implausible.[59]

For the anti-progressive interpreters, the no-differentiation stance of the Committee of Ten has often been seen as an admirable insistence on an equality of high intellectual standards. The usual progressive response has been to point out that Charles Eliot and the Committee of Ten conceived of the secondary schools as serving only a small minority of American youth. But, one anti-progressive commentator has retorted, "The Committee of Ten sought equal and democratic schooling precisely because its members expected a rapid rise in numbers of high school students, well beyond the 1894 percentages."[60] The

---

[57]For example, see John Francis Latimer, *What's Happened to Our High Schools?* (Washington: Public Affairs Press, 1958), 72; Paul Gagnon, "What Should Children Learn?" *Atlantic Monthly* 276 (December 1995):70; and Diane Ravitch, *Left Back: A Century of Failed School Reform* (New York: Simon & Schuster, 2000), 41–50.

[58]Latimer, 65. Ellwood P. Cubberley, *Public Education in the United States, a Study and Interpretation of American Education History* (Boston: Houghton Mifflin, 1934), 513–514.

[59]See Kliebard, *Struggle for the Curriculum*, 7.

[60]Paul Gagnon, "Letters," *Atlantic Monthly*, Mar. 1996, 14. Gagnon is replying to a letter from Gerald W. Bracey, a progressive interpreter of the Committee of Ten.

present writer has found no evidence of such prescience, either by the Committee of Ten or by its Mathematics Conference. But this much should be said in favor of the anti-progressive position: the Committee of Ten did sincerely believe that a rigorous academic education would benefit a wider segment of society than merely the college bound. This is clear from Eliot's pronouncements, and it is supported by the report of the Mathematics Conference as well. The Conference took a dim view of commercial arithmetic not because it was a vulgar concern of the many, but because the mathematicians were convinced that a broader mathematical training was in fact superior for business purposes.[61]

Questions of academic elitism and intellectual standards in relation to the Mathematics Conference are not successfully encompassed by either progressive or anti-progressive interpretations, and even Krug's "moderate revisionism" needs supplementation by attention to the professional development of the mathematicians. No anti-progressive defense of the Committee of Ten as a misunderstood egalitarian endeavor can obscure the fact that the Mathematics Conference was dominated by a small group of mathematical specialists of relatively similar backgrounds whose knowledge of what actually transpired in the majority of American schools was limited. That this group proceeded to recommend that the sons and daughters of practical Americans be diverted from extensive study of banking and insurance in favor of pondering the significance of $a \times b = b \times a$ can surely be characterized as academic elitism, in accord with the progressive view.

Yet the progressive complaint that the Committee of Ten represented college domination can also be seen to be too crude an analysis. What the case of the mathematicians shows is that the institutional and pedagogical arrangements sought by academic specialists were not naked grabs for power but complex negotiations between the internal imperatives of the specialties and external developments. For example, in championing concrete geometry and inductive methods of instruction mathematicians were able for a time to accommodate both the larger educational zeitgeist and their own evolving notions of how best to develop an intuitive grasp of their subject. Similarly, in championing algebra over arithmetic the mathematicians were briefly in accord with the advocates of educational efficiency. But such a balance of interests was fragile. The mathematicians were only willing to go so far in conforming to the preferences of the general educators, as we have seen in the case of the cube

---

[61]On Eliot's democratic views see Ray Greene Huling, "The New England Association of Colleges and Preparatory Schools," *School Review* 2 (1894): 617. On the failure to anticipate rising school populations see Hawkins, 246, 255; and Sizer, 132–133, 146.

root blocks. Moreover, these general educators, even Charles Eliot, were in the process of rethinking their own notions of the proper place for mathematics.

The case of commercial arithmetic was one where the central preoccupations of the mathematicians and popular conceptions of mathematical utility were already well out of alignment. For the mathematicians of the Conference, those who failed to understand that commercial arithmetic consisted of nothing but simple special cases of more general results were not worthy of serious engagement: the mathematical ignoramuses might as well have their special courses if they insisted; that is to say, they might as well be uncoupled from the great ongoing mathematical endeavor. Some mathematicians of the next generation were much more conscious than those of the Mathematics Conference that such attitudes held professional perils, but by then an even greater proportion of American mathematicians had been diverted from educational questions by "the esthetic quality which is the charm of the higher mathematics."

CHAPTER 5

E. H. Moore: Leader of a New Generation of American Mathematicians

Within a few months of the time that Simon Newcomb was guiding the Mathematics Conference of the Committee of Ten to its recommendations, Eliakim Hastings Moore of the University of Chicago, twenty-seven years younger than Newcomb, would take a prominent role in a different academic event of national, indeed international, scope. This event was the Mathematics Congress associated with the World's Columbian Exposition in Chicago in 1893. Although the Congress was not directly concerned with school education, its success ratified certain directions being promoted by American college and university mathematicians which would have significant implications for their evolving relationship with the secondary schools. Moore represented a species of mathematician different in significant respects from Newcomb, and analyzing these differences will provide insight into the changing attitudes of mathematicians to the problems of school education. The vocabulary used and the issues raised by Newcomb's Mathematics Conference continued to pervade the period of Moore's educational activity, but new concerns emerged as well.

To better understand Moore and the new educational attitudes, we will first describe the evolution from the mathematical generation of Simon Newcomb to that of Moore. We will then survey Moore's education and early career, followed by a discussion of the founding and growth of the institution to which Moore devoted most of his professional life, the University of Chicago. The chapter will conclude with an examination of Moore's initiatives for strengthening research mathematics in the United States, a development which in the short term would help amplify the voices of a few mathematicians to speak on educational matters, but in the longer term would distance most mathematicians from the schools.

5.1. Elite American Mathematical Activity from Newcomb to Moore

Roger Cooke and V. Frederick Rickey make a passing reference to Newcomb in their 1989 essay on William Story which is highly revealing. Cooke and Rickey, historians of mathematics trained as research mathematicians, allude in a footnote to the vexing topic of naming a successor to Sylvester at Johns Hopkins, clearly expecting their mathematical readers to be puzzled by the outcome:

> Story was passed over in favor of Simon Newcomb. This decision is unreasonable if one feels that the department head should be a pure mathematician, rather than an astronomer. Of course many consider Newcomb a mathematician, for he did serve as president of the AMS. Today, a recreational problem posed by Newcomb is of interest in combinatorics.[1]

This well illustrates the diminishment of Newcomb's standing as a mathematician since his death. Only the bare fact of his presidency of the American Mathematical Society remains visible, while the accomplishments and international recognition that led to that presidency have become dim indeed. Newcomb's vast output of publications, his mathematical textbooks, his meticulous compilations of tables of planetary motions, his investigations of series expansions and perturbative functions, his excursions into four-dimensional geometry, all this has been sifted and found to yield merely one "recreational problem." The changes in the mathematical profession thus manifested were already well under way during Newcomb's lifetime.

When Newcomb began his career there was often no clear distinction between mathematical astronomers and mathematicians. The celestial mechanical problems that attracted Newcomb derived from Isaac Newton's laws of motion and theory of universal gravitation, and had been pursued since Newton's day by a number of prominent investigators almost always referred to today as mathematicians, including Leonard Euler, Joseph-Louis Lagrange, Pierre-Simon Laplace, and Carl Friedrich Gauss. Newcomb, following the example of his teacher Benjamin Peirce, saw himself as very much in this tradition.

---

[1] Roger Cooke and V. Frederick Rickey, "W. E. Story of Hopkins and Clark," in Peter Duren, ed., *A Century of Mathematics in America* (Providence: American Mathematical Society, 1989), 3:49 (note 3). The problem referred to was posed by Newcomb in the *Philosophical Transactions of the Royal Society* 207 (Feb. 1908): 65. Given some rules for dealing out a deck of generalized playing cards it asks for the probability of the possible outcomes.

The fundamental problem was to identify the appropriate differential equations governing the motions of the bodies in the solar system (planets, moons, comets, asteroids) and to develop integration techniques to solve, or approximately solve, these equations so as to predict the motions of these bodies over long periods of time. It was early found that while the equations were tractable in the case of two bodies entirely isolated from other gravitational effects, the perturbing action of a third body created great mathematical difficulties. Since almost all the interesting gravitational interactions in the solar system involved more than two bodies the "three-body problem" loomed large, and the study of "perturbation" became central to the mathematical attacks on it.

During Newcomb's lifetime the intricacies of celestial mechanics opened up into an enormous field of investigation, well capable of supporting specialization. From performing celestial observations at the empirical end, to studying the qualitative behavior of the solutions of differential equations at the theoretical end, it was now possible to devote a career to a mere segment of the entire program, according to one's talents, interests, and opportunities. Those immediate predecessors with whom the young Newcomb most strongly identified, and whose work he spent the most time extending and correcting, were investigators such as Peter Andreas Hansen (1795–1874), Charles-Eugène Delaunay (1816–1872), and Urbain Jean Joseph Le Verrier (1811–1871), all of whom were far more interested in solving specific astronomical problems than in pondering the general mathematical questions that could be abstracted from these problems; none of these men is today usually classified as a mathematician, not withstanding the heavy use of mathematics in their work.[2]

Newcomb himself was most comfortable at the interface between theory and observation, organizing observations originally made by others and using them to test theories devised by others. Such testing of theories did indeed require considerable mathematical facility, but only rarely did Newcomb devote his time to theoretical derivations per se.[3] Thus did Newcomb become distinguished over the course of his career from more theoretically audacious astronomical investigators, including his sometime *Nautical Almanac* colleague G. W. Hill (1838–1914), who began his celebrated paper "Researches in the Lunar Theory" by proclaiming the advantages offered by a new infusion of abstract mathematical ideas into the subject:

---

[2] Newcomb discussed Hansen, Delaunay, and Le Verrier in his *Reminiscences of an Astronomer* (Boston and New York: Houghton Mifflin & Co., 1903), 315–321, 327–333.

[3] E. W. Brown, "Simon Newcomb," *BAMS* 16 (1910): 347–349.

> When we consider how we may best contribute to the advancement of this much treated subject, we cannot fail to notice that the great majority of writers on it have had before them, as their ultimate aim, the construction of tables: that is they have viewed the problem from the stand-point of practical astronomy rather than of mathematics.[4]

Even before publishing these words, Hill's investigations of lunar motions had led him to consider an infinite system of linear equations in infinitely many unknowns, a fruitfully innovative conception which posed substantial perplexities in the abstract. Hill's work was explicitly cited, and greatly elaborated upon, by Henri Poincaré (1854–1912), a prodigiously brilliant Frenchman who eclipsed Newcomb even more decisively on the mathematical side of astronomy.[5] Newcomb, while recognizing Hill's superior mathematical skill, nevertheless insisted in his autobiography on characterizing Hill's approach to astronomy in a way that is in truth more appropriate to himself:

> [Hill's] most marked intellectual characteristic is the eminently practical character of his researches. He does not aim so much at elegant mathematical formulae, as to determine with the greatest precision the actual quantities of which mathematical astronomy stands in need.[6]

An even younger astronomical theorist, E. W. Brown (1866–1938), candidly assessed Newcomb's limitations in a memorial essay (written when "computers" were people):

> Newcomb seems to have been troubled all through his life by the difficulty of mastering complicated analysis, and always turned by preference to a simple theory which could without much trouble be prepared for computers.[7]

One contemporary commentator who saw the work of both Newcomb and Hill as declining in mathematical prestige was Charles Sanders Peirce, son of Newcomb's old Harvard teacher. The younger Peirce, who had a complicated

---

[4]G. W. Hill, "Researches on the Lunar Theory," *American Journal of Mathematics* 1 (1878): 5. This paper was published in the inaugural issue of the journal, of which Newcomb was the cooperating editor in astronomy.

[5]Raymond Clare Archibald, *A Semicentennial History of the American Mathematical Society* (New York: American Mathematical Society, 1938; repr., New York: Arno Press, 1980), 117–124.

[6]Newcomb, *Reminiscences*, 222.

[7]Brown, "Simon Newcomb," 346.

and often contentious relationship with Newcomb over many years, wrote to Newcomb in 1907 to explain his recently published contention

> That the science of celestial mechanics by its own perfection-ment is now reduced to calculation. Of course, I did not mean numerical computation, but just that sort of art that there is in Delaunay's method. It is not a method for finding out any substantially new truth, but is a method for calculating a result according to well known principles ... That you and Hill and other theoretical astronomers find in the afternoon of life that their own successes have rendered their science uninteresting to most people,—even to most mathematicians is distressing; but it is a fact.[8]

Newcomb understandably did not readily accept Peirce's view, but he did recognize that America in the late nineteenth and early twentieth centuries was producing mathematicians whose primary interest was in pure abstractions, without reference to physical applications at all, and that in consequence his relation to the main lines of mathematical research was not as it once was. When he addressed the young New York Mathematical Society in 1893 he felt it necessary to begin with the following disclaimer:

> One who, like myself, is not a mathematician in the modern sense naturally feels that some apology is due for accepting the invitation with which this Society has honored me, to address it on a mathematical subject.[9]

To a great extent Newcomb welcomed the rise of pure mathematics as evidence that American science was at last being cured of the backwardness he had diagnosed in the 1870s. There is some evidence that he wished to have been able to play a more direct role here. His very first contribution to the *American Journal of Mathematics* (in fact the first paper in the inaugural issue) was a pure mathematical investigation of a question in higher-dimensional geometry. This was a subject he returned to again in his retiring address as president of the American Mathematical Society. Yet these remained mere excursions for Newcomb; unlike Benjamin Peirce, Newcomb never produced a

---

[8]C. S. Peirce to Newcomb, Oct. 31, 1907. Quoted in Carolyn Eisele, "The Correspondence with Simon Newcomb," in R. M. Martin, ed., *Studies in the Scientific and Mathematical Philosophy of Charles S. Peirce* (The Hague, Mouton Publishers, 1979), 85.

[9]Simon Newcomb, "Modern Mathematical Thought," *Bulletin of the New York Mathematical Society* 3 (1894): 95. This is the text of an address delivered on December 28, 1893.

substantial piece of pure mathematical exposition that might have endeared him to future generations of pure mathematicians despite his obvious bias toward astronomy.[10]

Newcomb's membership in the American Mathematical Society was a beneficial arrangement for both himself and the Society. For Newcomb it offered another platform, as the Johns Hopkins University had, to demonstrate his allegiance to mathematics and to promote his vision of science in America. Mathematics was central to this vision, for mathematization was a crucial outward sign that a science was progressing. For the young Mathematical Society Newcomb offered credibility; when he joined in 1891 he was a distinguished name indeed; he was head of the Hopkins mathematics department, he had a long list of publications and national and international honors to his credit, and his particular area of specialization was one to which some of the greatest names in mathematics had contributed.[11]

At the time Newcomb joined, the name of the organization was the New York Mathematical Society; it had been founded in 1888 by a small group of young enthusiasts associated with Columbia University.[12] A membership campaign was initiated in 1891, after a decision to publish a society bulletin, and Newcomb was among those whose membership was solicited. Indeed, Newcomb was among a select handful whose membership was solicited before the main appeal, in order that this appeal could contain his statement of support. Newcomb's reply to the initial contact from Society secretary Thomas Fiske was cautious:

> I belong to so many associations that I can hardly keep the run of them, or even be reminded of their existence except by the annual receipt of a small bill for dues. Hence I do not want to join another except for strong reasons. But your plan strikes me so favorably that I am willing to support it provided it is

---

[10] Simon Newcomb, "Note on a Class of Transformations Which Surfaces May Undergo in Spaces of More than Three Dimensions," *American Journal of Mathematics* 1 (1878): 1–4. Simon Newcomb, "The Philosophy of Hyperspace," *BAMS* 4 (1898): 187–195. Benjamin Peirce's major work of pure mathematics was *Linear Associative Algebra*, privately circulated in 1870 and published after his death, with editorial additions by his son C. S. Peirce, in *American Journal of Mathematics* 4 (1881): 97–229.

[11] W. W. Campbell, "Simon Newcomb," *Memoirs of the National Academy of Sciences* 17, 1st mem. (1924): 16.

[12] Archibald, *History of the AMS*, 3–5.

going to be a success ... Then the question will arise whether pure mathematics is not too narrow a field for the journal.[13]

We see here the burden of being a prominent professional figure during an age in which professional associations were being founded by the dozens.[14] We see further that by the 1890s Newcomb was very conscious of having a public reputation worthy of protection; the caution engendered by this consciousness would again be in evidence in the discussions surrounding the International Mathematical Congress of 1893, to be discussed later in this chapter. Finally, we see Newcomb's concern with the problem of specialization, in particular the "narrow field" of pure mathematics. Fiske took this warning to heart, clearly stating in the more widely distributed membership solicitation that the new journal was to be a "periodical review of pure and applied mathematics," and adding the further diplomatic note that

> the idea is not to enter into any competition with the American Journal of Mathematics, the Annals of Mathematics, or any other similar journal, but it is proposed to publish, primarily, historical and critical articles, accounts of advances in different branches of mathematical science, and reviews of important new publications.

Newcomb's favorable opinion was duly invoked in support, along with similar endorsements from Professor Henry B. Fine of Princeton and Professor William Woolsey Johnson of the Naval Academy.[15]

The membership drive was a success; during 1891 the membership rose from 23 to 210, and the *Bulletin of the New York Mathematical Society* commenced publication. By the end of 1893 Newcomb evidently felt that the organization had proved itself to such an extent that he accepted Fiske's invitation to address the meeting of December 28, 1893, the first time he had ever attended a meeting.[16] After the organization changed its name to the American Mathematical

---

[13] Newcomb to T. S. Fiske, Jan. 29, 1891, letterpress copy in SNP, Box 6.

[14] It has been estimated that at least 200 American learned societies were formed in the 1870s and 1880s. See Burton J. Bledstein, *The Culture of Professionalism* (New York: Norton, 1978), 86. By 1891 Newcomb had become a member or honorary member of at least 25 societies and associations. See Campbell, 15.

[15] New York Mathematical Society membership circular, signed by Thomas S. Fiske, Secretary, Mar. 23, 1891. Copy in SNP, Box 22.

[16] Newcomb to Fiske, Oct. 11, 1893, letterpress copy in SNP, Box 6. This is the talk that was published as "Modern Mathematical Thought." For AMS membership figures see Archibald, *History of the AMS*, 5.

Society (AMS), Newcomb served as the fourth president, during 1897–1898, succeeding G. W. Hill.

Although Newcomb's association with the AMS, with Johns Hopkins University, and especially with the *American Journal of Mathematics*, gave him the opportunity and the obligation to stay abreast of the latest developments in pure mathematics, he was not entirely comfortable with some of these developments, especially Georg Cantor's revolutionary work on infinite sets. Writing to C. S. Peirce in 1890 he dismissed Cantor's work, causing Peirce to reply with some heat that "Not to be acquainted with [Cantor's papers] is to be unprepared to enter into a modern discussion of the conception of infinity." Two years later Newcomb was still denying one of Cantor's most important demonstrations: "I have always held that infinity, considered in itself, could not be treated as a mathematical quantity, and that it is pure nonsense to talk about one infinity being greater or less than another." Thus, although Newcomb did oversee the publication of substantial quantities of pure mathematics while editor of the *American Journal of Mathematics*, there is certainly some likelihood that his conservative attitudes affected some of his editorial decisions. Peirce felt strongly that his own work was not treated fairly when submitted to Newcomb's journal, and even interpreted one comment of Newcomb as implying that the latter had actually rejected one of Cantor's papers for publication.[17]

In defense of Newcomb in his exchanges with Peirce it should be noted that Peirce was not the man most likely to sway Newcomb's mathematical opinions, especially on Cantor's work. Not only had Newcomb and Peirce had personal conflicts of long standing, but Peirce's defense of Cantor on infinities was couched within an exposition of his own defense of infinitesimals in calculus. Newcomb, a stout defender of "the doctrine of limits," as we have seen, could with much justice see himself here as a proponent of the most modern point of view, with Peirce apparently clinging to the discredited views of the eighteenth century.[18]

---

[17]Peirce to Newcomb, Dec. 1890. Newcomb to Peirce, Mar. 9, 1892. Quoted in Eisele, "Correspondence with Simon Newcomb," 65, 75, resp. I have found no evidence that Newcomb ever rejected a paper submitted by Cantor.

[18]See Simon Newcomb, "Remarks on the Doctrine of Limits," *The Analyst* 9 (1882): 114–115. For Newcomb infinitesimals were a mere terminological convenience, while for Peirce they were something more tangible. Since the advent of non-standard analysis about 1960 Peirce's views on infinitesimals have received more respectful treatment. See Joseph W. Dauben, "C. S. Peirce's Philosophy of Infinite Sets," *Mathematics Magazine* 50 (1977): 123–135.

When Newcomb engaged with another correspondent regarding Cantor's results in the late 1890s he did seem a bit more willing to accept the new ideas. The correspondent was E. H. Moore, a man much more representative of the growing mathematical community in the United States than the isolated Peirce, and a more patient and respectful expositor, as evidenced by the following letter to Newcomb:

> The Cantor theorem amounts essentially to this, that it is impossible to make a 1-1 correspondence between the numbers 1,2,3,4, ... and the numbers x of the interval
>
> $$a < x < b \text{ or } a \leq x < b \text{ or } a \leq x \leq b$$
>
> This theorem, I agree, appeals strongly to our geometric intuition, but still (even if only as a matter of form) we must prove it by reducing it to simpler thought elements if we can do so. And besides, we cannot properly <u>for analysis</u>, use the geometric intuition—the real numbers need <u>constructive definition</u> (in terms ultimately of positive integers) and one works always on basis of this (analytic) definition of the real numbers.[19]

This suggests that Newcomb had come around to accepting the truth of one of Cantor's key results, but was now proposing that it was so obvious as not to require proof. In politely but firmly rejecting Newcomb's proposal Moore was demonstrating the new level of abstraction that the younger generation of pure mathematicians was finding indispensable to its work. Moore was also declaring that geometric intuition, so central to the pedagogic ideas of both himself and Newcomb, had distinct limitations in the realm of the more advanced aspects of mathematics.

5.2. Moore's Education and Early Career

Eliakim Hastings Moore was born in Marietta, Ohio in 1862. During Moore's earliest days his father was an officer in the Union army, after which he resumed a career as a Methodist minister, at times itinerant. But despite these potentially chaotic circumstances, it is clear that in general Moore was the recipient of educational and social advantages quite unknown to the young Simon Newcomb. Partly this reflected Moore's uniquely fortunate circumstances, and

---

[19]E. H. Moore to Newcomb, Nov. 7, 1899, SNP, Box 32. Underlining in original.

partly the richer overall educational environment that was developing in parts of the United States.[20]

Moore spent much of his childhood in Athens, Ohio, where his grandfather and namesake was a man of prominence. Among young Moore's special advantages was a summer in Washington, D.C. as a congressional messenger, while his grandfather served as a member of Congress. By the early 1870s Moore and his family were living in Cincinnati. Here Moore was able to attend Woodward High School, graduating in 1879.[21]

It is evident that Woodward was a superior institution of its type. Physicist Carl Barus (1856–1935), who graduated from Woodward in 1874, reported that in his time the school was conscientiously "watched over" by such leading Cincinnati citizens as Judge Alfonso Taft, father of future United States president William Howard Taft, and that "Woodward was then virtually a junior college, with facilities as near a college as most Cincinnati lads dared to hope for." Indeed, Woodward had operated since its 1831 founding right along the blurry boundary between secondary and higher education. It was known as Woodward College from 1834 to 1851, after which time it ceased to be a private institution and became part of the public school system of Cincinnati.[22] That is to say, Woodward began as an institution hard to distinguish from the private academies founded in the late eighteenth century, but it evolved into an institution of a type that would be central to twentieth-century education in the United States: a public high school.

In Moore's day Woodward had a long-standing tradition of rigorous mathematics instruction. Barus was pleased to report having been "brought up on stiff original propositions and early taught self reliance and the habit of thought." Bestselling textbook writer Joseph Ray had been the chief mathematics teacher in Woodward's early days, and later served as principal of the school. By the 1870s Ray himself was deceased, but his textbooks remained at the core of the Woodward mathematics curriculum. Barus, writing 50 years later, when his classmate William Howard Taft was Chief Justice of the United State, reported his memory of classroom procedures at Woodward:

> I always recall Taft at the blackboard, struggling with what Dr. Ray in his book called the "theorems" of algebra. Truly

---

[20] G. A. Bliss and L. E. Dickson, "Eliakim Hastings Moore," *National Academy of Sciences Biographical Memoirs* 17 (1936): 83–102.

[21] Bliss and Dickson, 83. Parshall and Rowe, *Emergence*, 281.

[22] On Woodward High School see Jerry Dennis, "Joseph Ray," *Ohio Archaeological and Historical Quarterly* 46 (1937): 43; and Carl Barus, "Autobiography," CBP, 8, 24.

it should hearten the student, to think, that the man who
now confronts the most complicated problems of the nation, at
Woodward found in these wily theorems, foemen quite worthy
of his steel.

Barus also praised the advanced nature of some of Ray's treatments: "Much in Dr. Ray's admirable books on mathematics was hardly written for high school students."[23]

Yale University evidently welcomed Woodward graduates in the 1870s, including the Taft brothers, Horace and William Howard, and Horace's classmate E. H. Moore. During his brilliant undergraduate career from 1879 to 1883, Moore was especially guided by professor of mathematics and astronomy Hubert Anson Newton (1830–1896). Newton encouraged Moore to continue his studies at Yale past the bachelor's degree, with the result that Yale awarded Moore a Ph.D. in 1885, a then rare event in American education. Newton's own research interests had begun in pure geometry but had evolved more and more towards astronomy, especially the study of comets and meteors. His earlier student, the celebrated mathematical physicist J. Willard Gibbs (1839–1903), later speculated that the American intellectual environment had forced Newton's shift from pure mathematics to astronomy. In any event, Newton seems to have been agreeable to letting Moore pursue the abstract track he himself had abandoned. Moore's thesis was entitled "Extensions of Certain Theorems of Clifford and Cayley in the Geometry of n Dimensions."[24]

During his graduate phase at Yale Moore also took classes with Gibbs, who was uncompromisingly austere in presenting his work, whether in writing or in person. He later declared that only six students had ever been able to follow his lectures. One of the six was Moore.[25]

As he had done with Gibbs, Newton urged Moore to enlarge his education with a period of study in Europe. With Newton's financial assistance Moore studied the German language at Göttingen in the summer of 1885 and attended mathematical lectures in Berlin during the academic year 1885–1886. After returning from Germany Moore spent a year as an instructor in the preparatory

---

[23]Barus, 17, 25, 27. Punctuation as in original.

[24]Bliss and Dickson, 85; Parshall and Rowe, *Emergence*, 22; and Karen H. Parshall, "Eliakim Hastings Moore and the Founding of a Mathematical Community in America, 1892–1902," in Peter Duren, 2:157.

[25]See Clifford Truesdell, "Genius and the Establishment at a Polite Standstill in the Modern University: Bateman," in *An Idiot's Fugitive Essays on Science: Methods, Criticisms, Training, Circumstances* (New York: Springer-Verlag, 1984), 415.

department of Northwestern University, two years as a tutor at Yale, and then returned to Northwestern as an assistant professor. He had advanced to an associate professorship when he was called to join the newly founded University of Chicago on the south side of the city.[26]

5.3. Moore, Harper, and the University of Chicago

The rapid rise to prominence by the University of Chicago is an astonishing one, and the success of its mathematics department must surely be counted as one of its most remarkable features. The abundant financial resources provided by the University's patrons, headed by John D. Rockefeller; the eye for scholarly talent and the aggressive boosting of the first president, William Rainey Harper; the high ambitions and large organizational and mathematical skills of the first department head, E. H. Moore; and the expansion of nationwide teaching opportunities in mathematics: all combined to produce a most impressive result. Evidence will be offered in this and later chapters to support this last assertion, but one gross measure is striking: of all mathematics Ph.D.s conferred in the United States from 1862 through 1934, more than 18% were conferred by the University of Chicago; no other American university conferred more than 8%.[27]

With regard to the financial health of the University of Chicago, it must of course be acknowledged that all is relative to one's point of view. From the vantage of President Harper, for example, the university was in a nearly continuous state of excruciating financial crisis, requiring a never-ending search for more funds. But compared with two key academic predecessors with high ambitions, Johns Hopkins and Clark Universities, Chicago enjoyed steady support sufficient for it to avoid the painful contractions that those earlier institutions had to undergo. In contrast to Clark especially, the University of Chicago was notably successful in broadening its financial base by attracting substantial funds from local donors, supplementing the largess of the primary patron, Rockefeller. This is not to say that economic constraints played no part in the development

---

[26]Parshall and Rowe, *Emergence*, 23, 282.

[27]Richardson, 366. To give more precise figures, Chicago conferred 237 out of a total of 1286 mathematics Ph.D.s produced during 1862–1934. Harvard and Johns Hopkins, which both had a head start on Chicago, were tied for second place with 103 apiece during the same period.

of the University of Chicago, but only that these constraints were more subtle than at Johns Hopkins or Clark.[28]

The mathematics department was for the most part peripheral to the fundraising spectacle that marked the early University of Chicago. While elaborate courtship measures were undertaken to draw money for suitably impressive laboratories for the science departments, the Chicago authorities took advantage of Daniel Coit Gilman's insight that the mathematicians could get by with modest material facilities. The mathematics department was shunted into quarters in the Ryerson Physics Laboratory, and given distinctly limited appropriations for books, desks, and apparatus. It was not until 1930 that the mathematics department achieved a building of its own.[29]

Nevertheless, the mathematics department opened with an unusually strong faculty contingent by the standards of the time: one full professor, one associate professor, one assistant professor, two tutors, and an assistant. The decision to hire both an associate professor and an assistant professor proved to be a crucial step, and was materially aided by the providential arrival of a new donation just prior to President Harper's interview with Oskar Bolza, one of the candidates. But in one central respect the mathematics department faculty was obtained at a bargain price. Although the university was prepared to offer the previously unheard of sum of $7000 per year for well established scholars as department heads, it found no such individual to head the mathematics department. As a stopgap the university turned to the young E. H. Moore (just 30 years old when appointed) as acting head of the department at a salary of $3500 a year. The key to this appointment was the rapport between Moore and the not much older man who had been chosen as the president of the new university, William Rainey Harper (1856–1906).[30]

Harper was an intellectual prodigy as a child, and indeed seemed to race through his entire life at an exaggerated pace. A freshman at Muskingham College in Ohio at age 10, and a graduate at age 14, Harper proceeded to graduate study in classical languages at Yale in 1873. He received a Ph.D. in

---

[28]Richard J. Storr, *Harper's University: The Beginnings* (Chicago: University of Chicago Press, 1966), 341–358; and Willard J. Pugh III, "The Beginnings of Research at the University of Chicago" (Ph.D. diss., University of Chicago, 1990), 749, 763–764.

[29]Thomas Wakefield Goodspeed, *A History of the University of Chicago: The First Quarter Century* (Chicago: University of Chicago Press, 1916), 230–241, 301–307; Pugh, 554–555.

[30]Goodspeed, *University of Chicago,* 486–489; Parshall and Rowe, *Emergence,* 284–285; Oskar Bolza, *Aus Meinem Leben* (München: E. Reinhardt, 1936), 25; Thomas Wakefield Goodspeed, *William Rainey Harper* (Chicago: University of Chicago Press, 1928), 124–125.

1875, just before his nineteenth birthday, for a thesis entitled "A Comparative Study of Prepositions in Latin, Greek, Sanskrit, and Gothic." Harper ascended rapidly through a series of teaching positions, and filled his off hours and vacations with heavy instructional and administrative duties for the Chautauqua Summer Schools, all the while producing a blizzard of textbooks and specialized articles. By the time of the commotion that ended in the founding of the University of Chicago Harper was a professor of Semitic languages at Yale and Principal of the Chautauqua College of Liberal Arts. Not the least important of his career moves was his decision to become a Baptist while he was teaching at Denison University in Ohio. The University of Chicago was founded as a Baptist institution (John D. Rockefeller was a Baptist), and although there was no religious test for ordinary faculty members, it was required that its president be a Baptist.[31]

Observers since his own day have proposed different ways of evaluating Harper in relation to the continuing evolution of the American university. The interpretations have become more nuanced over time. Critics contemporary with Harper such as Thorstein Veblen and Upton Sinclair declared him to be an instance of the abject surrender of scholarship to business interests. In the 1960s Laurence Veysey portrayed Harper more kindly, treating him as representative of the "tendency to blend and reconcile" the multiple forces pressing on the late nineteenth-century universities. In the 1980s James P. Wind acknowledged the utility of Veysey's picture, but argued that it lacked a crucial element central to Harper's educational goals: Harper's devout Christianity. That is to say, Wind denied that Harper exemplified the "religion of convenience," in the phrase of Burton Bledstein. In Wind's persuasive account, Harper's emphasis on research was ultimately grounded in his religious ideals. It was by effectively making both the church and the university into laboratories that Harper proposed to reconcile reason and revelation and to master the forces of modernity. Lastly, in the 1990s, Julie Reuben further explored the tension between science and religion in Harper's presidency, noting his successful attempt to prevent his university from becoming a Baptist institution. Her illuminating study places Harper among those university leaders who sought to save religion as an integral part of higher education by making it more scientific.[32]

---

[31]Goodspeed, *Harper*, 8–10, 23, 26, 69, 73; James P. Wind, *The Bible and the University: The Messianic Vision of William Rainey Harper* (Atlanta: Scholars Press, 1987), 112–113.

[32]Wind, 4–5, 62–66, 152–159, 178; Laurence R. Veysey, *The Emergence of the American University* (Chicago: University of Chicago Press, 1965), 367–380; Bledstein, 198; Julie

Whatever its source, Harper's emphasis on research would prove significant for the trajectory of mathematics at the University of Chicago and indeed in the country as a whole. He was in important ways even more dedicated to the research role of the university than predecessors such as Charles Eliot and Daniel Coit Gilman; he personally participated in the research life far more whole-heartedly than these older men, and his scholarly accomplishments were more substantial. Coincidentally or not, Harper's enthusiasm for employing terms such as "research," "laboratory," and "inductive," surely outstripped the already noted efforts of these earlier university presidents. Few better than Harper can exemplify what Veysey has aptly referred to as the "giddy heights" reached by the "rhetorical allegiance to science" at the end of the nineteenth century:

> A century ago there was really no such thing as science. The laws of nature were still a secret. There had been much observation, but this was for the most part indefinite, imperfect, uncoordinated. The circle of scientific investigation has now, however, gradually extended itself, until it includes everything, from God himself to the most insignificant atom of creation.[33]

The church and the university, as just mentioned, were laboratories for Harper; even the Holy Land of Palestine was a laboratory, with God as the research director; the Old Testament was a laboratory notebook kept by a staff of laboratory assistants. Most significantly, it was by reference to research that educational and commercial activities could be reconciled:

> Every honest business enterprise has in it the essential elements of educational training. Every business enterprise is a school in which the manager is principal, the heads of departments are teachers, and the staff of employees the pupils. Nay more—it is a great laboratory in which men by working together secure results, the work and the results together exercising an educational influence.[34]

Anyone at the University of Chicago who used the word "laboratory" to characterize an educational proposal (we will see that E. H. Moore and John Dewey

---

A. Reuben, *The Making of the Modern University: Intellectual Transformation and the Marginalization of Morality* (Chicago: University of Chicago Press, 1996), 85–86, 96–101.

[33] William Rainey Harper, *The Trend in Higher Education* (Chicago: University of Chicago Press, 1905), 49. See Veysey, 173.

[34] Harper, *The Trend in Higher Education*, 41. See also Wind, 64.

were two who did), would have been appealing to one of Harper's deepest convictions.

Harper did not adhere to a strict definition of induction, but its essential feature for him was that data collection and organization should precede derivation of general principles. The word and the concept permeated his scholarly research methods, his teaching procedures, and his textbooks. Beginning with *Elements of Hebrew by an Inductive Method* of 1886, Harper was the author or co-author of six textbooks with "inductive" in the title. Induction also informed his approach to the Bible. The Bible reader should begin not with prior assumptions, but rather should examine the evidence of history, language, and personal experience.[35]

There is little evidence that Harper gave any great thought as to how mathematics fit into his view of science. Anecdotes that might suggest some special affinity between Harper and mathematics must be treated with caution. For example, the year after he got his Ph.D. from Yale Harper taught at a very small college in Tennessee. One student here later remembered Harper as "the finest teacher he ever saw, especially of Latin and mathematics." But this proves only to signify that Harper had mastered the standard classical curriculum. There is also a report, by a University of Chicago professor after Harper's death, that filling faculty positions in mathematics and Latin had been particular priorities of Harper in his initial days as President. This too is of doubtful significance; it sounds suspiciously like another instance of politic genuflection toward the classical curriculum such as we have seen with Gilman. Indeed, another faculty member recalled Harper's priorities as not having included mathematics, but rather being Semitics, classics, and philosophy.[36]

---

[35] Wind, 63, 70 and Storr, 57. Harper's "inductive" textbooks are William Rainey Harper, *Elements of Hebrew by an Inductive Method* (Chicago: American Publication Society of Hebrew, 1886); William R. Harper and William E. Waters, *An Inductive Greek Method* (New York: Ivison, Blakeman, and Co., c1888); William Rainey Harper, *Elements of Hebrew Syntax by an Inductive Method* (New York: C. Scribner's Sons, 1890); William R. Harper and Isaac B. Burgess, *An Inductive Latin Primer* (New York: American Book Company, c1891); William R. Harper and Clarence F. Castle, *An Inductive Greek Primer* (New York: American Book Company, c1893); William R. Harper and Isaac B. Burgess, *Inductive Studies in English Grammar* (New York: American Book Company, c1894).

[36] The Tennessee institution was the Masonic College of Macon, Tennessee, apparently no longer in existence. See Goodspeed, *Harper*, 30. The professor who claimed mathematics was a priority for Harper was Ernest D. Burton, professor of New Testament Greek, later the third president of the University of Chicago. See Pugh, 545. The professor who claimed it wasn't was James H. Tufts, professor of philosophy. See Goodspeed, *University of Chicago*, 211.

In designing the departmental structure of the university Harper seems to have regarded mathematics unproblematically as a science, or a tool of science. There is no indication, for example, that his elevation of induction gave him any qualms regarding deductive mathematics such as we have seen with T. H. Huxley. In his first published departmental plan for the University of Chicago he included a joint department of mathematics and astronomy, which would have been conventional for the time. When a patron emerged who wished to specifically support a graduate school of science, mathematics was included.[37]

But although Harper's understanding of mathematics may have been superficial, his knowledge of academic politics and institutions was not. His Chautauqua duties had included searching for and negotiating with teachers in all academic fields, providing him concentrated instruction on relations of power and prestige in late nineteenth-century American colleges and universities. As Daniel Lee Meyer has expressed it, "there were few in the United States by 1890 who could rival the breadth of his acquaintance with academic disciplines and schools of thought on the American educational scene."[38] Very likely Harper had acquired some insight into the relative standing of members of the mathematical community both nationally and internationally. His knowledge of such matters would have been further enhanced by his initial appointment of E. H. Moore, who was eminently well informed.

The previously cited advice of Daniel Coit Gilman had been passed on to Harper by one of his correspondents: "get a great mathematician and a famous Hellenist, and the rest will come." In 1892 Moore was not that great mathematician. He was, however, personally known to Harper from both Yale and Chautauqua, and although much has been made about the broad reach of Harper's faculty search, he began first with those he knew at Denison, Chautauqua, and Yale. Most likely Moore and Harper became acquainted at Yale during 1887–1889 when Moore was a tutor, with Harper subsequently hiring him to teach mathematics at Chautauqua during the summers. When Moore left for Northwestern in 1889 he felt he could no longer spare the time for such elementary teaching as desired by Chautauqua, but he and Harper kept in touch. Harper was thus in a position to know several things about Moore: the high regard for him among the scientists and mathematicians at Yale; his basic competence as a teacher of elementary mathematics; and his ability to function comfortably within an educational organization. Harper may not have fully

---

[37]Daniel Lee Meyer, "The Chicago Faculty and the University Ideal, 1891–1929" (Ph.D. diss., University of Chicago, 1994), 70, 356; Goodspeed, *University of Chicago*, 173–176.

[38]Meyer, 58.

realized what a remarkable talent he had in Moore, but his choice was probably better informed than any one else's could have been at the time, rather than being due to ineffable intuition or a blind stroke of luck. In making Moore "Acting Head" of the Mathematics Department Harper showed the judicious distinctions he was capable of making. He was under no illusions that Moore had already proved himself to be a major scholar in a class with those being hired for other departments, but at the same time he evidently considered Moore competent to carry out the bulk of the administrative duties. In 1896 Moore was awarded the full title of head professor.[39]

For his part, Moore was attracted to the University of Chicago by Harper's repeated assurances that the emphasis would be on advanced instruction and original research. The clear model in the minds of both men was the German university with its scholarly specialists. Harper had not encountered German scholarship at first hand, but had been inspired in its ways by his Yale mentor William Dwight Whitney. Moore pushed the German connection explicitly, and with a specific focus. Immediately after his appointment he began urging on Harper the desirability of hiring the progeny of Felix Klein to assist him in the department. Harper clearly took this message to heart. Within a few weeks he was negotiating with two of Klein's students who had emigrated to the United States, Oskar Bolza (1857–1942) and Heinrich Maschke (1853–1908). Bolza particularly noted how eager Harper was to hire a student of Klein. In the end, both Bolza and Maschke joined Moore in the initial configuration of the Mathematics Department. The two Germans were at first skeptical of how well the department would function under the untried Moore. They were both older than Moore, and moreover, "accustomed to the absolute freedom of the German universities." But Moore soon won their confidence with his mathematical brilliance, his ambitious projects, and his tactful administrative hand.[40]

The story of the remarkable research accomplishments of Moore, Bolza, Maschke and their students is detailed by Karen Parshall and David Rowe. In brief, the Chicago mathematics department in its first fifteen years managed fully what the Johns Hopkins University had given only intimations of doing:

---

[39]On Gilman's advice see Storr, 69. On Moore and Harper's early relations see Goodspeed, *University of Chicago*, 211; Pugh, 743–744; and Moore to Harper, September 20, 1890, UCPP, Box 46, Folder 26. On Moore's promotion to head professor see Bliss and Dickson, 86.

[40]Parshall and Rowe, *Emergence*, 283–291; Wind, 32–34. For Bolza'a views see Bolza, 24, 26. My translation.

to produce a self-reproducing class of research mathematicians who could compete with any in the world. Several features of this achievement are especially notable. First, the overwhelming emphasis of the Chicago mathematicians was on pure mathematics. Applied mathematics and mathematical astronomy as practiced by Simon Newcomb did exist at Chicago, but were of little consequence to the research interests of Moore, Bolza, Maschke, and their students. Moore, Bolza, and Maschke had strong backgrounds in the physical sciences, but when given the freedom to pursue research as they wished each chose topics with little current applicability. This pent up desire to do pure mathematics was clearly more than an American phenomenon, as it was especially well exemplified by Maschke, who had found himself working in industry just prior to accepting the position at Chicago. "I beg you, dear sir," he wrote to Harper in his acceptance letter, "to consider me as very much obliged to you for giving me this excellent opportunity to return again to pure science to which I have aspired ever since I was engaged in more practical work."[41]

Conversely, the research programs of the mathematicians were of minimal interest to the astronomers and physical scientists at Chicago. This is not to say that there was no interaction between the pure mathematicians and others at Chicago. Forest Ray Moulton of the department of astronomy, whose primary research was very much in the Newcomb tradition, contributed papers to the Chicago section of the American Mathematical Society, founded by Moore, and to the *Transactions of the American Mathematical Society*, edited by Moore, in the latter case apparently making an effort to venture into pure mathematics.[42] The Chicago pure mathematicians do not seem to have reciprocated. It would be very difficult to construe any of Moore's research papers as being applied mathematics. There is little trace of any substantive input by the pure mathematicians to research outside their field at Chicago.

The triumph of pure mathematics at Chicago during Moore's tenure, and the political problems this created on occasion, is exemplified in a revealing way by a document written by one or more of the Chicago mathematicians during the last decade of the Moore era, in 1924. Entitled "Pure and Applied Mathematics at the University of Chicago," this 13-page paper was part of an effort to

---

[41] Maschke to Harper, June 17, 1892, WRHP, Box 14, Folder 10. For the research record of the Chicago department see Parshall and Rowe, *Emergence*, 372–401 and Saunders Mac Lane, "Mathematics at the University of Chicago: A Brief History," in Peter Duren, 2:127–154.

[42] See Tropp, "Moulton," 553; and Moulton, "A Simple non-desarguesian Geometry," *Transactions of the AMS*, 3 (1902): 192.

convince the university administration that the mathematics department was deserving of improved facilities. The paper began by noting that since the city of Chicago was the center of an industrial and business empire which was "enormously indebted to science for its prosperity," the university was obligated to "support the researches on which the continued progress of our race depends." Since, moreover, mathematics was "the basis of all science," it followed that the university ought to do the right thing by mathematics "by the erection of a proper building for it and the creation of an adequate endowment." The writer of this document had a figure in mind as the proper amount. "It is not often that $2,000,000 can be invested to better advantage in education work."[43]

While the general argument of this proposal is clear enough, the supporting details are a bit obscure. Two pages were devoted to describing the accomplishments of the Chicago physicists A. A. Michelson and Robert Millikan. Here it was baldly asserted that Millikan's work on measuring the charge of the electron was "important in the development of wireless telephony," but it was left to the reader's imagination how this work was dependent on mathematics in general and Chicago mathematics in particular. The "kindred Departments of Mathematics and Mathematical Astronomy" were described as "working closely with the Department of Physics, and in the same building," a happy circumstance which it was the express purpose of the document to abolish.

In recounting the accomplishments of mathematics at Chicago the researches of Forest Ray Moulton on "the vital problem of increasing the range and accuracy of artillery fire" during the World War were strikingly moved to the front rank. The pure mathematicians were described as winning many honors, but the writer was evidently hesitant to explain what they had done to deserve them. Listing some of the specialized studies initiated by the Chicago pure mathematicians (general analysis, the arithmetic of algebras, modular invariants, and projective differential geometry) created a cloud of obscuring terminology, as the writer was only too aware, but he persevered:

> To the layman these researches in pure Mathematics may seem far removed from the sphere of practical affairs. But the value of Mathematics in the numerous instances in which it has furnished the solution of problems of other sciences otherwise insoluble, to say nothing of its value in developing the power of thought, is sufficient answer to this criticism. Two incidents

---
[43]"Pure and Applied Mathematics at the University of Chicago," EHMP, Box 4, Folder 9, pp. 2–3, 10, 13.

in which Mathematics at the University has served a practical purpose will suffice to illustrate this point:

> The first is the story of a physicist. He was also a mathematician of some attainment, but the solution of fifteen simultaneous equations which resulted from a piece of his work was beyond him. He worked two months with no success. Then he took the problem to Professor Dickson, of the Department of Mathematics, and in two hours he had the solution.

The second story involved Professor Moulton rushing to the assistance of "Mr. Frank Vanderlip," who had run up against an unspecified "need of assistance from higher Mathematics." The one area of departmental achievement in which the writer felt confident enough to be specific was not regarding production of knowledge but of people. The department's prowess in graduating Ph.D.s was noted. Their present occupational categories were listed, with 86% found to be engaged in college and university teaching.[44]

About the same time that this document was being written within the mathematics department, a former Chicago faculty member, Robert Herrick, expressed through the thoughts of a fictional alter ego a cynical view of how the relationship between the university and its environment had evolved:

> The world was finding out every day how it could make profit out of the labors of scholars and ruthlessly drafting them into the universal processes of gain. Few could resist. And the worst was that the higher authorities in the university itself aided and abetted the prostitution, advertising the utility values of the wares they had to offer.[45]

Certainly there is some measure of confirmation of this in the mathematics department document, which did indeed raise utility to a preeminent place. One can see a hazy allusion to the vestiges of the mental discipline thesis ("developing the power of thought", but it was quite subordinate, and with no linkage to

---

[44]Ibid., pp. 5–6, 8–9. Capitalization as in original. Frank Vanderlip (1864–1937), unidentified in the document, was a prominent banking executive. See Eugene A. Agger, "Vanderlip, Frank Arthur," *DAB*, (supp. 2):677–679.

[45]Robert Herrick, *Chimes* (New York: Macmillan, 1926), 70. Herrick joined the Chicago English department in 1893 and stayed 30 years. *Chimes* is a thinly disguised fictional depiction of the University of Chicago as he knew it, with President Harper transformed into President Harris, and Herrick himself represented by Professor Clavercin. See Wind, 149–152. I cannot detect in *Chimes* any fictional portraits of the mainstays of the mathematics department.

the traditions of the classical curriculum. A striking feature of this attempt to advertise the utility values of mathematics was the slim documentation. One must presume that if the pure mathematicians had any strong evidence to offer of sustained intellectual contact between themselves and the physical sciences or the business world that they would have offered it up on this occasion, but all they could produce were vague anecdotes. Not only was it unclear what these anecdotes had to do with "the continued progress of our race," but the ability to solve fifteen simultaneous equations was forced to stand in for the entire range of accomplishments in pure mathematics by the Chicago faculty and students for over 30 years. Thus by the close of the Moore era pure mathematics had become vigorous enough to demand increased resources from the outside lay public, but apparently resigned to evasion in the face of explaining its workings to that public. In the case of the particular endeavor represented by this document the stratagem seems to have been effective, or at least not damaging: the University of Chicago mathematics department moved into its new quarters in Eckert Hall in 1930.

But while the mathematics department's orientation was decisively purist, its members contributed to a wide variety of subspecialties within pure mathematics. The interests of Moore, Bolza, and Maschke were not identical, and Moore's interests were especially broad. Bolza recalled that he and Maschke

> saw with increasing admiration how Moore in swift succession embraced an astonishing series of the most various mathematical disciplines, constantly drawing in the most difficult problems, always striving after rigor and generality, and throughout leaving his traces behind in the form of deeply penetrating, valuable publications.[46]

It is furthermore evident that research at Chicago was highly collaborative. Two of the most successful products of the program described Moore's approach to bringing his advanced classes to the leading edge of the research frontier:

> Frequently he came to his class with ideas imperfectly developed, and he and his students studied through successfully or failed together in the study of some question in which he was at the moment interested.[47]

---

[46]Bolza, 26. My translation. Moore's biographer G. A. Bliss divided Moore's publications into five categories: (1) geometry; (2) groups, numbers, algebra; (3) theory of functions; (4) integral equations, general analysis; and (5) miscellaneous. See G. A. Bliss, "The Scientific Work of Eliakim Hastings Moore," *BAMS* 40 (1934):501.

[47]Bliss and Dickson, 87.

The Mathematical Club which Moore had almost immediately established kept all participants aware of what others in the department were doing, and the seminars concentrated the attention of an entire class on a particular region of investigation, with several publications possibly resulting. Thus, for example, although Oswald Veblen is usually credited with the first rigorous proof of the celebrated Jordan curve theorem, Veblen's paper on the subject suggests a culmination of a group effort. He cited closely related work of two of his fellow students, N. J. Lennes and G. A. Bliss, Moore's 1901–1902 seminar on Foundations of Geometry, and more generally "the discussions of the subject that have taken place under the leadership of Professor Moore."[48]

It seems clear that the versatile and varied mathematical talents developed by Chicago graduate students benefited them greatly when they moved on to faculty positions in their own right, at Chicago itself and at other institutions. Probably the most prominent of Moore's students were Leonard Eugene Dickson (1874–1954), Oswald Veblen (1880–1960), George David Birkhoff (1884–1944), and Robert Lee Moore (1882–1974). These four between them covered a wide range of research subspecialties in geometry, analysis, algebra, and topology; they became major figures at the University of Chicago, Princeton, Harvard, and the University of Texas, respectively, where they produced many additional doctoral students in their turn; they each became president of the American Mathematical Society; and they each became a member of the National Academy of Sciences.[49]

The Chicago doctoral students benefited not only from their own talents but also from the vigorous support of E. H. Moore on their behalf. Moore was very sensitive to prospective job openings and opportunities for honors for American mathematicians, and he was especially eager that Chicago students get their

---

[48]Oswald Veblen, "Theory on Plane Curves in Non-Metrical Analysis Situs," *Transactions of the AMS* 6 (1905): 83.

[49]On Dickson see Della Dumbaugh Fenster, "Role Modeling in Mathematics: The Case of Leonard Eugene Dickson (1874–1954)," *Historia Mathematica* 24 (Feb. 1997): 7–24; Parshall and Rowe, *Emergence*, 379–381; and Ronald S. Calinger, "Dickson, Leonard Eugene," *DSB*, 4:82–83. On Veblen (nephew of Thorstein) see Parshall and Rowe, *Emergence*, 383–384; and Saunders Mac Lane, "Veblen, Oswald," *DSB*, 13:599–600. On Birkhoff see Parshall and Rowe, *Emergence*, 390–392; and Marston Morse, "Birkhoff, George David," *DSB*, 2:143–146. On R. L. Moore see John Parker, *R. L. Moore: Mathematician and Teacher* (n.p.: Mathematical Association of America, 2005); Parshall and Rowe, *Emergence*, 384–387; and Albert C. Lewis, "Moore, Robert Lee," *DSB*, 18:651–653. Altogether, Moore supervised 30 doctoral students at the University of Chicago. See Bliss and Dickson, 88.

share.⁵⁰ His aggressive approach was sharply different from Simon Newcomb's much more cautious attitude as head of the Hopkins department. Moore was likewise much more confident that his moves had the potential to influence the future course of American mathematics as a whole. He was proud of his students, and convinced that they would elevate any department to which they became attached. By the time he had been running the Chicago department for 15 years he was secure in believing that it had few rivals in the United States. In 1907 when Oswald Veblen was debating whether to take a position at Yale, Moore mused on "the thought of making Yale stand on a par with Harvard and Chicago."⁵¹ Then, a short time later, when Veblen chose to stay at Princeton with assurances of increased emphasis on graduate education, Moore was again enthused:

> This means Princeton is really entering the lists in the way of graduate work, and that you and the others of the Chicago circle should be in and effectively active at the beginning of the movement is a great source of joy to me.⁵²

Such claims to distinction for Moore's department as a producer of mathematical talent had become firmly established outside Chicago by 1919, when R. G. D. Richardson of Brown University wrote to Moore as follows:

> It was very disappointing that we were not able to get a Chicago man for our Brown position ... Shouldn't something be done to insure a succession of teachers of mathematics and also to provide for a proper distribution?⁵³

5.4. Moore's Initiatives for Enhancing American Mathematical Research

As the preceding pages have suggested in passing, Moore expended his energies not merely on behalf of his department of mathematics in isolation,

---

⁵⁰The topics of honors and jobs are frequent in the surviving correspondence between Moore and Oswald Veblen. See OVP, Box 8.

⁵¹Moore to Veblen, May 22, 1907, OVP, Box 8.

⁵²Moore to Veblen, May 26, 1907, ibid. Also at Princeton at this time were E. H. Moore's student R. L. Moore, Oskar Bolza's student G. A. Bliss, and Moore's brother-in-law John Wesley Young, a doctoral grandson of Bolza. See Archibald, *History of the AMS*, 201, 240; R. D. Beetle and C. E. Wilder, "John Wesley Young—In Memoriam," *BAMS* 38 (1932): 604; and Henry S. Tropp, "Young, John Wesley," *DSB*, 14:559.

⁵³Richardson to Moore, July 6, 1919, EHMP, Box 3, Folder 4.

5.4. MOORE'S INITIATIVES FOR AMERICAN MATHEMATICAL RESEARCH

but in the service of a larger vision of how American mathematics should develop as a whole. With respect to the politics of American mathematics Simon Newcomb had for the most part been content to lament and applaud from a dignified distance, occasionally participating in an action initiated by someone else, "providing it is going to be a success."[54] It was Moore's generation of mathematicians who were willing to take the pioneering risks, and Moore himself was one of the boldest. "Aggressive" was a word that seemed to come readily to mind when his contemporaries sought to describe Moore and his circle, and always as a positive attribute.[55] Moore established this reputation during his first ten years at Chicago. His Chicago colleague Oskar Bolza later noted that there occurred a series of events during these ten years "of the greatest importance for the development of mathematics in America," and that the "true propelling force" behind all these events was "E. H. Moore mit seinem aggressiven Enthusiasmus und seinem Organisationstalent," a characterization that readily leaps the language barrier.[56]

The first of these events referred to by Bolza was the Mathematical Congress held in Chicago in 1893, in conjunction with the World's Columbian Exposition. The organizers of this exposition, building upon a precedent set by the Paris fair of 1889, arranged to have "congresses" convened during the period of the exposition to discuss various topics of interest to humankind. It will be recalled that the Committee of Ten Report was originally to have been unveiled at the International Congress on Education in July 1893. The Mathematical Congress was one of several Congresses on Science and Philosophy scheduled to meet the week of August 21.[57]

Moore almost immediately saw this occasion as a great opportunity to advance American mathematics. Two aims especially animated his actions: to bring Felix Klein to the United States, however briefly, and to defend the interests of pure mathematics. As early as June 1892 President Harper (likely at Moore's behest) had invited Klein to lecture at the university during the

---

[54]Newcomb to T. S. Fiske, Jan. 29, 1891, letterpress copy in SNP, Box 6.

[55]Bliss and Dickson use forms of "aggressive" at least five times in their biographical sketch of Moore; three times on page 86 and twice on page 92.

[56]Bolza, 31–32. My translations.

[57]Parshall and Rowe, *Emergence*, 303. Reid Badger, *The Great American Fair: The World's Columbian Exposition and American Culture* (Chicago: Nelson Hall, 1979), 99. Henry S. White, "A Brief Account of the Congress on Mathematics, Held at Chicago in August, 1893," in E. Hastings Moore et al., eds., *Mathematical Papers Read at the International Mathematical Congress Held in Connection With the World's Columbian Exposition Chicago 1893* (New York: Macmillan and Co., 1896), vii.

summer of 1893 in parallel with the fair and the congresses.[58] This apparently could not be arranged; in December 1892 Moore urged Harper to see if Klein might be persuaded to come to the University of Chicago for the fall quarter of 1893, dwelling on the tremendous boost it would give to the department: "The stay of Klein at Chicago would make us for the time being the focus of mathematical activity for the whole country," resulting in an influx of graduate students both immediately and over the longer term.[59]

Simultaneously Moore was joining forces with another young University of Chicago dynamo, astro-physicist George Ellery Hale, to make local arrangements for the World's Congress in Mathematics and Astronomy. The national effort on behalf of this congress, which was to include mathematics, general astronomy, and astro-physics, was being conducted by a committee of the American Association for the Advancement of Science (AAAS), headed by none other than Simon Newcomb.[60] In writing to Newcomb, Moore expressed his basic aim:

> Permit me to express the hope that in the mind of the committee the interests of pure mathematics be recognized as coordinate with those of the other two branches. This is especially important because we have in America so few men in pure mathematics, and most of the papers ought to be by foreigners, leaders of investigation in England and the continent. A request from your committee to these men for papers would be apt to be favorably received, for the members of the committee have international standing. I am anxious that the congress in pure mathematics be as successful as those in astronomy.[61]

In his reply, Newcomb sought to dampen Moore's enthusiasm, and to distance the AAAS from responsibility for what he clearly saw as a dubious undertaking. "I see no hope of a separate congress of pure mathematics," he asserted, and he doubted that many European mathematicians would be able to come.[62] Newcomb did follow up on Moore's request for aid by writing to at least one prominent European mathematician of his acquaintance, Gösta Mittag-Leffler (1846–1927), but his promotion of the congress in this letter was decidedly half-hearted. Newcomb expressed the hope that Mittag-Leffler might

---

[58]Parshall and Rowe, *Emergence*, 304–305.
[59]Moore to Harper, Dec. 23, 1892, UCPP, Box 17, Folder 2.
[60]Newcomb to Hale, Oct. 17, 1892, letterpress copy in SNP, Box 6.
[61]Moore to Newcomb, Nov. 3, 1892, SNP, Box 32.
[62]Newcomb to Moore, Nov. 5, 1892, letterpress copy in SNP, Box 6.

be able to come as an official representative of Sweden, but he warned him that "the success of the project is quite doubtful."[63]

Moore and the Chicagoans pressed on regardless. They were rewarded with one piece of good fortune not of their making but of which they were well prepared to take advantage. Germany had elected to use the Chicago exposition as a major world forum for advertising its industrial and scientific accomplishments. From this emerged a plan to stage a German university exhibition at the fair, and also to send Felix Klein as an official representative of the Prussian Ministry of Culture. Klein, who saw this as an opportunity to further his own agenda as a leader of German mathematics, then arranged with Moore and the local committee not only to attend the Mathematical Congress, but also to stay an additional two weeks to deliver a series of lectures. As it turned out these lectures were held not at the University of Chicago but north of the city at Northwestern, with Klein's former doctoral student Henry S. White as host.[64]

Moore had proposed to Harper in May 1893 that the University of Chicago offer Klein a suite of dormitory rooms for the colloquium, but Harper vetoed the proposal. By the time the colloquium got underway in September, Harper (ever the harried executive) had evidently forgotten his earlier decision, and seems to have asked Moore why the colloquium was being held at Northwestern rather than at the University of Chicago! Moore diplomatically reminded Harper of the latter's strictures against "creating a precedent likely to be troublesome during the World's Fair."[65] Whatever aggravation Moore may have felt at any of this was easily outweighed by his enthusiasm for having Klein in the vicinity, even for this limited time. Indeed, in this same letter to Harper of September 1893 Moore was willing to hand Chicago mathematics into Klein's hands entirely, if it could be arranged:

> Perhaps some day you will be ready to establish head-professorship and Klein will be willing to accept: which would be the thing for our department.[66]

---

[63]Newcomb to Mittag-Leffler, Nov. 7, 1892, letterpress copy in SNP, Box 6.

[64]Parshall and Rowe, *Emergence*, 303–305, 331.

[65]Moore to Harper, Sept. 5, 1893, UCPP, Box 46, Folder 26. The nationwide depression of 1893 created grave financial difficulties for the University of Chicago by drying up many expected contributions. Harper was in a near constant state of high anxiety until the end of August. See Storr, 244–250.

[66]Moore to Harper, Sept. 5, 1893, UCPP, Box 46, Folder 26. Underlining in original.

Acting Head-Professor Moore was not yet confident that he would become the thing himself.

Newcomb turned out to be quite right in one of his predictions regarding the Mathematical Congress; only four of 45 mathematicians attending were from outside the United States. Indeed, even the eastern United States was only sparsely represented. Newcomb himself, preoccupied with *Nautical Almanac* politics such as the Morrison affair (see Chapter 3) was unable to get to Chicago, although he had been invited to speak not only by mathematicians and astronomers but by economists as well. Nevertheless, Klein's presence alone made the Mathematical Congress a significant event, which would prove precedent setting. Moreover, Moore maneuvered to produce what was effectively a pure mathematical congress. The mathematicians and astronomers met jointly for about 90 minutes for some general speech-making (including one by Klein). Then, "on motion of Professor E. H. Moore the Congress divided into two sections, one of Mathematics and the other of Astronomy and Astro-Physics."[67]

For the rest of the week the mathematicians heard a series of lectures and paper presentations on a wide range of topics, dominated by pure mathematics. Klein spoke not only on his own behalf but also presented to the congress a number of papers he had brought with him by leading mathematicians back in Germany, including David Hilbert, Hermann Minkowski, Adolf Hurwitz, and Arthur Schoenflies, names well known to many mathematicians more than 100 years later. In his own remarks Klein was much concerned with promoting mathematical unity, both intellectual and social. He saw hopeful signs in the increasingly central role being played by such unifying concepts as "function" and "group," and in the movement to form new organizational structures, national and international. He lauded the founding of the New York Mathematical Society in particular. Such organizations were needed because "What was formerly begun by a single master-mind, we must seek to accomplish by united efforts and cooperation."[68]

Klein also expounded upon educational reforms he was proposing in Germany. He mentioned the desirability of giving future teachers exposure to physical and astronomical applications of mathematics and the general need for more

---

[67]"The Congress on Mathematics, Astronomy, and Astro-Physics," *Astronomy and Astro-Physics* 12 (1893): 743. See Parshall and Rowe, *Emergence*, 309–312. On Newcomb's preoccupation see Newcomb to Mittag-Leffler, July 5, 1893, letterpress copy in SNP, Box 6.

[68]Felix Klein, "The Present State of Mathematics," in Moore et al., 134–135.

5.4. MOORE'S INITIATIVES FOR AMERICAN MATHEMATICAL RESEARCH

cooperation among mathematicians, physicists, and engineers.[69] These ideas of Klein would later be echoed by E. H. Moore.

The Chicago mathematicians acquitted themselves well, with Moore providing "unquestionably the most important contribution to mathematical research to issue from the Chicago Congress," in the judgment of Parshall and Rowe. Moore reported to Harper that "Our Congress was a very good success, and must have done our Chicago department good."[70]

Two days after the end of the congress many of the attendees reconvened in Evanston to listen to Klein's series of survey lectures on modern mathematics, and for abundant informal interaction with Klein. His lectures, delivered in English, inspired what Klein described as "scientific enthusiasm" in the audience. For those best prepared, notably Moore and his University of Chicago colleagues, this two week colloquium was an intense mathematical experience, with some of the sessions lasting five hours.[71]

Moore quickly built upon the success of the congress and the Evanston Colloquium. In April 1894 Moore, who had joined the New York Mathematical Society during its first major membership campaign in 1891, proposed that the society publish the proceedings of the congress. The conservative Newcomb opposed this move on the reasonable grounds that the society "is in no way responsible for the work of the Chicago Congress." Moore's purpose, however, was precisely for the society to take on this responsibility retroactively, thus making it more of a national society. He made his case sufficiently persuasive that his publication scheme was adopted in June. His larger purpose was symbolically ratified later that summer when the society officially changed its name to the American Mathematical Society. This name change was another one of the key events noted by Bolza, and due at least in part to Moore's activism.[72]

Bolza also saw Moore's hand behind the 1896 proposal of Northwestern's H. S. White that the AMS sponsor regular colloquia somewhat along the lines of Klein's Evanston lectures of 1893: concentrated cycles of lectures by leading experts. Another Moore initiative was the 1897 founding of a "Chicago Section"

---

[69]Felix Klein, "Über die Entwicklung der Gruppentheorie während der letzen zwanzig Jahre," in Moore et al., 136. For reasons that remain obscure, the title of Klein's talk corresponds not at all to the published abstract. See Parshall and Rowe, *Emergence*, 325.

[70]Parshall and Rowe, *Emergence*, 324–325. Moore's paper was entitled "A Doubly-Infinite System of Simple Groups." Moore to Harper, Aug. 28, 1893, UCPP, Box 46, Folder 26. Underlining in original.

[71]Parshall and Rowe, *Emergence*, 333–334.

[72]Newcomb to Thomas Fiske, May 21, 1894, letterpress copy in SNP, Box 6. Archibald, *History of the AMS*, 7. Bolza, 31. Bliss and Dickson, 91.

of the AMS, a further scheme for giving a national patina to ongoing local mathematical activity.[73] This, like other of Moore's organizational ideas, was also designed to counter what he saw as the condescending attitude of many mathematicians of the northeastern United States toward the "west." Some 30 years later he would remind his former student Oswald Veblen of the bad old days:

> You realize how the North Center has developed in the last 40 years. I am older. I know how the North East formerly wrapped robes of—shall I say? - Pharisaism [sic] about itself and ignored the signs of promise in the North Center.[74]

Finally, Bolza noted Moore's role in founding a new journal, the *Transactions of the American Mathematical Society*. By the late 1890s a number of members of the AMS were complaining that there were too few American outlets for research articles by American mathematicians. The first solution proposed was that the AMS should take over and expand the *American Journal of Mathematics*, run by Johns Hopkins University. Negotiations were conducted with Simon Newcomb, editor of the *American Journal*. These negotiations failed, partly because of general philosophical differences and partly because of personal conflicts. The final result was the founding of an entirely new journal, under the auspices of the AMS, with Moore as editor-in-chief. Moore wrote to Newcomb that his goal in these negotiations had been, and continued to be, "to utilize this organization [i.e., the AMS] to the utmost in the furtherance of the interests of the science." He looked forward to having "two high grade journals in relations of friendly rivalry."[75] Moore's vigilant editorship of the *Transactions* set a high standard for that publication.

The capstone was placed upon Moore's first decade of institution building by his election as president of the AMS in 1901. He was much younger than any of his predecessors and more devoted to pure mathematics. He was also the first AMS president to have earned a Ph.D.[76] On his retirement from this position in December 1902 Moore delivered an address to a meeting of the society, as had become customary. His address surprised many of his listeners by being primarily devoted not to research issues but to pedagogy. Moore's attention

---

[73] Bolza, 31; Archibald, *History of the AMS*, 67–68, 74–76.

[74] Moore to O. Veblen, Nov. 3, 1928, OVP, Box 8.

[75] Moore to Newcomb, Feb. 6, 1899, SNP, Box 22. See Bolza, 31–32; Parshall and Rowe, *Emergence*, 411–414; and Archibald, *History of the AMS*, 56–60.

[76] Parshall and Rowe, *Emergence*, 415; Archibald, *History of the AMS*, 110, 112, 117, 124, 139.

had not in fact been entirely devoted to promotion of mathematical research in his first ten years at Chicago, and it is to these other interests that we now turn.

CHAPTER 6

The Development of E. H. Moore's Pedagogic Program

E. H. Moore himself never fully explicated his interest in pedagogy. Near the end of his most important pedagogical statement, his 1902 speech to the American Mathematical Society (AMS), he made a tantalizing feint toward this topic, but then veered off into standard speaker's etiquette: "I might explain how I came to this question of pedagogy of elementary mathematics. I wish, however, merely to express my gratitude to many mathematical and scientific friends ..."[1] This suggests that the explanation of Moore's interest in pedagogy was not something easily encapsulated in a sentence or two.

To make a beginning at explaining the genesis of Moore's pedagogical ideas we will look first at his relationships with several individuals with whom he interacted at the University of Chicago, both above him and below him in the organizational hierarchy. These relationships help us understand how mathematics education was viewed within the university, and beyond. We will then examine some pedagogical ventures made by Moore and his mathematics department colleagues during his first decade at Chicago, including a textbook and participation in several educational committees at the regional and national level. Finally, the meat of the chapter is devoted to describing Moore's laboratory method of mathematics instruction, which he expounded most fully and enthusiastically in the above mentioned address to the AMS at the end of 1902.

Moore's intellectual life was thoroughly focused on mathematics, and he did not embed his pedagogy within a publicly articulated comprehensive worldview, as Simon Newcomb had done.[2] At the same time, the professional import

---

[1]Eliakim Hastings Moore, "On the Foundations of Mathematics," *Science* 17 (Mar. 13, 1903): 415.

[2]Newcomb's worldview is detailed in Albert E. Moyer, *A Scientist's Voice in American Culture: Simon Newcomb and the Rhetoric of Scientific Method* (Berkeley: University of California Press, 1992).

of Moore's pedagogical thought was much less diffuse compared to Newcomb. Because Moore was so vitally concerned with the health of the mathematical enterprise, especially with establishing opportunities for research in pure mathematics, his pedagogical efforts were more closely tied to his professional agenda as a mathematician. Moore, from the time he took command of the mathematics department at Chicago, sought to advance the cause of mathematical research in the United States. He recognized, however, that production of research results would not suffice to justify a university department of mathematics to the wider society, or even to the wider university. Nor could the university research mathematicians content themselves with merely reproducing their own kind, a species whose proportion among university graduates would never be large. To thrive as Moore desired, the mathematicians must service a much larger clientele than themselves. Consequently he sought a curriculum that would satisfy the needs of consumers of mathematical knowledge, while at the same time retaining the interest and allegiance of the mathematicians producing that knowledge. In this grand compromise was born his "laboratory method" of instruction.

Moore's instructional ideas were not resoundingly original. He borrowed from various sources, as will be further discussed in the next chapter. What does make Moore distinctive, especially among mathematical educators of his time, was his perception that pedagogy, the theory and practice of classroom activity, could not be divorced from developments outside the classroom, especially the political relationships among the professional groups concerned with education. Moore saw that ideas for reforming instruction would not sell themselves, but required negotiation among and within these professional groups. College and university mathematicians were his special concern. This was not merely because he knew them better than other stakeholders, but because he sincerely believed that these mathematicians would need to change some of their fundamental attitudes in order to make his proposed reforms succeed. Whatever one might say about the merits or demerits of Moore's program within the classroom, it was a delicate political exercise outside the classroom, and he understood it to be so.

6.1. The Place of Pedagogy in Moore's Department of Mathematics

It should be acknowledged that there is one appealingly simple explanation for Moore's attraction to pedagogy which, if true, he might not have been inclined to proclaim loudly. In this view, which is intimated in the dissertation of Willard J. Pugh III in 1990, the causal chain goes back to the relations between John D. Rockefeller and William Rainey Harper. Rockefeller, according to Pugh, had originally envisioned a relatively modest Baptist college in Chicago, while Harper wanted a great university, with research the paramount endeavor. Harper, through low cunning or high-minded persuasive skill, managed to put much of his vision in place initially, but Rockefeller, holding the financial power, eventually reigned in Harper's taste for research extravagance and encouraged more attention to undergraduate teaching. Harper, pinched by Rockefeller's manipulation of the purse strings, felt obliged to modify his views regarding teaching versus research in his university, and in turn pressed his faculty to devote more effort to pedagogy. Moore was one of those thus pressured.[3]

This view cannot be entirely dismissed. There does seem to be a rough correlation between University of Chicago financial crises, Harper's pushing of the teaching mission, and Moore's pedagogical excursions. But there were other factors at work as well. It is important to assess more carefully the nature of Harper's aims, Moore's aims, and the changing educational environment in which they functioned.

As for Rockefeller's aims, they are difficult to assess. Most of his negotiation concerning the University of Chicago was done through intermediaries. Some individuals involved in the founding were strongly committed to establishing a college rather than a university, but it is unlikely that Rockefeller was among them. It would be more fair to say his views were unformed. There can be no doubt that Harper often pushed ahead faster than Rockefeller and his advisers thought fiscally prudent, and that the resulting calls for restraint were sometimes interpreted by Harper and his advisors as questioning the value of research. But one must be cautious about detecting fundamental shifts in Harper's ideas based on isolated quotations from his correspondence, which is Pugh's method.

Harper operated in an atmosphere of perpetual crisis. Many of these crises were brought on by himself, as he pressed for advantage in one direction or another, or tried to respond to the consequences of earlier decisions. Unlike Charles Eliot at Harvard, Harper was creating almost everything from scratch, and he opted for a far faster pace of development than Gilman at Hopkins. His many pronouncements and decisions were not always well thought through, and

---

[3]Willard J. Pugh III, "The Beginnings of Research at the University of Chicago" (Ph.D. diss., University of Chicago, 1990), 742, 757, 763, 766.

cannot always be mapped to his deepest convictions; we have seen a minor instance with the Klein colloquium mix-up. There were periods of loud complaint by the research-minded faculty that opportunity and facilities for investigation were not what they had been promised when Harper lured them to Chicago, but such incidents should not be taken as indicating a major shift in Harper's thinking about research.[4]

More persuasive pictures of Harper are provided by James Wind and Richard Storr, who acknowledge that Harper was pushed hither and yon by the exigencies of administration, but conclude that his fundamental designs did not change. These designs of Harper were immense. Building on his Chautauqua experience, and in the service of his Christian mission, according to Wind, Harper proposed to extend the reach of the new university well beyond the Hyde Park campus.

One means of reaching out was through University Extension, an ambitious program of off-campus courses and correspondence study programs. Another was to create a web of connections to other educational institutions through so-called affiliation. Colleges, professional schools, and secondary schools were all offered the opportunity to affiliate with Harper's rising university. Curricula and examinations would interlock, books and apparatus would be loaned, qualified students would be able to move readily among the component institutions. In seeking to bring secondary schools within its influence, Harper remarked, "The University confesses frankly its desire to affect the work of these institutions as to secure more thoroughly prepared students for college and university work." Thus affiliation of academies and high schools was in part a version of the accreditation method for entry into college which had become popular in the Midwest.[5]

One should also note Harper's debt to emerging trends in American educational thought. His desire to bring a new order to the chaos of the educational world was closely allied to the movement that spawned the Committee of Ten. Indeed, Storr sees Harper as attempting to use the University of Chicago to achieve the very same goals that Eliot was aiming at through the Committee. And one can also readily see Harper's imperial scheme as a prime example of the proliferating interest in educational efficiency. As one of his lieutenants put it, the mission of the University of Chicago "is to help make every educational

---

[4]Richard J. Storr, *Harper's University: The Beginnings* (Chicago: University of Chicago Press, 1966), 20–21, 267–68; Pugh, 750–762.

[5]Storr, 196–222 and James P. Wind, *The Bible and the University: The Messianic Vision of William Rainey Harper* (Atlanta: Scholars Press, 1987), 113.

institution within the circuit of its possible influence stronger and more efficient than it could be without the co-operation of a great central educational clearing house."[6]

The overriding point is that Harper's concern with education was a broad one from the beginning of his involvement with the University of Chicago, and that while he certainly changed tactics many times, his overall strategy remained largely consistent. In particular, it is an oversimplification to see Harper as originally placing research ahead of teaching and later transposing these priorities. He did not propose a purely graduate institution and then find himself forced to retreat to include lower level instruction; such a description is far more appropriate to G. Stanley Hall at Clark University. Rather, Harper from early on saw the University of Chicago as encompassing the widest possible range of teaching levels. Further, as we have already noted, research for Harper was not a discrete activity that could be readily separated from other parts of the educational endeavor. Rather, research ought to be at the foundation of all education. Harper even proposed that Sunday Schools, for example, ought to institute seminar methods, with pupils given prizes for the depth of their scholarly work.[7]

Moore's interest in pedagogy likewise cannot be solely attributed to direct pressure from Harper. Whatever Harper's particular directives, the need to handle the large numbers of undergraduate students appearing in the mathematics classrooms at Chicago was in itself a powerful incentive to become involved with instructional issues. Enrollments began at a relatively high level, and increased rapidly. All undergraduates in the arts and sciences were required to take some mathematics.[8] Moore was faced with the continuing challenge of marshaling the individual talents of his graduate fellows and faculty to meet these demands, while endeavoring to facilitate mathematical research as much as possible. Ideally all his doctoral students and faculty would have been able and committed researchers like himself, but he recognized when this was not so and compromised accordingly. Three cases are especially revealing.

The first case is Jacob William Albert Young (1865–1948), who had the title of tutor in the original Chicago mathematics department of 1892, having just received a Ph.D. from Clark University. Negotiations to obtain Young began

---

[6]Storr, 214–215.

[7]Daniel Lee Meyer, "The Chicago Faculty and the University Ideal, 1891–1929" (Ph.D. diss., University of Chicago, 1994), 314–315 and Wind, 100.

[8]See Meyer, 133. Storr, 118–119. Also see *Circular of the Departments of Mathematics, Astronomy and Astrophysics, Physics, Chemistry*, University of Chicago, (1908): 7.

as early as April 1892.⁹ In an early letter to Moore, Young emphasized his determination "to continue my own mathematical work." He recognized that if he were hired as a tutor in mathematics at Chicago he would be teaching primarily undergraduate courses, but added that

> I should be glad to offer a course for the graduate students beginning the subject in which I am specially interested,—the <u>theory of groups</u>,—or also an introductory course in the <u>theory of numbers</u>, which has been an important auxiliary in my work, and to which I have paid quite a little attention.[10]

These words were well calculated to appeal to Moore, who was at that time developing a strong interest in the theory of groups himself. But Young, it would turn out, did not take his place beside Moore in exploring the research frontier. One finds, for example, in looking over the early records of the department's Mathematical Club, that after an initial flurry of presentations by Young on group theory, closely related to his dissertation, he tended more and more to concentrate on exposition of pedagogical topics.[11] By 1896 Moore acknowledged the reality of the direction Young had taken, writing to Harper to argue the case for promoting Young to an assistant professorship:

> He is a vigorous and in every way good teacher. His papers before the Mathematical Club are invariably well thought out and clearly presented. He is not so deeply and actively interested in the investigational side of mathematics as I should wish. But, as I told you, he is developing much interest in the pedagogical and historical side of the subject.
>
> Under the new reorganization of the faculties he is certainly exactly the man to take charge of the work of the department in the Academic College, and you must share my satisfaction that we have such a man available.[12]

---

[9] See W. E. Story to W. R. Harper, Apr. 6, 1892, WRHP, Box 15, Folder 23. J. W. A. Young is not to be confused with J. W. Young (John Wesley Young, Moore's brother-in-law mentioned previously).

[10] Young to Moore, May 9, 1892, WRHP, Box 15, Folder 23. Underlining in original, although whether it was done by Young or by Moore is unclear.

[11] MCR, Boxes 1–2. On Moore's interest in group theory see Bliss, "Scientific Work of E. H. Moore," 504.

[12] Moore to Harper, Jan. 31, 1896, UCPP, Box 17, Folder 2. The four year program of undergraduate instruction had from the first been divided into two components: students in their first two years were considered to be in the "Academic College" while those in their

This shows both Moore and Harper adjusting to their environment. The reorganization referred to had its origin not with Rockefeller or Harper, but rather from within the faculty itself. Much of the agitation came from the more junior faculty members, who were most likely to be responsible for the more elementary classes, and who sought greater autonomy in this endeavor.[13] Probably some senior faculty welcomed it as an opportunity to further distance themselves from lower level instruction. Doubtless within Moore's eagerness to put Young into what he considered the most appropriate career track one can discern the continuation of the call for stratification of instructional duties found earlier in the century in Benjamin Peirce, but we shall also see that there was more to Moore's position than this.

Young's status was more fully formalized in 1897 when he was given the title of Assistant Professor of the Pedagogy of Mathematics. He continued to devote most of his attention to mathematical pedagogy until his retirement from the University of Chicago in 1926 as an associate professor.[14]

Moore's stance is further revealed by his attitude toward Herbert Ellsworth Slaught (1861–1937). Slaught was among the first class of graduate students given a fellowship to study in the mathematics department. He was not fresh out of college (indeed he was slightly older than Moore), nor had he lately been engaged with higher mathematics at all; rather he had been principal of a private academy in New Jersey. This caused Moore to hesitate, but Slaught made a good personal impression, and strong recommendations had arrived on his behalf.[15] Moore was led to conclude that

> in Mr. Slaught's case we shall be abundantly justified in giving him a fellowship in view of the excellence and independence of his work in the elementary teaching and in view of his evident force of character and scholarship.[16]

Slaught arrived in Chicago with a strong pedagogical orientation, which was reinforced by the fact that he felt compelled to augment his fellowship with additional lower-level teaching whenever he could, since he had a family to support. These extra teaching responsibilities delayed Slaught's obtaining the

---

last two years were in the "University College." Later these terms were altered to "Junior College" and "Senior College." Goodspeed, *University of Chicago*, 138.

[13] Meyer, 206–209.

[14] *University of Chicago Register* (1897–1898): 278; Saunders Mac Lane, "Mathematics at the University of Chicago: A Brief History," in Peter Duren, 2:133.

[15] G. A. Bliss, "Herbert Ellsworth Slaught—Teacher and Friend," *AMM* 45 (1938): 6–7.

[16] Moore to Harper, Mar. 31, 1892, WRHP, Box 14, Folder 15.

Ph.D.[17] Although Moore was a strong believer in the doctorate as the crucial credential for academic advancement, he soon recognized Slaught's value to the department, as illustrated in the following extract from a letter to Harper:

> The general principle that no man without mathematical doctorate go higher than assistantship I believe in and hold to. If there are to be exceptions, Slaught certainly deserves to be an exception, for he is a thoroughly good teacher, and he is doing excellent work in supervising the entrance-examination reading.[18]

The question at issue was whether Slaught should be advanced from "assistant" to "tutor," the original system of faculty ranking at the University of Chicago being, in ascending order: docent, reader, assistant, tutor, instructor, assistant professor, associate professor, professor, head professor.[19]

Slaught's pre-doctoral pedagogical work gave him occasional contact with Harper independent of Moore. In 1898 the two men discussed the choice of geometry textbooks for the schools of the city of Chicago and for the Chicago Manual Training School, with Slaught giving voice to ideas likely to please Harper, such as the advisability of "the inductive plan."[20]

Yet with all the evidence apparently available to Moore that Slaught's primary interest was pedagogy, he nevertheless seems to have retained some illusions regarding Slaught's potential as a researcher, until Slaught finally took his degree in 1898. At that point he had to admit to Harper that "Slaught's doctorate-examination was a great disappointment and surprise to Bolza, Maschke and myself." In this letter Moore went on to say that "In the domain of mathematics, as teachers, critical scholars, and investigators, Bolza and Maschke are today, in my judgment, far superior to what Hancock, Boyd, or Slaught will ever become."[21] Slaught stayed at Chicago, but after 1898 decisively put mathematical research aside and concentrated on pedagogy and administration, with much encouragement and assistance from Moore. Slaught wrote several textbooks and was much involved in the founding of professional

---

[17]Bliss, "Slaught," *AMM*, 8.

[18]Moore to Harper, Jan. 31, 1896, UCPP, Box 17, Folder 2.

[19]Meyer, 121.

[20]Slaught to Harper, Feb. 5, 1898, UCPP, Box 17, Folder 2. The Manual Training School became part of the University of Chicago in 1897. See Storr, 134. Harper devoted much attention to the public schools of Chicago. See Julia Wrigley, *Class Politics and Public Schools: Chicago 1900–1950* (New Brunswick: Rutgers University Press, 1982), 92–104.

[21]Moore to Harper, June 28, 1898, UCPP, Box 17, Folder 2.

associations to promote both secondary and college teaching of mathematics, as will be discussed in more detail in due course.

Through Slaught we can readily glimpse how rapidly the University of Chicago became committed to the training of teachers. This may be an instance where the structure of the university created conditions that outpaced even Harper's vision. The intellectual significance and administrative complications attending the Department of Pedagogy (later Department of Education), the Laboratory School, and the School of Education at the early University of Chicago have long fascinated historians of education.[22] Through the efforts of John Dewey, Francis Parker, and others, the university became engaged in training secondary school teachers directly; and also indirectly, through training administrators, normal school teachers, and professors of pedagogy. But another less conscious form of professional teacher training emerged as well, through the students who obtained graduate degrees in the arts, literature, and science with the intention of achieving or advancing in college teaching positions. Harper's institution of the "four-quarter plan" was especially conducive to this result.

The quarter system can be seen in large part as a quintessential Progressive Era innovation, designed to increase efficiency and decrease waste by using the university facilities equally all year long. In particular, a student could take the same courses with the same credits during the Summer Quarter as during any other. This naturally attracted ambitious students who were preoccupied with their own teaching duties during the standard academic year. In addition, the mathematics department (and no doubt other departments as well) was frequently able during the summer to employ visiting professors of some attainment who were looking for additional income or a change of scene.[23] In making itself attractive to a clientele of present and prospective college teachers the University of Chicago was tapping into what Robert Kohler has called "one of the great growth industries of the late nineteenth century."[24]

---

[22] See, for example, Robert L. McCaul, "Dewey and the University of Chicago," *School and Society* 89 (Mar. 25, 1961): 152–157; (Apr. 8, 1961): 179–183; and (Apr. 22, 1961): 202–206.

[23] Storr, 296–303, 310; Goodspeed, *University of Chicago*, 155–156; and Karen Hunger Parshall and David E. Rowe, *The Emergence of the American Mathematical Research Community, 1876–1900: J. J. Sylvester, Felix Klein, and E. H. Moore* (American Mathematical Society, 1994), 369.

[24] Robert E. Kohler, "The Ph.D. Machine: Building on the Collegiate Base," *Isis* 81 (1990): 645.

Slaught became well informed on the role the university was playing in producing college teachers, as from 1903 to 1914 he was in charge of the university's Board of Recommendations, a job placement service for the university's graduates. Writing to Moore about 1920 Slaught summed up the "work of our department in supplying teachers to the schools, colleges and universities during the past 25 years":

> I can only give a rough estimate of the number of bachelors who are or have been teachers of mathematics in elementary schools or high schools. On the other hand the information is exact and complete as to our doctors and approximately so as to our masters, practically all of whom have become teachers for the most part in higher institutions ... [T]he standing of the University ... and the early established principle of conservative recommendations ... soon led to a continuous demand far greater than the supply for the University as a whole and for our department in particular. There has never been a year in which it would not have been possible to place in good positions at least twice as many doctors and masters and other advanced students as we were able to supply.[25]

Unlike J. W. A. Young, Slaught rose to full professor, retiring in 1931, and it is clear that he was a more central figure in the Chicago mathematics department and in the university generally. "Every graduate department needs a man like Slaught if it is fortunate enough to find one," a Chicago graduate later opined.[26] The implication being that when research is the primary goal, many of the non-research demands on such a department may be best handled by specialized labor. Slaught was just the right sort of specialist, on easy terms with the researchers and loyal to the research agenda, but not himself a participant.

---

[25] Slaught to Moore, n.d., EHMP, Box 4, Folder 4. Comparison with other correspondence in the Moore papers suggests a date around January 1920, when Moore was apparently writing to several members of the department to assess the department's accomplishments and future needs. On the Board of Recommendations see Goodspeed, *University of Chicago*, 375–377 and Gilbert A. Bliss, "Herbert Ellsworth Slaught," *University of Chicago Magazine* 30 (1937): 24.

[26] W. L. Duren, Jr., "Graduate Student at Chicago in the Twenties," in Douglas M. Campbell and John C. Higgins, eds., *Mathematics: People. Problems, Results* (Belmont, CA: Wadsworth International, 1984), 1:181. On Slaught's full professorship see Bliss, "Slaught," *University of Chicago Magazine*, 22.

A third figure of interest in the early University of Chicago mathematics department was Harris Hancock (1867–1944). He had been a peripatetic student, touching down at the University of Virginia and Johns Hopkins before flitting across Europe, where he came in contact with a number of important European mathematicians, whose names he was forever after dropping into his correspondence. For example: "Although what I add has nothing to do with the point in question, I may write that I later took the Dr. Sc. at Paris with Professors Poincaré, Darboux and Picard." Probably these contacts initially impressed Moore and led to his recommending that Hancock be hired to join the mathematics department for the opening of the university. But Hancock got off to a rocky start by not having a Ph.D. in hand as he had promised, causing him to be hired at a lower level than he had hoped. Hancock seems to have tried to bluff his way through this impasse, suggesting that he was eligible for Ph.D.s at multiple institutions, and writing to Harper that, "As a dernier resort in case I can't do any better, I can get the Ph.D. degree at Hopkins whenever I want it."[27]

Hancock never did get a Hopkins degree, and although he eventually took a leave of absence to finish his Ph.D. in Berlin, his relations with Moore and Harper were never afterwards smooth. The correspondence is suffused with Hancock's pleadings for promotion and his requests for letters of recommendation so as to obtain a better position elsewhere. Harper eventually was exasperated enough to declare "I wish Hancock would die or something happen to him. His case wearies me." In addition to being a namedropper and poseur, Hancock at one point descended to denouncing Bolza and Maschke as "foreigners" and would later reveal himself as a shameless anti-Semite. Moore meanwhile had concluded that Hancock did not have the makings of a creative mathematician, although he might well "do considerable compilation-work." Finally in 1900 Hancock left Chicago to head the mathematics department at the University of Cincinnati, where he remained more than 30 years.[28]

---

[27] For an overview of Hancock's career see Parshall and Rowe, 364–365 and Pugh, 563–568. Hancock's namedropping letter was to Frank Goodnow, March 31, 1915, JHUPP. Moore's initial good impression is found in a letter to Harper, Mar. 2, 1892, WRHP, Box 14, Folder 15. Hancock's boast about his Hopkins accomplishments is from a letter to Harper, Oct. 21, 1891, UCPP, Box 37, Folder 12. Underlining in original.

[28] Harper's wish for Hancock's demise is from a letter to Moore, Feb. 10, 1896, quoted in Pugh, 565. Moore noted Hancock's attack on Bolza and Maschke in a letter to Harper, June 28, 1898, UCPP, Box 17, Folder 2. For Hancock's anti-Semitism see Hancock to Oswald Veblen, Apr. 20, 1929, OVP, Box 6. Moore's judgment of Hancock's mathematical prowess is found in a letter to Harper, Jan. 31, 1896, UCPP, Box 17, Folder 2.

Parshall and Rowe interpret the Hancock saga as demonstrating the high standards insisted on by Moore in his department.[29] No doubt this is true. It also casts some suspicion on the rigor of the German doctoral programs, at least where American students were concerned. Moreover, comparison with Young and Slaught shows that proclaiming oneself a researcher was neither necessary nor sufficient to survive in a department serving such a varied clientele as that of the University of Chicago. Both Young and Slaught carved out places for themselves at Chicago without making any pretensions of doing mathematical research. Hancock was a more ambiguous type; Moore considered him a pseudo-researcher, and as such there was no long-term home for him at Chicago. But also significant to the portrait of university mathematics ca. 1900 was how long Moore and Harper were willing to put up with this man, despite their evident annoyance with his behavior. Hancock's promotions were delayed, but he was never fired, and when he left Chicago he proceeded directly into the chairmanship of another department. Together these facts point yet more strongly to the imbalance between demand and supply for college level teachers of mathematics at this period. Even a troublesome character like Hancock was a valuable commodity.

6.2. Initial Instances of Moore's Interest in Pedagogy

Herbert Slaught, who was at pains to ensure that Moore's pedagogic interests not be overlooked in toting up the latter's accomplishments, saw the first manifestation of this interest as being Moore's 1894 encouragement to Benjamin Franklin Finkel to found the *American Mathematical Monthly*. According to Slaught, both Moore and Finkel hoped the journal would serve high school teachers of mathematics. But this market would prove to be about a decade from maturing. Instead, the *Monthly* became the preferred forum for college teachers of mathematics desirous of reading and writing less advanced, more expository mathematical articles than were found in the *American Journal of Mathematics* or the *Bulletin of the American Mathematical Society*. With frequent help from Moore, Slaught, and other Chicago mathematicians, the *Monthly* managed to survive for 20 years until its clientele formed a separate professional association, the Mathematical Association of America (MAA); the

---

[29]Parshall and Rowe, 365.

6.2. INITIAL INSTANCES OF MOORE'S INTEREST IN PEDAGOGY 155

Monthly then became the official journal of the MAA, and has so remained to the present day.[30]

Further evidence of Moore's early involvement with pedagogical matters is provided by his evaluating a textbook for a publisher in 1894. The book was a geometry text by Wooster W. Beman and David Eugene Smith and the publisher was D. C. Heath & Company. Moore declared that the Beman-Smith book had good potential, but complained that it did not have enough of "the modern spirit." He felt the authors were deferring too much to the secondary schools: "Of course present high school conditions must be recognized. The authors would make still better book if they recognized less, and felt freer."[31] The conscious quest for striking the correct balance among competing demands would characterize Moore's pedagogical excursions.

In 1897 Moore made a modest entry into the textbook field himself, editing an arithmetic book designed for fifth through eighth grades. The designation of "editor" conveys, among other things, that the project received less than Moore's undivided attention. Given his heavy responsibilities in building up his department at Chicago, where elementary arithmetic was not one of the course offerings, it is hard to see how it could have been otherwise. Very likely Moore merely picked and chose from materials assembled by the publisher, perhaps adding some original material here and there. Moore's recent promotion in faculty rank was clearly considered valuable by the publisher; "Head Professor of Mathematics, The University of Chicago" was prominently displayed on the title page. But Moore probably would have received only a modest monetary reward for his involvement.[32]

The one passage in the book that is unambiguously the work of Moore himself is the "Editor's Preface," which is thus worthy of being quoted at length:

---

[30]On the early history of the *Monthly*, see Herbert Slaught, "Eliakim Hastings Moore, An Appreciation," *AMM* 40 (Apr. 1933): 193, and B. F. Finkel, "The Human Aspect in the Early History of the *American Mathematical Monthly*," *AMM* 38 (1931): 305–320.

[31]Moore to W. S. Smyth, Nov. 7, 1894, DESP. Underlining and syntax as in original. It is not clear whether Moore's criticism had anything to do with the fact that the Beman-Smith geometry eventually was published not by Heath but by Ginn. Woodruff W. Beman and David Eugene Smith, *Plane and Solid Geometry* (Boston: Ginn & Co., 1895).

[32]Eliakim Hastings Moore, ed., *Grammar School Arithmetic by Grades* (New York: American Book Co., 1897). During this period editors of textbooks would have received a single flat fee for their labors, usually from $75 to $400. See John Tebbel, *A History of Book Publishing in the United States*, vol. 2, *Expansion of the Industry 1865–1919* (New York: Bowker, 1975), 573. In Moore's case his fee would likely have been no more than 10% of his yearly salary, possibly much less.

The inductive or laboratory method is applied throughout. By carefully graduated questions and hints, and with the use of graphs and models, the pupil is led, for himself, first to recognize the number relations of concrete quantities, and then to create abstract quantity endowed with the corresponding number relations.

The book gives no rules: the teacher should give no rules. As soon as the pupil has correctly analyzed a particular problem, he recognizes all problems of the same type and analyzes them at sight; and if the type of problem is one of fundamental importance, he devises for himself a definite process for the solution of all such problems.

Thus by the inductive method the pupil is made from the very beginning self-reliant and independent of book and teacher. He gains the clear insight into arithmetical relations and processes, and the accuracy and facility in their practical application, so necessary to success in his subsequent career, be it in high school or in business.[33]

In large part Moore was here in harmony with the ideas of the Mathematics Conference of the Committee of Ten, but with some interesting extensions which would prove to be characteristic. The agreement on the importance of induction is clear. Moreover, as appropriate for one associated with William Rainey Harper, Moore explicitly equated inductive and laboratory. The laboratory method would become a hallmark of his later remarks on pedagogy. The Mathematics Conference had written approvingly of using objects and models in proceeding from the concrete to the abstract. Likewise Moore, who added one specific feature: graphs. This too would receive more extended treatment by Moore. As for Moore's animadversions against rules, it will be recalled that they had become suspect well before the Committee of Ten. The Mathematics Conference attitude, eminently moderate-revisionist, had been that "So far as possible, rules should be derived inductively, instead of being stated dogmatically."[34] Moore might seem to be more radical, apparently calling for banishing rules entirely. But he did retain something he termed "a definite process," surely very like a rule.

---

[33]Moore, *Arithmetic*, preface. I have reproduced all of the preface except the first paragraph.

[34]"Report of Mathematics Conference," in James K. Bidwell and Robert G. Clason, *Readings in the History of Mathematics Education* (Washington, D.C.: NCTM, 1970), 131.

6.2. INITIAL INSTANCES OF MOORE'S INTEREST IN PEDAGOGY 157

What is more striking than Moore's precise attitude toward rules is the set of qualities he ascribed to his hypothetical grammar school student. This "self-reliant" paragon, who "recognizes all problems of the same type and analyzes them at sight," and who "devises for himself" suitably abstract methods for solving problems of "fundamental importance," would seem to have more in common with a productive research mathematician than with the general run of grammar school students. Yet the last sentence of his preface shows that Moore was perfectly well aware that this text was to serve even those proceeding directly into business after grammar school. There was thus a tension here between Moore's most vital experience of mathematics, namely with research, and his recognition that for most students mathematics would be primarily a tool. The attitude that Moore would adopt in later writings to resolve this tension was to affirm that although the specific abstruse knowledge possessed by the research mathematician might be useless for most people, the general approach by which the mathematician obtained this knowledge was in fact widely applicable.

The text of *Grammar School Arithmetic*, however, offered a more prosaic means of resolving the tension between mathematics and business. Some relatively innovative geometric and rule-free treatments, promised by the preface, were simply interspersed with standard sections on business topics. In the section on multiplication, for example, we find a dot diagram accompanying the pair of equations $4 \times 3 = 12$ and $3 \times 4 = 12$, very much as recommended by the Mathematics Conference of the Committee of Ten. No such diagram is found in the widely used text marketed under the name of Joseph Ray.[35] But going beyond the Mathematics Conference, and consistent with Moore's claim to avoid rules, no generalization to $a \times b = b \times a$ was presented in his text, no commutative law expressed; presumably the student was meant to induce it. Fractions too were introduced using pictures of rectangles, again highly consistent with Newcomb and the Mathematics Conference, and differing from the non-geometric treatment previously standard, such as found in Ray.[36]

Entirely inconsistent and conservative, however, was the coverage of commercial arithmetic. Almost every single business topic specifically belittled by the Conference can be found in Moore's book, including banking, insurance, discount, equation of payments, exchange, and customs. Duodecimals

---

[35]Compare Moore, *Grammar School Arithmetic*, 26 with Joseph Ray, *Ray's New Practical Arithmetic* (Cincinnati: Van Antwerp, Bragg & Co., 1877), 41.

[36]Compare Moore, *Grammar School Arithmetic*, 59–63 with Ray, *New Practical Arithmetic*, 131–164.

and compound proportion do seem to have disappeared from Moore's text, but we have seen that this recommendation of the Mathematics Conference was no radical departure from existing trends. The handling of cube root in the Moore book is the most interesting of all. Careful perusal of the book reveals an eight-page "Supplement" on cube and cube root bound into the end of the book but unreported in the table of contents. The treatment is the traditional one using the block diagram discussed in a previous chapter, and culminates in the description of a procedure under the heading "RULE" in explicit violation of Moore's preface.[37] Although this is a matter of speculation, there is surely reason to believe that Moore had agreed with the Mathematics Conference that cube root ought to be deleted, but that the publisher, feeling constrained by tradition, insisted on putting it back in.

In summary, Moore's *Grammar School Arithmetic* was clearly produced independently of the recommendations of the Mathematics Conference of the Committee of Ten, although with significant areas of agreement. In many ways Moore's book was more accommodating to traditional business applications of arithmetic than Simon Newcomb and the Conference; in other ways the book was even more radically inductive than the Conference. By virtue of Moore's position as editor, these features cannot be traced indisputably to him, but we shall see that in some respects they did in fact correspond to important elements of his thought.

The reception of this textbook seems to have been mixed. Moore's friend Benjamin Finkel obligingly declared that

> The treatment of arithmetic as given in this book is a definite departure from the old ruts, and we believe that the timely appearance of this work will go far towards correcting many of the vicious and unwholesome methods pursued in many schools.

But D. E. Phillips feared the book was not consistent with the general level of teacher training:

---

[37]On business topics compare Moore, *Grammar School Arithmetic*, 6 with "Report of Mathematics Conference," 133–134. On cube root see Moore, *Grammar School Arithmetic*, 337–344. The last section reported in Moore's table of contents is on the metric system, pages 329–336. The specific copy of *Grammar School Arithmetic* I first examined came via interlibrary loan, but I made no note of the loaning institution. I subsequently examined a second copy at American University's Artemus Martin Collection, and found this copy apparently identical to the first except that it had no supplement on cube and cube root. The word "rule" is definitely not used in the section on square root on pages 272–276, nor elsewhere in the book, as far as I can determine.

6.2. INITIAL INSTANCES OF MOORE'S INTEREST IN PEDAGOGY

We can well see how such a book would be all right in the hands of competent teachers, but there is a gap between the subject matter therein presented and the child's knowledge that must be supplied some way.

There appears to have been at least one other review of an even more negative character. Harris Hancock, as part of a campaign to turn president Harper against Moore, presented Harper with a clipping of a review of the book; Hancock claimed that Harper would see that "it reflects no credit on the University of Chicago."[38]

Moore's venture into the schoolbook field followed closely upon some muttering by president Harper in 1896 that the university could not afford for its faculty to be too exclusively devoted to research at the expense of teaching. Coincidentally or not, Moore's next flurry of activity in the pedagogical line also followed hard upon a scolding by Harper, this time indeed directed at Moore himself. Disagreeing with what he took to be Moore's position, "that the whole corps of instructors should be made up of instructors who are primarily successful producers," Harper in 1899 in particular questioned whether L. E. Dickson ought to be appointed to the Chicago faculty, since he was "a teacher of low merit." Dickson was in fact appointed to an assistant-professorship in 1900, where he would stay for the rest of his long and distinguished research career, but that year also saw both Moore and Dickson participating in a special session of the Mathematical Club led by J. W. A. Young on teaching college algebra.[39]

Meanwhile other forces were afoot to encourage pedagogical activity among the Chicagoans. For one thing, the purely bureaucratic momentum of the Committee of Ten had not yet played itself out. In 1895 at a meeting of the National Educational Association (NEA) a pertinent question was asked: "What Action Ought to be Taken by Universities and Secondary Schools to Promote the Introduction of the Programs Recommended by the Committee of Ten?" Discussion led to the creation of a Committee on College Entrance Requirements (CCER),

---

[38] I have not discovered any sales figures for the book. Finkel's capsule review is in *AMM* 4 (1897): 232, signed B.F.F. Phillips' assessment is included in D. E. Phillips, "Number and Its Application Psychologically Considered," in John Dewey, *The Early Works, 1882–1898*, vol. 5 *1895–1898, Early Essays*, ed. Jo Ann Boydston (Carbondale: Southern Illinois University Press, 1972), lxxviii. Phillips's article was first published in *Pedagogical Seminary* 5 (Oct. 1897), 221–278. Hancock's aspersions are in a letter to Harper, June 24, 1897, UCPP, Box 37, Folder 12. I have been unable to determine what review Hancock was referring to. See also Pugh, 565–566.

[39] Pugh, 557, 763, and Notes on "Conference on Teaching College Algebra," Dec. 7, 1900, MCR, Box 2, Folder 4, pages 95–110.

with Dr. A. F. Nightingale, superintendent of the Chicago high schools, named as chairman. There were twelve other committee members, including mathematician Henry Fine of Princeton, whom we have met as a member of the Mathematics Conference of the Committee of Ten. Among the early decisions made by the CCER was that ideas should be solicited from specialist associations. Accordingly the American Mathematical Society was approached: "The request was too late for the general association, but the Chicago branch was empowered to appoint a committee to study the matter and to report. Professor J. W. A. Young, of the University of Chicago, was chairman." The reasons for it being "too late" for the general body of the AMS to act are obscure. Since the end result was to produce a committee run from Chicago with a subordinate of Moore in charge it is hard to avoid suspecting that some measure of E. H. Moore's aggressive style was involved. In any event, since he was a vice president of the AMS at the time, and had moreover been the force behind the creation of its Chicago Section, the maneuver could hardly have been taken without his approval and encouragement, although he does not seem to have played any direct part in the subsequent deliberations.[40]

The report of the CCER was published in July 1899. Like the Committee of Ten Report it contained both a general report by the overseeing committee and a series of more detailed reports by specialists. The general report elaborated upon a number of the themes sounded by the Committee of Ten, and retained a moderate revisionist flavor. Two large issues were mentioned but left unplumbed: the place of commercial studies in the schools, and whether the preferred method of admission to college should be by certification of schools or by student examinations.[41]

The report of the CCER's specialist subcommittee on mathematics, chaired by J. W. A. Young, was in general agreement with the Mathematics Conference of the Committee of Ten. Like the Mathematics Conference the Young committee recommended regular oral drill, and called for stress on neatness and accuracy in written work. The Young committee also agreed on the desirability of using graphic illustrations in arithmetic, on emphasizing the equation in algebra, and on using models in geometry.[42]

---

[40] *Report of the Committee On College Entrance Requirements* (Chicago: University of Chicago Press, 1899), 5–6, 9,11; Archibald, *History of the AMS*, 77; and Edward A. Krug, *The Shaping of the American High School 1880–1920* (New York: Harper & Row, 1964), 137–145.

[41] *Report of CCER*, 47.

[42] Ibid., 136, 138, 140, 143.

6.2. INITIAL INSTANCES OF MOORE'S INTEREST IN PEDAGOGY

The Young committee explicitly embraced correlation, a notion that was becoming popular among educators, especially due to the efforts of those who considered themselves followers of the German pedagogue Johann Friedrich Herbart. Herbart had died in 1841, but his ideas became most potent in the United States some decades later, reaching a peak of popularity in the years immediately after the Committee of Ten Report. Herbart was interpreted as favoring the centrality of the content studies (history, literature, science) over the formal studies (grammar, mathematics), thus attracting many of those who chafed at the primacy of mental discipline. Further, Herbartians advocated that the formal subjects be taught in close relation with the content subjects. In an era when the crowding of the curriculum was of great concern, this concept of correlation of subjects had broad appeal, even to those not enthused by the full Herbartian program. As applied to mathematics by the Young committee, correlation yielded the following precept:

> The subjects arithmetic, geometry, algebra should be treated as branches of one whole—mathematics—and each of these subjects freely applied in illustrating and broadening the others.

Such thinking had been implicit in the Mathematics Conference, as had been noted by Eliot when he commended it for its "interlacing of subjects."[43] It is likely that such notions would have appealed to many mathematicians even without the inspiration of Herbart. Much research progress was being achieved at this time by unifying disparate portions of mathematics through such concepts as function and group. This had been one of Felix Klein's major themes at the Chicago Congress and the Evanston Colloquium.

The Young committee was also notable for its optimistic view of the prospects for cooperation among the different varieties of mathematical educators. Befitting a progressive era initiative, "waste of teaching energy" was identified as the major problem involved. For example, waste occurred whenever a student studied the same material in college that had already been studied in high school. Declared the Young committee: "Much can be done to diminish this waste by close relations between the teachers of the three divisions [grade school, secondary school, college], and the comparison of results and adaptations of work to mutual needs." In addition the committee seems to have felt

---

[43]The Young committee quote is from *Report of CCER*, 138. Eliot commended "interlacing" in *Report of Committee of Ten*, in Theodore R. Sizer, *Secondary Schools at the Turn of the Century* (New Haven: Yale University Press, 1964), 234.

that mathematics was still an unchallenged component of the curriculum, and sensed little potential conflict with advocates of other school subjects. It recommended that all secondary school students take mathematics throughout the four year course, but did not attempt to justify this in any way.[44] Very few mathematical educators would be able to retain such relaxed attitudes in the future.

In the summer of 1900, at the Seventh Summer Meeting of the AMS, both Moore and Young participated in a session on the undergraduate mathematical curriculum. Moore's paper advanced a number of ideas that he would emphasize again in his 1902 address: that college instructors should have a firm grasp of "the fundamental modern ideas and methods" of mathematics; that the best way to view mathematics was as a body of results deduced logically from "basal notions and axioms"; that "intuition" nevertheless retained an important place in mathematical investigation; and that different levels of logical rigor were appropriate for different stages of the mathematics curriculum.[45]

Young's comments at this session provide a revealing coda to his committee report of the previous year by alluding to some of the professional interests at stake. His committee had strongly endorsed the principle that all high school mathematics teachers should have "a good college course, with special attention to mathematics, either by electives during the course or by graduate study."[46] Young said precisely the same to his AMS colleagues at the meeting, but with the added framing comment that such words were a "truism to say to this audience." Moreover, "[i]t is equally well known to this audience that this standard is by no means generally reached." The college-level mathematical community, as has already been suggested, owed not a little of its new vitality to the fact that more high school teachers were indeed coming out of the colleges, and the continued growth of this segment of the educational market should thus have been of concern to alert members of the AMS. Young also endorsed "special training in the art of teaching mathematics," but since "mature" students would most benefit from this he recommended that "[t]he best place for the pedagogic training and instruction is therefore *after* the college course; either in the university or by some form of apprenticeship in the secondary schools."[47]

---

[44] *Report of CCER*, 145, 147.

[45] W. H. Maltbie, "The Undergraduate Mathematical Curriculum," *BAMS* 7 (1900): 15–16. The remarks of Moore and the other session participants were published only in abstract. This is the only documentation known to me of the paper Moore gave on this occasion.

[46] *Report of CCER*, 146.

[47] Maltbie, 22, 23. Italics in original.

The University of Chicago of William Rainey Harper was well equipped to provide such training. The independent normal schools were conspicuously absent from Young's ideal educational world.

Following the CCER's avoidance of the issue of examination versus accreditation for college admission, proponents of both methods moved quickly to support their views. As noted earlier, this issue frequently distinguished educators along geographic lines. This can be seen among the mathematicians in particular. For example, we find Thomas Fiske of Columbia engaged like many eastern educators on behalf of the examination system, while Moore of Chicago participated in the accreditation system favored by many in the midwest. There is little evidence that these actions of Fiske and Moore indicated deeply held commitments as to the best way to organize college admissions; more likely these two skillful academic politicians were operating pragmatically within the particular circumstances in which they found themselves.

Throughout much of the 1890s Charles Eliot of Harvard and Nicholas Murray Butler of Columbia had pushed for colleges to cooperate to form a single board of examiners to administer tests to those seeking to enter college. In 1899 they finally succeeded in persuading the Association of Colleges and Preparatory Schools of the Middle States and Maryland to sponsor such an undertaking. Butler was named the first secretary of the resulting College Entrance Examination Board (CEEB) in 1900, and the first examinations were conducted in July 1901. In that same year Butler was named president of Columbia and chairman of the executive committee of the CEEB. He persuaded his colleague Fiske, professor of mathematics at Columbia, to succeed him as secretary. Thus did Fiske, the principle founder of the American Mathematical Society, commence his long relationship with the CEEB. The initial cooperating institutions included Johns Hopkins and Columbia, and were primarily from New York and Pennsylvania. Eliot was unable to convince the Harvard faculty to join until 1904.[48]

Accreditation, on the other hand, was pioneered by state universities in the midwest, California, and the south. Interested secondary schools could apply to be inspected by colleges or universities to which they wished to send an appreciable number of their graduates. Movement to bring uniformity to the process came to be centered in the North Central Association of Colleges

---

[48]Krug, *High School*, 146–151; Edward Krug, *Salient Dates in American Education, 1635–1964* (New York: Harper & Row, 1966), 102; John A. Valentine, *The College Board and the School Curriculum* (New York: College Entrance Examination Board, 1987), 3–16; Archibald, *History of the AMS*, 151.

and Secondary Schools which was founded in 1895.[49] In 1901 this association formed a Commission on Accredited Schools, with dean Harry Pratt Judson of the University of Chicago as chairman and including superintendent Nightingale of the CCER. The commission published a first report in March 1902, with little attention to mathematics. Then in April 1903 it published a number of improvements suggested by specialist committees, in particular a three-page supplement by a three-member committee on mathematics led by E. H. Moore that emphasized proper attitudes to be inculcated in the students. The specific rhetoric and references used leave little doubt that Moore himself was the principal author.[50]

6.3. Moore's Championship of the Laboratory Method of Instruction

Moore's pedagogical activism ratcheted up in 1902, as his term as president of the American Mathematical Society wound down, and is worth examining closely, step by step. The theme for the University of Chicago Mathematical Club in the summer quarter that year was "Methods of Teaching Mathematics." Moore himself led off the series with a paper entitled "What is Mathematics?" In this paper, evidently a forerunner to his AMS address at the end of the year, Moore distinguished the mathematician from the scientist by citing the former's primary interest in "the deductive logical organization of the scientific structure on the basis of a well-defined system of well-defined axioms," but warned of mathematical instruction becoming too isolated from the sciences:

> In particular, the teacher of mathematics in all but the most highly specialized courses must guard against the supposition that the students in his classes are or should be, or should be made to be primarily of mathematical bent. He should, on the contrary, endeavor to teach mathematics in constant and intimate relation with its applications.

The fact that his own teaching was primarily of "highly specialized courses" seems to have given him no hesitation about speaking on this subject. In the

---

[49]Krug, *High School*, 152.

[50]"Report of the Commission on Accredited Schools," *Appendix to the Proceedings of the Eighth Annual Meeting of the North Central Association of Colleges and Secondary Schools* (1903): 181–184.

6.3. MOORE'S CHAMPIONSHIP OF THE LABORATORY METHOD

same talk Moore cited the pedagogic importance of "the correlation of subjects usually held to be distinct," for instance algebra and geometry.[51]

The next session of the Mathematical Club featured a paper on the subject of correlation of algebra, geometry, and physics by Edith Long of the Lincoln High School in Lincoln, Nebraska. This was a precursor of a paper she would soon publish in the *Educational Review* that would be cited by Moore in his 1902 address. Long stated that her high school was inspired to experiment with correlation "by the report of the Chicago section of the American Mathematical Society" of 1899; i.e. the report chaired by J. W. A. Young appended to the CCER report. In Long's account algebra and geometry were taught together, with problems from physics and physical geography used "for the sake of a broader concrete basis for the algebra and of a more interesting application of the geometry." Solution by rote was discouraged: "Mathematical operations must be performed in harmony with laboratory operations." With the aid of experiments with balances, much attention was given to "the equation, its meaning, its formation, and the meaning of each step taken in its solution." Long noted that one criticism given of this method of instruction was that it handicapped students who needed to be preparing for college and university entrance examinations. She confidently denied this, stating that "the student who has taken this kind of training is much better fitted to solve the problems of the examination paper than the student who has had nothing but pure algebraic theory and exercises."[52]

Less than three weeks after Long's Mathematical Club presentation Moore wrote his first letter to Harper sketching out his plans for adopting at the University of Chicago "the laboratory method of instruction in mathematics." This letter informed Harper of some presumably cost-free initiatives which had already been taken, and requested new resources for the short and long term to carry these initiatives to "full fruition." The letter was well calculated to appeal to Harper. In addition to the deployment of "laboratory," announced in the first sentence of the letter, Moore suggested that the idea he proposed was just coming to the forefront of mathematics education reform throughout Europe and North America, and that the University of Chicago had "the opportunity

---

[51] MCR, Box 2, Folder 4, pages 310–311, June 27, 1902.

[52] The quotes are all from Edith Long, "Report of experience in the correlation of algebra, geometry and physics in the Lincoln High School, Nebraska," MCR, Box 2, Folder 4, pages 314–321, July 11, 1902. Compare Edith Long, "Correlation of Algebra, Geometry, and Physics," *Educational Review* 24 (1902): 309–311. The point about the advantages to college bound students is not made in the *Educational Review* article.

to take the lead in the systematic realization of the notion." He surmised that "the method will appeal with especial force to our natural constituency as being evidently the proper method."[53] Judging from the rest of the letter and from Moore's other pronouncements, "our natural constituency" most likely referred to the majority of students who took mathematics for essentially utilitarian purposes. Most certainly it did not refer to the small minority destined to become research mathematicians. This latter group would need special handling to convince them of the merits of the laboratory method; this would be a major aim of Moore's address to the AMS in December 1902.

In the July 29 letter Moore went on to propose that secondary and introductory college courses in pure mathematics should be taught so that

> The pure theory will be illuminated (not merely by the development of (its theoretic applications but also) (and this is where the laboratory method in the narrower sense enters) by the constant use of numerical computations, graphical illustrations, models and mechanisms of all kinds.
>
> The student will learn not merely verbally or even logically but also by the use of all the intuitions and senses. And what he learns in this way he will really understand and control for later use in whatever field ... <u>provision should be made for the actual construction</u> <u>by the students of many of the simpler drawings, models and mechanisms.</u>

(The puzzling proliferation of parentheses seems to indicate Moore's hurried attempts to appropriately qualify his thoughts in the act of writing.) He further asserted that the introductory courses he envisioned would satisfactorily support

> the more advanced and special courses in pure mathematics, much as at present, and also a body of courses specializing along the line of the theory and construction of mechanisms. This second direction of specialization would surely be very attractive to a considerable percentage of students.[54]

This last passage supports the comment about "our natural constituency" above. Moore was maneuvering to retain control of the training of the mathematical specialists, while at the same time reaching out to attract new students

---

[53]Moore to Harper, July 29, 1902, UCPP, Box 17, Folder 2. Underlining in original, most likely by Moore, who often employed such emphasis in his letters.

[54]Ibid. Underlining, double-underlining, and parentheses in original.

not currently as interested in mathematics as in his view they naturally should be.

Moore told Harper that he had been mulling over such ideas "for a number of years," and that he had just that summer taken the step of "establishing a Calculus Laboratory in the fourth floor of Ryerson." He hoped such provisional arrangements would suffice "until the new building for Mathematics and Astronomy comes to provide adequate facilities." In the meantime he needed one laboratory assistant immediately, and more later. If the plan was implemented correctly at the University of Chicago Moore had hopes that it would "speedily receive recognition and adoption wherever possible in all our dependent territory," referring to the network of affiliated educational institutions that Harper was busy drawing into his empire.[55]

Despite Moore's best efforts, fiscal realities proved troublesome. As earlier noted, Moore did not obtain his desired new building for mathematics for another 28 years. Nor did he receive his requested laboratory assistant for the fall quarter of 1902. A letter he wrote to Harper on October 11, 1902 indicates that the man he requested had not in fact been hired, and that on his own responsibility he had drafted Arthur Lunn, associate in applied mathematics, to assist with the laboratory course in calculus, leaving some junior college instruction previously assigned to Lunn temporarily uncovered. Evidently Harper had put Moore off by pleading the intransigence of the board of trustees, but Moore remained optimistic and committed to his course of action: "I am bold enough to confess my belief that I can make you see the matters so clearly that seeing with you the Trustees will have a way of providing for this instruction." Indeed, Moore's enthusiasm for the laboratory method was such that after only nine days of the experiment with a class of 24 students he declared it "will work a revolution in mathematical instruction," a revolution that he hoped to convince Harper should "take place under the auspices of the University of Chicago." One of the great beauties of the method, according to Moore, was that it could be molded to the individual needs of the students, serving those intending to teach mathematics as well as those going into electrical engineering, mining engineering, and medical research, all of which were represented in the present class. Moore's sweeping vision contemplated a unified mathematics instruction which would

> join hands with the representatives of the newer primary education, on the one hand, and that on the other hand in this

---
[55]Ibid.

way we shall train up students whose mathematical insight will be such that they can go forward powerfully either into pure mathematics or theoretical applied mathematics, or any of the lines of the physical or technical sciences, involving mathematics, and that furthermore we shall remove ultimately the stigma upon mathematics that there are those quite incapable of comprehending mathematics.[56]

In the same letter Moore politely added a new request, that the needs of "proper laboratory facilities" for mathematics be taken into account in the ongoing planning for "laboratory provisions for the manual training school." He also proposed that the laboratory instruction in mathematics could at the same time function as a highly desirable form of teacher training. That is, the prospective mathematics teacher should first take the laboratory course as a student, and then later participate in the same course as an assistant to the professor:

This suggestion will, I suppose, appeal to you at once as feasible and desirable, and I am sending a copy of this letter to Mr. Dewey with the hope that he may see a way whereby the department of mathematics may be permitted to arrange its work in such a way as to train up a strong body of thoroughly initiated teachers of laboratory mathematics.[57]

This is the first reference to John Dewey of which the present writer is aware in the correspondence of E. H. Moore. Note that Moore was citing Dewey not as a source of ideas, but rather as a powerful figure within the University of Chicago whose cooperation would be necessary for Moore's proposals to be fully implemented. That is to say, anxieties of professional jurisdiction were here preeminent. At bottom Moore was proposing to take greater control of the teaching of mathematics throughout the educational system; the new teachers must be "thoroughly initiated" as he saw fit. But Moore recognized that in making such a move he would encounter other interested parties, Dewey and his department of pedagogy in particular.

For the next several months Moore maintained vigorous activity on behalf of just such jurisdictional calculations, attempting to garner the cooperation of a variety of educators. He was not turned aside by his many other responsibilities.

---

[56]Moore to Harper, Oct. 11, 1902, UCPP, Box 17, Folder 2.

[57]Ibid. On the Chicago Manual Training School, and its relationship with the University of Chicago, see Storr, 134–135.

As he traveled east in late October 1902 he deputized graduate student Oswald Veblen to keep the pot boiling on the pedagogical front:

> Will you please collect a sizable bundle (three or four) of my papers (R45 [room number in Ryerson Hall]) (Geometry, Functions) and send (<u>large</u> envelopes R45) to Prof. ? ? Perry (Perry's Calculus for Engineers) and write him in my name requesting him to send reprints or references to his writings referring to a reform in pedagogy of mathematics, especially in engineering schools—with references also to the other writers and books on the subject.[58]

This is Moore's first reference to John Perry known to the present writer. His acquaintance with Perry was evidently vague at this time; the question-marks he interpolated would seem to indicate that he was unclear either on Perry's first name or on whether Perry was entitled to be addressed as professor. There is a hurried, even slapdash, flavor to Moore's efforts here, a consciousness that he was plunging into pedagogy more fully than he ever had before and that he needed to become acquainted with the literature expeditiously.

Moore's request for information from Perry appears to have succeeded. On November 20 Moore circulated a letter to teachers of mathematics in which he cited John Perry in support of his own ideas, with extensive references to Perry's recent writings. The letter called for reorganization of *"the mathematics and physics in the secondary school into a thoroughly coherent four years' course"* and encouraged teachers while awaiting this reorganization to immediately adopt the tools of the laboratory method: "laboratory record books with cross-section paper, colored chalks and inks, much work of numerical and of graphical character, the cultivation of the laboratory spirit." He proposed that the mathematics teachers attend the November 28 meeting of the Central Association of Physics Teachers, and closed with the thought that mathematics teachers should be thinking of founding a similar organization for themselves.[59]

In early December Moore wrote to Harper on the ostensible subject of the University Secondary School, then in process of being created as an amalgamation of the Chicago Manual Training School and the South Side Academy. A new building was planned, and Moore wished to convey his desire for adequate "laboratory facilities" for the work of the mathematics department within the

---

[58] Moore to Veblen, Oct. 28, 1902, OVP, Box 8. Underlining and question marks in original.

[59] Moore to "My Dear Colleague," Nov. 20, 1902, UCPP, Box 17, Folder 2. Italics in original.

school, placing on record his recommendations for optimal room dimensions and his estimates of how many students the school would ultimately serve. He stated that all his computations were based "on the supposition that the work of physics and mathematics will be brought into the closest correlation in the Secondary School." He noted further that Dewey was calling a pedagogic conference of all teachers in the university who taught mathematics or science at the junior college level or below. Moore indicated he would be promoting his reform ideas at this conference, ideas that Harper would better appreciate if he "were to find it possible to read the literature connected with the Perry movement in England, as cited in the circular letter which is enclosed." Moore closed this letter by urging Harper to explore bringing Perry to Chicago for the next summer quarter.[60]

The culminating event of all this activity was Moore's presidential address to the American Mathematical Society, delivered in New York City on December 29, 1902.[61] There can be no doubt that he took this opportunity seriously indeed and put much thought into his speech. He wrote to Simon Newcomb in March of 1903 that during the fall of 1902 he had neglected other matters because "I was in the pressure as to my Mathematical Society address." The speech as it was published is a carefully crafted piece of work. One immediately notices a liberal sprinkling of footnotes; on closer study one finds the piece studded with fascinating rhetorical choices and exhibiting a remarkable structure as well.

The focus of Moore's concern was on certain undesirable "chasms" and "branchings"; there was a great need for "bridging," "correlation," and "unification"; and the gist of the solution he prescribed was "indirection." Indeed, the entire talk may be described as an exercise in indirection. In the first paragraph Moore explained that his title, "On the Foundations of Mathematics," carried a double meaning to be revealed later:

> The American Mathematical Society gives its retiring president the privilege of speaking on whatever he may have at heart. Accordingly, this afternoon I propose to consider with you some matters of importance—indeed, perhaps of fundamental importance—in the development of mathematics in this

---

[60]Moore to Harper, Dec. 3, 1902, UCPP, Box 17, Folder 2. On the University Secondary School see Goodspeed, *University of Chicago*, 354.

[61]Moore, "Foundations of Mathematics," 401.

6.3. MOORE'S CHAMPIONSHIP OF THE LABORATORY METHOD

country; and it will duly appear in what non-technical sense I am speaking 'On the Foundations of Mathematics.'[62]

There are two implications of Moore's reference to his privileged position in the AMS that seem borne out by the reception of his talk: that his words would carry more weight coming from him than from most of his colleagues; and that he did not feel compelled to restrict himself to the expectations of his audience. By the time of his talk the 40 year old Moore had been awarded an honorary Ph.D. from the University of Göttingen, been elected to membership in the National Academy of Sciences, and had become an Associate Fellow of the American Association for the Advancement of Science. In a survey conducted in 1903 by J. McKeen Catell for the first edition of his *American Men of Science*, Moore was ranked first for achievement among 80 leading American mathematicians.[63]

Moore's talk was composed of two major subsections, entitled "A View" and "A Vision." As Moore was doubtless aware, his audience would be expecting that as the leading pure mathematician in the country he would survey the state of pure mathematics, just as his predecessor R. W. Woodward had surveyed applied mathematics two years before. The initial portion of "A View" seemed to promise just such a survey. He pointed out that although the idea of basing mathematical investigation on "defining the objects of consideration ...by a body of properties" was of ancient lineage, its true importance had not been appreciated until the nineteenth century. There had thus arisen a new "abstract mathematics," with Guiseppe Peano and David Hilbert being among the most influential recent practitioners. The latter's *Grundlagen der Geometrie* of 1899 was especially significant. Moore also called attention to the valuable contributions of his own student, Oswald Veblen. At the center of these endeavors stood a "system of undefined symbols" and a collection of "undemonstrated or primitive propositions, or postulates," from which one proceeded to deduce other propositions "from the postulates by a finite number of

---

[62] Moore, "Foundations of Mathematics," 401. Moore's address was published in the *BAMS* as well, but all my references are to the version published in *Science*. For Moore's admission of neglecting other matters to prepare his talk, see Moore to Newcomb, Mar. 16, 1903, SNP, Box 32.

[63] Archibald, *History of the AMS*, 108. Catell's 1903 survey was based on responses from ten "outstanding leaders" in mathematics chosen by him. I do not know who these individuals were. See Stephen Sargent Visher, *Scientists Starred 1903–1943 in American Men of Science* (Baltimore: Johns Hopkins University Press, 1947; repr., New York: Arno Press, 1975), 3.

logical steps." Here then was the technical sense of the phrase "foundations of mathematics."[64]

At this point Moore began to voice some skeptical sentiments regarding the developments that he had just sketched. He accused abstract mathematicians of "losing sight of the evolutionary character of all life-processes," and proclaimed that "All science, logic and mathematics included, is a function of the epoch— all science, in its ideals as well as in its achievements." He asserted his "feeling that the carrying out of the program of the abstract mathematicians will be found impossible," but maintained that the quest was nevertheless of great value. Moore was impressed (as we have earlier noted that Sylvester was) by the fact that mathematical research employed intuition and induction as well as deduction:

> and if this is true with respect to the research of professional mathematicians, how much more is it true with respect to the study, which should throughout be conducted in the spirit of research, on the part of students of mathematics in the elementary schools and colleges and universities.[65]

Here was where Moore first gave notice that his talk was veering away from research and toward pedagogy, but with the important qualification, consistent with the views of William Rainey Harper, that the ideal pedagogy should be permeated with "the spirit of research."

Moore proceeded to briefly sketch the historical development of mathematics. In particular, the just completed nineteenth century had seen the "critical reorganization of the foundations of pure mathematics." The problem was that such critical pondering could be done without any thought whatsoever of applications to other sciences: "There has thus arisen a chasm between pure mathematics and applied mathematics." It was one of the great merits of Felix Klein that he had actively sought the "bridging" of this chasm. John Perry in England was also to be commended for his attempted rapprochement between theory and practice "with respect to the teaching of elementary mathematics." Moore endorsed Perry's specific proposals to emphasize "squared paper" and "graphical methods," as well as Perry's general philosophy that by building

---

[64]Moore, "Foundations of Mathematics," 401–403, 408.

[65]Moore, "Foundations of Mathematics," 403–404. In support Moore cited a paper of Poincaré, but did not acknowledge that Poincaré's formulation was precisely the reverse of his own: "Si [l'intuition] est utile à l'étudiant, elle l'est plus encore au savant créateur." Henri Poincaré, "Du rôle de l'intuition et de la logique en mathématiques," *Compte Rendu du Deuxième Congrès International des Mathématiciens* (1902): 124.

upon practice with "experiment, illustration, measurement" the student could be made "*familiar*" with abstract ideas and "thoroughly *interested*" in their use. Moore admitted that he was calling for a "diminution of emphasis on the systematic and formal sides of the instruction in mathematics," and that "many mathematicians" would fear "irreparable injury to the interests of mathematics." But Moore stood firm, maintaining that the Perry approach would ultimately result in more students developing the all too rare "feeling that mathematics is indeed itself a fundamental reality of the domain of thought, and not merely a matter of symbols and arbitrary rules and conventions."[66]

Moore concluded his "View" by observing that although the American Mathematical Society had been primarily concerned with "promoting the interests of research," he was gratified to see that lately it had "recognized that those interests are closely bound up with the interests of education in mathematics." As evidence he cited four recent developments: the mathematics committee that had worked in conjunction with Nightingale's Committee on College Entrance Requirements; a committee formed by the Chicago Section of the AMS to report on requirements for the Master of Arts degree in mathematics; an AMS committee formed to cooperate with the NEA and the Society for the Promotion of Engineering Education to formulate mathematics requirements for admission to colleges and technological schools; and the role of AMS founder Thomas Fiske as secretary of the College Entrance Examination Board. Moore did not acknowledge that only the last of these four instances was independent of his own efforts, nor that this last instance was highly indebted to unusual local circumstances. Rather, he boldly took all these examples as demonstrating a general AMS interest in "the pedagogy of elementary mathematics," giving him license, he asserted, to devote the rest of his address to this subject.[67] (He made clear throughout his talk that by "elementary mathematics" he meant all mathematics up through introductory calculus.)

The second section of Moore's talk, "A Vision," turned out to be specifically "a vision of the future of elementary mathematics in this country." He

---

[66]Moore, "Foundations of Mathematics," 405–408. Italics in original.

[67]Moore, "Foundations of Mathematics," 408. The first committee on Moore's list has already been discussed. The second committee, on master's requirements, consisted of three persons, one of whom was Moore's Chicago colleague Oskar Bolza. "Report on the Requirements for the Master's Degree," *BAMS* 10 (1904): 380–385. The third committee cited by Moore consisted of five members, one of whom was Moore's Chicago colleague, J. W. A. Young. "Report of the Committee of the American Mathematical Society on Definitions of College Entrance Requirements in Mathematics," *BAMS* 10 (1903): 74–77.

began by inviting pure mathematicians to look outside their profession, "to determine how mathematics is regarded by the world at large." Mathematicians, he suggested, needed to change their ways in order to "win for [mathematics] the very high position in general esteem and appreciative interest which it assuredly deserves."[68] Here then was the non-technical sense of the "foundations of mathematics."

Moore acknowledged the by then standard stratification of the school system into primary school, secondary school, and college, but noted that there were movements afoot to have the secondary schools absorb both the higher primary grades and the first two years of college; in other words, to make them more like the German gymnasia and the French lycées. Moore was here alluding to notions strongly advocated by his chief, William Rainey Harper. He further extended the implicit compliment to Harper by lauding the desirability of having secondary schools "closely related to strong colleges and universities."[69]

But whatever might be done about unification at the institutional level, there was an intellectual issue as well:

> The fundamental problem is that of the *unification of pure and applied mathematics*. If we recognize the branching implied by the very terms 'pure,' and 'applied,' we have to do with a special case of *the correlation of different subjects* of the curriculum, a central problem in the domain of pedagogy from the time of Herbart on. In this case, however, the fundamental solution is to be found rather by way of indirection—by arranging the curriculum so that throughout the domain of elementary mathematics the branching be not recognized.[70]

Moore here framed his subject within language familiar to his specialized audience of mathematicians, the language of mathematical problem-solving, in which improved understanding may often be obtained by recognizing that a particular problem is a special case of a more general one. At the same time Moore gently introduced some notions of non-mathematical educators. His allusion to Herbart was not a deep one. The German pedagogue was widely associated with the concept of correlation, and it is likely that many American

---

[68]Moore, "Foundations of Mathematics," 408.

[69]Ibid., 409. Compare W. R. Harper, "The High School of the Future," *School Review* 11 (Jan. 1903): 1–3.

[70]Moore, "Foundations of Mathematics," 409. Italics in original.

6.3. MOORE'S CHAMPIONSHIP OF THE LABORATORY METHOD

educators knew little about Herbart beyond this.[71] Moore's peculiar use of "indirection" is more interesting, though perhaps of doubtful significance. As we shall note later, it was a word used by John Dewey during his Chicago period.

For the primary schools Moore recommended that mathematics should always "be directly connected with matters of a thoroughly concrete character." He encouraged making models, drawing graphs of magnitudes associated with natural phenomena, and deducing conclusions about the phenomena from the behavior of the graphs. But he specifically disavowed "allowing the mathematics to enter only implicitly in connection with the other subjects of the curriculum." He indicated his awareness that "some" saw this implicit method as an ideal way to teach mathematics, but tantalizingly failed to explain to whom he might be alluding.[72]

Moore's professional anxieties were most clearly expressed in his discussion of the secondary schools. He lashed out at the conventional practice of teaching algebra, geometry, and physics each in their own "water-tight compartment." It was engineers, he asserted, who most clearly recognized the absurdity of this state of affairs; mathematicians, with a few exceptions, were often all too complacent about it. In a fascinating footnote he asked two pointed rhetorical questions:

> Why is it that one of the sanest and best-informed scientific men living, a man himself not an engineer, can charge mathematicians with killing off every engineering school on which they can lay hands? Why do engineers so strongly urge that mathematical courses in engineering schools be given by practical engineers?

To combat these problems Moore proposed *"to organize the algebra, geometry, and physics of the secondary school into a thoroughly coherent four years' course."* These words were almost identical to Moore's circular letter of November 20 cited above, with the difference that in the earlier letter the words were not illuminated by allusions to jurisdictional competition which Moore evidently felt were more significant for his AMS audience.[73]

---

[71]See Krug, *High School*, 98–105.

[72]Moore, "Foundations of Mathematics," 409.

[73]Ibid., 410. Italics in original. I have not discovered the "scientific man" to whom Moore was referring, nor can I offer a definite reason as to why Moore should have left him anonymous. It does suggest to me that Moore perceived some taboos related to the professionalization issues he was broaching. It would be most interesting to learn whether Moore's footnote comment on engineers threatening to teach mathematics courses was included in his

Moore was quite willing to defer to the engineers on much of the content of his proposed secondary school course, but with an all important proviso to protect the jurisdiction of the mathematicians:

> Let the instruction in the course, however, be given by men who have received expert training in mathematics and physics as well as in engineering, and let the instruction be so organized that with the development of the boy, in appreciation of the practical relations, shall come simultaneously his development in the direction of theoretical physics and theoretical mathematics.[74]

Here was the quest for the holy grail of mathematical pedagogy: a course that would "simultaneously" satisfy the needs of the entire hierarchy of mathematical users. We have seen this quest anticipated by the Mathematics Conference of the Committee of Ten. But in Moore's version, only ten years on, there was no mention whatsoever of the value of mathematics as a mental discipline.

At this point Moore expanded upon his endorsement of John Perry, joining with him to entirely reject the vaunted thoroughness of early nineteenth-century educators, who had aimed never to introduce new topics of study until all previous topics had been mastered in full detail. Moore held that Perry was "quite right" to suggest that mathematics instruction ought to begin by unproblematically assuming the truth of various properties and principles: "Sufficient unto the day is the precision thereof." In this way the engineer could rapidly attain to the mathematics truly of most use, while the mathematician could advance to the leading edge of research before subjecting the foundations to searching criticism and delicate analysis. To support his position Moore made a footnote reference to John Dewey's ideas on teaching geometry, and also proposed remarkably radical ideas of his own that he perhaps thought would appeal to his mathematically highbrow audience. In Moore's vision of high school geometry no axioms would be laid down by the teacher:

> [W]hy should not the students be directed each for himself to set forth a body of geometric principles, on which he would proceed to erect his geometric edifice? ... The various students would have different systems of axioms, and the discussions thus arising naturally would make clearer in the minds of all

---

spoken remarks. The footnote was included in the printed version of his talk as it appeared in both *Science* and the *BAMS*.

[74]Ibid.

precisely what are the functions of the axioms in the theory of geometry.

This is another case where Moore appears to have made an exceedingly optimistic extrapolation from his advanced teaching experience. Indeed, it is easily conceivable that his suggestion of having students develop competing axiom systems had been a feature of his advanced graduate seminar on "Foundations of Geometry." But Moore insisted that even for high school students his proposal was "thoroughly practical and at the same time thoroughly scientific." In another concession to his audience Moore admitted that no one could claim truly to understand euclidean geometry without also understanding non-euclidean geometry. But, he added, it would be a grave mistake to prematurely burden the average secondary-school student with such complications.[75]

Moore proceeded to describe further details of his proposal for a "laboratory method" of mathematics instruction, mentioning both his broad goal of developing the "true spirit of research" in all students, and such practical details as the desirability of assigning two consecutive class periods for the instruction. Besides recommending graphs, models, and physical measuring apparatus, he stressed the importance of deriving all important results in more than one way, so that "the student is made thoroughly independent of all authority," thus reiterating an aim he had espoused in his 1897 arithmetic textbook. Moore also explained how problems in practical mathematics could be gradually transformed to yield insight into fundamental theoretical properties, and to awaken interest in abstract concepts and desire for formal proof. Thus problems involving measurement of physical quantities could become exercises in the study of approximation and allowable error bounds, eventually leading to the theory of limits and irrational numbers. Study of the areas under specially selected curves could yield the notion of uniform convergence. The laboratory method also offered manifold opportunity for individually tailored instruction and cooperative learning, leading Moore to expand upon his claim for this method as the best one for all students: "for students in general, and for students expecting to specialize in pure mathematics, in pure physics, in mathematical physics or astronomy, or in any branch of engineering."[76]

When it came to implementation of his proposals, Moore advocated *"Evolution, not Revolution."* Even small steps in the preferred direction would be beneficial. Introduction of the laboratory method into the beginning college

---
[75]Ibid., 410–411.
[76]Moore, "Foundations of Mathematics," 411–413.

courses, that is into the "junior colleges," was especially to be sought for its leverage value, since some of these students would then propagate the method by going on to teach in the secondary schools. He reiterated the teacher training proposals he had vouchsafed to Harper. Moore noted in passing that he conceived of the normal schools as mainly for training primary school teachers.[77]

Moore announced that "Teaching must become more of a profession." This goal would require better training for teachers, as well as more freedom and more responsibility. Most significantly, it entailed that "closer relations should be established between the teachers of the colleges and the teachers of the secondary schools." The AMS could best do its part by "enlarging its membership by the introduction of a large body of the strongest teachers of mathematics in the secondary schools." Possibly two divisions could be established, one for research and one for pedagogy.[78]

Moore called attention to what he saw as the exciting developments along desirable lines taking place in the Midwest, citing in particular the work of high school teacher Edith Long in Nebraska, the Central Association of Physics Teachers, and initiatives being undertaken at various institutions in and around Chicago. He reiterated his call for "indirection"; that in elementary mathematics no distinction be recognized "between pure mathematics and its various applications." He hoped that the twentieth century would at last bring the mathematics of the seventeenth century "[t]o young students during their impressionable years." He concluded by affirming that the key to all these desirable developments was to adopt the laboratory method.[79]

---

[77]Ibid., 413–414. Italics in original.
[78]Ibid., 414.
[79]Moore, "Foundations of Mathematics," 415–416.

CHAPTER 7

Moore's Pedagogy in Relation to Contemporary Educational Thinkers

We can further illuminate E. H. Moore's ideas in relation to contemporary educational thought by looking more closely at five educators we have already met, with whom he had personal interactions: William Rainey Harper, John Dewey, John Perry, Felix Klein, and Simon Newcomb. The degree of his connection with these educators varied considerably, but each case reveals important features of his pedagogical thinking.

Looking at University of Chicago president William Rainey Harper and philosopher/psychologist John Dewey makes clear that Moore's use of laboratory as word and concept did not arise from a vacuum, but surely derived in part from the general atmosphere at Chicago. The case of Dewey further reveals the fraught relationship between university mathematicians and more general educators minded to make pronouncements on mathematics education. The casual arrogance of some mathematicians, secure in their abundant specialized knowledge of their subject, and drawing from this firm conclusions about their jurisdictional rights across the whole range of schooling, has been a feature of educational controversies of the late twentieth and early twenty-first centuries. Similar phenomena were already in evidence by the 1890s, and Dewey had felt the effects. Moore took a more conciliatory line towards non-mathematicians venturing into mathematics education than many mathematicians, and was rewarded with one brief mutually supportive interchange with Dewey. It is unlikely, however, that Moore took any specific ideas from Dewey; use of the same words by the two men might suggest such a substantive alliance, but this evaporates on close inspection.

With English engineering educator John Perry there is no doubt; Moore explicitly acknowledged a debt. Moore adopted several of Perry's ideas, ranging from the specific tool which came to be called graph paper, to the general notion that university mathematics instruction should focus on future engineers.

Moore embraced Perry despite the latter's sometimes bumptious style, and despite the suspicion generated among some of Moore's mathematical colleagues owing to Perry not being a mathematician. These were not problems with the German mathematician Felix Klein, another influence acknowledged by Moore. Klein had immense prestige among the young and growing American community of mathematicians, both for his own research accomplishments and for his skill in smoothing the way for other researchers. Moore tried to emulate Klein in many respects. Examination of Klein's career brings to the fore notable sociological features of the mathematics profession which help explain the lukewarm commitment to mathematics education often exhibited by mathematicians, and in particular some of the resistance experienced by Moore with regard to his pedagogical program.

With this background we conclude the chapter by returning to consider Moore in relation to Simon Newcomb and the Committee of Ten. This examination shows that while there was similarity between Newcomb and Moore with regard to proposed classroom procedures, they were imbedding these procedures within differing political goals and with differing perceptions of the challenges facing American mathematics education.

Long enduring provincialism within both the history of mathematics and the history of education has meant that figures such as Klein and Dewey or Harper and Perry, are rarely discussed in close proximity. We attempt here to see what results when such barriers are broken down.

7.1. William Rainey Harper

In his AMS address Moore conscientiously noted sources and supports for many of his pedagogical ideas, with one prominent exception: William Rainey Harper. Why he should have neglected to name Harper is unclear; possibly he simply thought that Harper's name would hold no special magic for mathematicians. His address included several references to publications and initiatives of University of Chicago faculty, certainly an implicit acknowledgement of Harper's encouragement of pedagogical work. In any event we have already noted that there can be little question of Moore's debt to Harper. This may or may not mean that Moore literally did not conceive certain thoughts until they were suggested by Harper; the historical record may never be able to resolve this. Some of Harper's dearest notions were analogous to notions Moore might easily have encountered elsewhere. What is clear is that within the University of Chicago Moore actively sought to frame his ideas in ways designed to appeal to Harper and that outside the university Moore spoke out with a confidence that surely derived in part from the consciousness that the like-minded Harper was behind him. The fact that Moore became much less pedagogically active after Harper's death in 1906 seems unlikely to be pure coincidence.

The central place of the word "laboratory" is the most important instance where Moore allied himself with Harper. We have seen that the word had been percolating through pedagogical thought for some while, and that it had been seized by mathematical educators even before Harper became president of the University of Chicago. The use of the word in the pedagogical endeavors of the Chicago science departments may seem unproblematic.[1] But Harper's vigorous proclamations added considerably to the aura surrounding this word, making it almost politically indispensable within his university. While Chicago physical and biological scientists complained about their facilities, other departments saw funds being unfairly lavished on science laboratories. Willard Pugh has found that the agitation of the arts and humanities faculty became especially intense after the Decennial celebration of June 1901, an occasion during which the present and future glories of the university were loudly trumpeted. Under the circumstances, classicists and economists thought it only fair that they too be given laboratories for graduate research.[2] It is hard to believe that Moore's concurrent activity was an entirely independent development.

Other Harper themes echo through Moore's pedagogical endeavors. Moore's invocation of the spirit of research has been mentioned, as well as his endorsement of the affiliation scheme, with the great university at the top of the educational hierarchy bringing order to the lower ranks. The desirability of inculcating independence of authority in the student is yet another Harper

---

[1] For example, two productions of the Chicago physics department during this period were the following: *College Course of Laboratory Experiments in General Physics* by Stratton and Millikan in 1898; and *A Laboratory Course in Physics for Secondary Schools* by Millikan and Gale in 1906. See Alfred Romer, "Robert A. Millikan, Physics Teacher," *The Physics Teacher* 16 (Feb. 1978): 82.

[2] Willard J. Pugh III, "The Beginnings of Research at the University of Chicago" (Ph.D. diss., University of Chicago, 1990), 752–753. See also Daniel Lee Meyer, "The Chicago Faculty and the University Ideal, 1891–1929" (Ph.D. diss., University of Chicago, 1994), 328–329. On the Decennial celebration see Thomas Wakefield Goodspeed, *A History of the University of Chicago: The First Quarter Century* (Chicago: University of Chicago Press, 1916), 399–405.

notion.[3] In all these cases Moore could well have received additional stimulation from other quarters, but Harper's versions would have been consistent and insistent.

One can also observe that Moore adapted ideas associated with Harper to a more mathematical context. For example, Moore's knowledge of the research process had been forged through his own wide-ranging mathematical inquiries, his graduate teaching, and his reading of contemporary mathematicians, especially major figures such as Klein and Poincaré. He certainly would have had no need of Harper to convince him of the importance of induction; Moore had a far more sophisticated understanding than Harper of the roles of induction, intuition, and deduction in mathematics and in scientific inquiry generally, and his proposals reflect that understanding.

Similarly, Moore's revolt from authority reflects not only Harper, Protestantism, and American individualism, but also Moore's experience of mathematics at the dawn of the twentieth century. One could not simply declare the *truth* of geometry. According to Poincaré, in an article Moore cited in the published version of his 1902 address, it was fruitless to ask whether space was *really* euclidean or non-euclidean: "we choose this geometry rather than that geometry, not because it is more *true*, but because it is more *convenient*." Even a given set of axioms for geometry or for abstract groups (two of Moore's favorite topics at the time) could be recast in numerous equivalent forms. He remarked in one of his research papers of 1902 that "the canons of relative simplicity of equivalent definitions by sets of postulates are not well established." The conventionalism of Poincaré and Moore's own research experience would thus have given him ample support for anti-authoritarian sentiments, whatever he may have taken from Harper.[4]

7.2. John Dewey

Moore's relationship with John Dewey is both more interesting and more inconclusive than his relationship with Harper. Dewey's name naturally attracts

---

[3]For Harper's disdain for authoritarian teaching, see James P. Wind, *The Bible and the University: The Messianic Vision of William Rainey Harper* (Atlanta: Scholars Press, 1987), 100 and Julie A. Reuben, *The Making of the Modern University: Intellectual Transformation and the Marginalization of Morality* (Chicago: University of Chicago Press, 1996), 96–97.

[4]Moore quoted Poincaré in Eliakim Hastings Moore, "On the Foundations of Mathematics," *Science* 17 (Mar. 13, 1903): 405, as found in Henri Poincaré, "On the Foundations of Geometry," trans. T. J. McCormack, *Monist* 9 (1898): 42. Italics in original. Moore's remarks on defining the group concept are from Eliakim Hastings Moore, "A Definition of Abstract Groups," *Transactions of the American Mathematical Society* 3 (1902): 488.

attention well beyond the bounds of mathematics education, and it would be gratifying to make some definitive statements in this case. Some commentators have discerned a causal link from Dewey to Moore. William Duren refers to "John Dewey, whose ideas Moore supported by proposing a mathematics laboratory." Karen Parshall states: "A follower of his Chicago colleague John Dewey (1859–1952), Moore renounced the standard lecture method of teaching mathematics in favor of a laboratory method in the Deweyan sense." The present writer suspects that Duren and Parshall have made incautious leaps on the basis of the word "laboratory." This word was so pervasive at Chicago that it cannot be used to establish a special connection between Dewey and Moore. A third scholar, Sidney Ratner, in an essay promisingly titled "John Dewey, E. H. Moore, and the Philosophy of Mathematics Education in the Twentieth Century," lauds "the parallels between their ideas," but his subsequent discussion reveals little.[5] Yet there clearly was some connection between these two head professors, and although the evidence is murky, sketching it nevertheless casts some revealing light on the interaction between mathematicians and less specialized educators.

In 1895, about a year after Dewey came to Chicago, he and James McLellan published *The Psychology of Number*, a book that aroused critical comment from Princeton mathematician Henry Fine in the journal *Science*. Fine, who had served with Simon Newcomb on the Mathematics Conference of the Committee of Ten, had declared Dewey and McLellan mistaken in grounding elementary arithmetic on measurement and ratio. Much more fundamental, said Fine, was counting, "one of the simplest of intellectual acts." He asserted that "Counting is not measuring and number is not ratio." Rather, counting was characterized by the procedure mathematicians had come to call "one-to-one correspondence," for example the matching of one's fingers with another set of objects. Fine proceeded to make the following remarkable comparison:

> The number of things in a group is not its measure, but, as Kronecker once said very happily, its "invariant," being for the group in relation to all transformations and substitutions what

---

[5]W. L. Duren, Jr., "Graduate Student at Chicago in the Twenties," in Douglas M. Campbell and John C. Higgins, eds., *Mathematics: People. Problems, Results* (Belmont, CA: Wadsworth International, 1984), 1:181. Karen H. Parshall, "Eliakim H. Moore and the Founding of a Mathematical Community in America, 1892–1902," in Peter Duren, ed., *A Century of Mathematics in America* (Providence: American Mathematical Society, 1989), 2:169. Sidney Ratner, "John Dewey, E. H. Moore, and the Philosophy of Mathematics Education in the Twentieth Century," *Journal of Mathematical Behavior* 11 (1992): 105.

the discriminant of a quantic, say, is for the quantic in relation to linear transformations, unchangeable.[6]

Fine was here using the word "group" in an informal sense, merely any collection of things. His invocation of "the discriminant of a quantic," on the other hand, was unashamedly technical. "Quantic" has now disappeared from mathematical language, usually replaced by "polynomial" or "form"; for example the binary quadratic form

$$ax^2 + 2bxy + cy^2,$$

whose discriminant is

$$b^2 - ac.$$

Fine was making the point that no matter in what order one counts one's fingers, for example, one will always come up with ten, a fact considered unworthy of attention by the overwhelming majority of people, but capable of arousing great interest among mathematicians. But observe also that Fine, who had just remarked on the intellectual simplicity of counting, was now alleging that the most basic notion of number (how many fingers, how many apples) could be substantially clarified by comparing it with concepts known only to an elite few. Note Fine's deliciously casual "say"; any respectable mathematician, Fine was suggesting, could offer many other like comparisons, but invariants of quantics would do very happily.[7]

Fine's excruciating condescension is hard to miss; this was a pointed assertion of the dominion of mathematicians in matters of pedagogy, and Dewey duly felt himself to have been skewered. His reply, in the form of a letter to the editor of *Science*, was suffused with defensiveness. He made a diffident attempt to grapple with Fine on the technical issues, remarking that if only he had known the phrase "one-to-one correspondence" he surely would have used it; to Dewey the phrase seemed to confirm "the implicit presence of the ratio idea in every number." But Dewey admitted that he was not a mathematician, and he furthermore noted that most school children were not destined to be mathematicians, so it was really not quite fair for the mathematicians to take the

---

[6]H. B. Fine, "Review of *Psychology of Number*," in John Dewey, *The Early Works, 1882–1898*, vol. 5 *1895–1898, Early Essays*, ed. Jo Ann Boydston (Carbondale: Southern Illinois University Press, 1972), xxiv–xxv.

[7]Invariant theory was a very active area of research in the nineteenth century, recast in a much more abstract manner by David Hilbert beginning in the 1880s. See Morris Kline, *Mathematical Thought from Ancient to Modern Times* (New York: Oxford University Press, 1972), 925–932.

high-handed attitude exemplified by Fine. Dewey called for a rapprochement between mathematicians and psychologists regarding mathematical instruction. Conceding the likelihood that he had made "blunders on the mathematical side," he hoped the mathematicians would "venture a little blundering" on the psychological side.[8]

Dewey would venture no more public remarks specifically on mathematical pedagogy until the time of Moore's AMS address. In the meantime he busied himself with many projects, including the establishment of his "Laboratory School":

> The conception underlying the school is that of a laboratory. It bears the same relation to the work in pedagogy that a laboratory bears to biology, physics, or chemistry. Like any such laboratory it has two main purposes: (1) to exhibit, test, verify, and criticize theoretical statements and principles; (2) to add to the sum of facts and principles in its special line.[9]

But although Thomas Chamberlain in geology, John Coulter in botany, Charles Whitman in zoology, Jacques Loeb in physiology, and Albert Michelson in physics have all been recorded as having taken an active role in the Laboratory School, no evidence has yet surfaced indicating any special contact with the mathematicians. Other evidence of interaction between Dewey and the Chicago mathematicians is sparse but tantalizing. In 1899 Dewey made brief acknowledgement of the pedagogical work of J. W. A. Young. Sidney Ratner was told by Moore's student Oswald Veblen in the 1950s that Veblen "in the early 1900s" took Dewey's seminar in logic and emerged from it convinced that Dewey was "the greatest teacher he had ever had."[10] But how this experience may have been transmitted to Moore is unknown.

---

[8]Dewey, "Letter to the Editor of *Science*," in *1895–1898, Early Essays*, 426, 429.

[9]John Dewey, "The Need for a Laboratory School," in Dewey, *1895–1898, Early Essays*, 437.

[10]For the role of University of Chicago science professors in Dewey's Laboratory School see Katherine C. Mayhew and Anna C. Edwards, *The Dewey School* (New York: Appleton-Century-Crofts, 1936) 10; and Arthur G. Wirth, *John Dewey As Educator: His Design for Work in Education (1894–1904)* (New York: John Wiley & Sons, 1966), 51–52. Dewey mentions J. W. A. Young in John Dewey, *Lectures in the Philosophy of Education: 1899*, ed. and with an introduction by Reginald Archambault (New York: Random House, 1966), 360. The Veblen quotes are from Ratner, 108. Veblen was a graduate student at Chicago from 1900 to 1903.

In December 1902 there appeared in the *Educational Review* an article by J. J. Sylvester's old Johns Hopkins student George Bruce Halsted entitled "The Teaching of Geometry." Halsted championed the recent advances in geometry:

> It is time now that the creation of non-Euclidean geometry, and the subsequent work of the masters on the foundations and meaning of geometry and science, so fruitful for the theory of knowledge, ... should be made fruitful for the general teaching of elementary geometry. ... There must be a textbook of rational geometry really rigorous. ... We must from the beginning bring up ourselves and our pupils on, not only the truth, but the whole truth. ... How soon the recent researches of Hilbert and others on the foundations of geometry must take their place in elementary text-books on plane and solid geometry cannot be said. But that is purely a matter of time unless some reactionary tendency sets in—and there are some who think it is already beginning to set in.

Halsted proceeded to criticize in technical detail recent expositors of what he saw as an insufficiently rigorous approach to teaching geometry. He cited John Perry as the leader of this "reactionary tendency."[11]

It should be evident that Halsted was touching on topics central to Moore's AMS address, and taking positions quite contrary to Moore. Indeed, Moore's cautions on pushing advanced geometric research concepts too rapidly into the curriculum, and his championing of John Perry, seem expressly designed as replies to Halsted. In particular, Moore and Halsted both cited Eduard Study's claim that understanding non-euclidean geometry is a prerequisite to full understanding of euclidean geometry, but to different effect.[12]

But it was Dewey rather than Moore who replied most directly to Halsted. In the April 1903 issue of the *Educational Review* there appeared a paper on "The Psychological and the Logical in Teaching Geometry," inspired, according to its author, John Dewey, by "Professor Halsted's able and suggestive article in the December number."[13] Halsted had remarked, evidently to Dewey's agreement, that "One advance which has been safely won, and may be rested on, is

---

[11] George Bruce Halsted, "The Teaching of Geometry," *Educational Review*, 24 (December 1902): 456–460.

[12] Moore, "Foundations of Mathematics," 411; Halsted, 470. Halsted was apparently not present to hear Moore's address. See list of attendees in *BAMS* 9 (Mar. 1903): 281.

[13] John Dewey, "The Psychological and the Logical in Teaching Geometry," *Educational Review* 25 (April 1903): 387.

that there should be a preliminary course of intuitive geometry which does not strive to be rigidly demonstrative, which emphasizes the sensuous rather than the rational."[14] We have seen that such a geometry course had been promoted by the Mathematics Conference of the Committee of Ten and others, including Moore.

However, Dewey saw a problem in Halsted's apparent desire to make the transition from the "sensuous" geometry to the "rational" geometry in one sudden leap. He found Halsted's proclamation about bringing the student up on "the whole truth" as far too rigid:

> I should say that to bring up the pupil from the beginning on the whole truth is simply impossible ... *Towards* the whole truth with all out heart; *on* it, no, because it is a meaningless requirement. The need and demand for teaching arise from the fact that the whole truth is not there to build upon.

Dewey felt that there needed to be a bridge from the experience of the student (the psychological) to the abstractions of mathematics (the logical), and he suggested a need to recognize that some students would never fully traverse this bridge. He admitted logical rigor as a fine thing for those who could appreciate it, but

> What of those whose interest in this mode of instruction is restricted? Those to whom the game of absolute logic does not appeal, those who are not called to ascend to the higher levels of science, or ever perhaps to reach much control of geometry as a pure tool? Mathematicians as mathematicians are not called upon to reckon with this class; but those who are concerned with teaching must take them into account.

For such students as those just described, Dewey argued, insistence on full logical rigor was foolish. He proceeded to defend the "Perry Movement" as recognizing for mathematics the "need of modulation in transition from the more intuitive to the more demonstrative phases of the subject."[15]

Even from these few extracts one can find a number of points of contact between Dewey's paper and the published version of Moore's address. Moreover, these two documents contain a pair of reciprocating footnotes, reproduced below in their entirety, with emphasis added to highlight the parallels:

---

[14]Halsted, 456.

[15]Dewey, "Teaching Geometry," 390, 394, 398. Italics in original.

> In an article shortly to appear in the *Educational Review*, on 'The Psychological and the Logical in the Teaching of Geometry,' Professor John Dewey, calling attention to **the evolutionary character of the education of an individual**, insists that there should be no abrupt transition from the introductory, intuitional geometry to the systematic, demonstrative geometry.[16]

and

> Since this article was written there has appeared in print (*Science*, March 13, 1903) the identical address of Professor E. H. Moore, entitled "On the Foundations of Mathematics," which dwells upon **the evolutionary character of all mathematics** as a reason for not making too fixed separation between various branches of mathematics or between pure and applied mathematics. I wish to record my indebtedness to Professor Moore for various suggestions.[17]

This is indeed fascinating. Clearly for at least a brief period Moore and Dewey conferred together, and this occurred just when it is known that Moore was inviting Dewey to meetings of the Mathematical Club concerned with pedagogy.

Unhappily this is as far as the record currently known to the present writer will take us. There remain many questions to which only informed speculation can be offered in response. One may surmise that Dewey, a previous contributor to the *Educational Review*, and surely a regular reader of that periodical, would likely have encountered the Halsted article on his own. Perhaps Dewey drew Moore's attention to the Halsted piece and the two then conferred on an appropriate response. Dewey would have needed no particular spur from Moore. The absolutist tenor of Halsted's article was clearly at odds with Dewey's deeply held beliefs in any case. But Dewey would likely have been desirous of some mathematical support for his positions; the Halsted piece contained technical details and literature references that Dewey would probably have found unfamiliar. John Perry, for example, was not a figure who seems to surface elsewhere in Dewey's oeuvre. Moreover, we have seen that Dewey had been burned in his last venture into mathematics for not consulting the experts. In his reply to Halsted he observed that "a psychologist encroaches on such a field only at his

---

[16] Moore, "Foundations of Mathematics," 411. Emphasis added.
[17] John Dewey, "Teaching Geometry," 387. Emphasis added.

deadly peril," but this time he came armed with testimony such as the following: "I am told that even in the very highest phases of mathematical inquiry there are still some matters in which even the trained mathematician finds it advisable to resort to intuitive constructions."[18] There can be little doubt that Moore was the informant here alluded to.

It is also a fact that Dewey's explicit reference to Moore appeared on the first page of his article as a footnote to the title. Indeed, it was the only footnote in Dewey's article, whereas Moore's footnote reference to Dewey was one of 27 footnotes. This encourages the conclusion that the deviation from parallelism in their published references to each other is significant: note that Dewey included a second sentence acknowledging "suggestions" from Moore; Moore's lack of such a sentence gives his acknowledgement of Dewey's work a more casual character. As regards their pronouncements on mathematical pedagogy circa 1902–1903 I therefore surmise that Dewey was likely more indebted to Moore than the other way around.

These speculations on the extent of the Dewey-Moore consultations are buttressed by considering the main features of their respective educational proposals, yielding the following general conclusion: Moore found enough correspondence between his ideas and those of Dewey to make Dewey a useful ally, but it is doubtful that much of Moore's thought derives from Dewey or that he was in any sense a "follower" of Dewey. Let us consider their respective use of some key pedagogical terminology.

Moore's use of "correlation," for example, was highly expedient. He referenced Herbart, but there was little substance to this reference. For Moore, correlation was an idea with wide currency which could be easily interpreted to support his particular proposals for mathematics. Most significantly, Moore's correlation proposals derived from his endeavor to protect the profession of mathematics by tying it more closely to science and engineering. For Dewey, in contrast, correlation did not stand alone as a worthy educational goal to be applied in various contexts; rather, it was subordinate to his overarching aim to connect education with the wider society:

> [T]he social life of the child is the basis of concentration, or correlation, in all his training or growth.[19]

and
---
[18]Dewey, "Teaching Geometry," 390, 398.
[19]Dewey, "My Pedagogic Creed," in *1895–1898, Early Essays*, 89. Originally published in 1897.

Relate the school to life and all studies are of necessity correlated.[20]

Such broad social conceptions of education were not present in Moore's writings.

Similar remarks may be made about the concept of student interest. Moore, as did many educators of the time, sought to cultivate it. One of his avowed aims was "connecting the abstract mathematics with subjects naturally of interest to the boy."[21] This has some resemblance to the conventional misinterpretation of Dewey as the champion of child indulgence, but it reflects little of the subtle critique of interest that Dewey in fact expounded. During his Chicago years Dewey specifically disparaged the conception of interest as something external to the student, to which one had "recourse to adventitious leverage to push it in, to factitious drill to drive it in, to artificial bribe to lure it in."[22]

This was but one instance of Dewey's distaste for dualisms of all kinds. The child and the curriculum, the psychological and the logical, interest and effort, all these for Dewey were falsely seen as paired off in struggle. Underlying all these false dualisms was the industrial revolution's shattering of the previous organic unity between education and life. No more did children learn what they needed to know as a natural part of growing up; the school had to compensate for this change in the social structure.[23] Moore, in contrast, placed his pedagogical thoughts within no large socio-economic frame, and his wariness of dubious dualisms was restricted to those connected with his own profession: pure versus applied mathematics, teaching versus research.

Another term used by both Dewey and Moore was "evolutionary." Since this term was used conspicuously in their parallel footnotes one must acknowledge some degree of intellectual interchange here, at least an agreement that they both considered the notion important. It is doubtful that it is more than this. Philip Wiener has commented on the notion of "*evolutionism* as a generalization invading every field of study" during the late nineteenth century,

---

[20]John Dewey, *The School and Society*, introduction by Leonard Carmichael (Chicago: University of Chicago Press, 1956), 91. Originally delivered as a lecture in Chicago in 1899; published 1900.

[21]Moore, "Foundations of Mathematics," 407.

[22]Dewey, *The Child and the Curriculum*, introduction by Leonard Carmichael (Chicago: University of Chicago Press, 1956), 27. Originally published in 1902. See also Dewey, "Interest in Relation to Training of the Will," in *1895–1898, Early Essays*, 114–117. Originally published in 1896 in the yearbook of the National Herbart Society.

[23]Dewey, *School and Society*, 9–12. On Dewey's anti-dualism within his educational thought see Morton White, *Social Thought in America: The Revolt Against Formalism*, 2nd ed. (Boston: Beacon Press, 1957), 7, 94–102.

and Arthur Wirth has cited "the general interest in evolutionary theory on the Chicago campus" in particular. In Dewey's case it is well known that evolution was deeply embedded in his thought.[24] Once again Moore's use of the concept seems more expedient, something grafted on from outside.

Finally, let us examine the manner in which Dewey and Moore used the word "indirection." There is some possibility that this represents a genuine appropriation by Moore from Dewey, and in any case there are revealing features. It is true that the notion of indirect proof, or *reductio ad absurdum*, is a very old one in mathematics. Hippocrates in the 5th century B.C. is often credited with inventing it. Moore most certainly used the method and the terminology in his mathematical work.[25] Nevertheless, his reference to indirection in his AMS address seems distinct from the mathematical usage, and much closer to the usage of Dewey, who certainly used the term during his Chicago period to designate an educational concept he considered valuable:

> The process of learning, in other words, conforms to psychological conditions, in so far as it is *indirect*; in so far that is, as attention is not upon the *idea of learning*, but upon the accomplishment of a real and intrinsic purpose—the expression of an idea.[26]

Moore, characteristically, was less interested in the psychology of indirection than in its usefulness as a tool of professional politics, namely, to solve "[t]he troublesome problem of the closer relation of pure mathematics to its applications."[27] The pure mathematical professional venture, according to Moore, would be best accomplished if pedagogical attention was not exclusively upon the ideas of pure mathematics but upon the expression of these ideas in applications; that is, upon the accomplishment of purposes more readily seen to be

---

[24]Philip P. Wiener, *Evolution and the Founders of Pragmatism* (Cambridge, MA: Harvard University Press, 1949; repr., New York: Harper Torchbooks, 1965), 6. Italics in original. Wirth, 100. See also Morton White, *Revolt Against Formalism*, 20–21.

[25]See, for example, Eliakim Hastings Moore, "A Simple Proof of the Fundamental Cauchy-Goursat Theorem," *Transactions of the AMS* 1 (1900): 499–506. On Hippocrates and indirect proof see Morris Kline, 40–41.

[26]Dewey, "Plan of the University Primary School," in *1895–1898, Early Essays*, 229. Italics in original. This piece was probably written about 1895 but never published during Dewey's lifetime. For other use of "indirection" by Dewey during his Chicago period see ibid., 232; and Dewey, "Educational Psychology: Syllabus of a Course of Twelve Lecture-Studies," in *1895–1898, Early Essays*, 308, 324. Dewey at this time also wrote that history for the educator "must be an indirect sociology." Dewey, *School and Society*, 151. See Wirth, 139.

[27]Moore, "Foundations of Mathematics," 416.

real and intrinsic by the ordinary student. There was no trace here of proof by contradiction.

Education historian Julie Reuben has noted that during this same time period indirection was a strategy also attractive to educators concerned with moral and religious instruction, the theory being that "Studying religion scientifically would indirectly stimulate students to be religious."[28] Whether or not there was any causal influence either way, I do think there was a substantial analogy between religious and mathematics education at this time. Many leaders in both fields had come to feel that presenting their respective subjects directly to students in unadorned fashion was likely to be counterproductive, and therefore both had recourse to indirect strategies. Both tried to use the cultural prestige of science for this purpose.

Dewey left Chicago for Columbia University in 1904, so whatever face-to-face interaction he may have established with Moore could not be continued.[29] There do not seem to be any extant letters between the two men, either from Dewey's Chicago period or later, to further elucidate their relationship.

7.3. John Perry

Only about a year before the Halsted-Dewey exchange in the *Educational Review*, that journal had been the venue for printing one of John Perry's polemical statements on mathematics education.[30] Perry (1850–1920) had been stirring up controversy in England for some years, but this was apparently the first time his work had been published on the other side of the Atlantic. His remarkable career began in Belfast, where he served an apprenticeship in a foundry and acquired a bachelor of engineering degree from Queen's College, studying with James Thomson. There followed a variety of positions in education and industry, including a brief stint in Glasgow assisting Thomson's more famous brother William (later known as Lord Kelvin), and three years at Tokyo's Imperial College of Engineering in the 1870s. From 1882 to 1896 he taught at Finsbury Technical College in London, after which he became Professor of Mathematics and Mechanics at the Royal College of Science, South Kensington. In both Japan and England Perry sought to create a new engineering education which would synthesize the best elements of academic laboratories and industrial workshops. He found the role of mathematics in the

---

[28]Reuben, 101.
[29]Meyer, 164–165.
[30]John Perry, "The Teaching of Mathematics," *Educational Review* 23 (Feb. 1902): 158–181.

education of engineers to be especially problematic. In particular, by the late 1890s Perry had injected himself forcefully into an already vigorous debate in Britain regarding the dominance of Euclid in mathematical training.[31]

Perry felt he had achieved considerable success teaching mathematics to engineers, and he concluded that his methods should be made widely available to all students of mathematics. He especially insisted that mathematics should be taught as an inductive science; for instance, propositions from Euclid should be tested by careful measurements using "squared paper," a device pioneered by Perry which became ubiquitous.[32]

Perry achieved his biggest splash with a symposium he organized on "Teaching of Mathematics" in September 1901 at the meeting of the British Association in Glasgow. This symposium, held before a joint session of the Mathematics and Physics Section and the Education Section, began with his own forceful exposition of his views and was then followed by a round of comments and discussion, all later incorporated into a book. The result in Britain was the long-sought victory of the anti-Euclid forces in the secondary schools, with some corresponding modifications at the university level as well.[33]

In addition to the individuals actually present in Glasgow, Perry sought responses to his views from other educators interested in mathematics, not confining himself to Britain. At the suggestion of A. R. Forsyth, then Sadlerian professor of pure mathematics at Cambridge, Perry sent a copy of his Glasgow address to the American David Eugene Smith, then recently appointed to a position at Teachers College of Columbia University. Smith, who had been quickly building a reputation in the world of mathematical education by his textbooks and his publications on the history of mathematics, sent Perry his response to be included with the published account of the Glasgow meeting.

Smith agreed with Perry in general, offering the supportive comment that from his observation of education in England, France, Germany, and the United States he judged England to be doing the worst job of teaching mathematics

---

[31]Graeme J. N. Gooday, "Perry, John," *Oxford Dictionary of National Biography* (Oxford: Oxford University Press, 2004), 43:832–833; William H. Brock and Michael H. Price, "Squared Paper in the Nineteenth Century: Instrument of Science and Engineering, and Symbol of Reform in Mathematical Education," *Educational Studies in Mathematics* 11 (1980): 373–376. On the earlier discontent with Euclid in Britain see W. H. Brock, "Geometry and the Universities: Euclid and His Modern Rivals 1860–1901," *History of Education* 4 (1975):, 21–29.

[32]John Perry, ed., *Discussion on the Teaching of Mathematics* (London: Macmillan and Co., 1902), 2; Brock and Price, 372–375.

[33]Brock, 30.

(he rated Germany best). It is likely that it was via Smith that Perry's Glasgow address came to be printed in February 1902 in the *Educational Review*, edited by the president of Columbia, Nicholas Murray Butler. Later in the year another address by Perry was published in *Science*, and then the expanded version of the Glasgow meeting, *Discussion on the Teaching of Mathematics*. It was this book to which both Moore and Halsted made most reference at the end of 1902.[34] The "Perry movement" had arrived in the United States.

It should be evident that Perry, although he did teach mathematics among his other duties, and although he is sometimes referred to as a mathematician by later scholars, was not a mathematician in the same sense that Moore and Felix Klein were mathematicians. From the point of view of Moore's American audience of pure mathematicians, Perry was certainly a far riskier ally than Klein. David Eugene Smith for one had detected "a tone running through [Perry's] address that grates on one who loves mathematics for its own sake." Indeed, Perry ridiculed the "affectation" of pure mathematicians who claimed their studies were entirely useless and all the better for it. In a 1903 letter Perry used words that many pure mathematicians might have interpreted as sheer philistinism:

> Our mathematicians are spending their time now in proving all sorts of things doable which the plain man knows to be doable which the plain man does without getting preliminary permission. I spent two weeks when I was very young in proving that an arbitrary function can be expanded in Fourier's Series. Good Lord! But after all, perhaps this was good work. What I complain of is the carrying of this idea into all our mathematical physics. Things assumed as axiomatic by Kelvin and Stokes are now being proved.[35]

---

[34]Perry's side of the correspondence is found in Perry to D. E. Smith, Oct. 10, 1901, and Perry to Smith, Nov. 30, 1901, both in DESP. Part of Smith's reply is quoted in Perry, *Teaching of Mathematics*, 89. The further Perry address is John Perry, "Address to the Engineering Section of the British Association," *Science* 16 (Nov. 14, 1902): 761–782. Cited in Moore, "Foundations of Mathematics," 407. See Moore, "Foundations of Mathematics," 406 and Halsted, 470 for their references to Perry's book.

[35]Perry to Victor C. Alderson, Mar. 14, 1903, quoted in Alderson, "Five Cardinal Points in the Perry Movement," *School Mathematics* 2 (1904): 194. Lack of punctuation in the original. Perry is referred to as a mathematician by Brock, 30. Smith comment on Perry's tone is found in Perry, *Teaching of Mathematics*, 91. See ibid., 3 for Perry's ridicule of pure mathematicians.

It is not likely that Perry's plain man would have been greatly enthralled by many of the products of Moore's Chicago department of mathematics.

Nevertheless, Moore seems to have satisfied himself that Perry was jeering not at creative mathematicians but at mind-numbing instructional practices, and that Perry too held mathematics to be "a fundamental reality of the domain of thought."[36] As Perry expressed it, "My engineering friends think that I have an exaggerated notion of the importance to all men of possessing a love for mathematics." The talented few could often develop such love for the subject despite bad teaching; the big problem was with the average student. Perry's faith was that such students could be brought to love mathematics by being introduced to it as an experimental science where they were expected to make discoveries for themselves. This entailed avoidance of "the scholastic devil" such as had plagued medieval philosophy; that is, avoidance of nit-picking logical rigor in favor of rapidly conferring the capacity to apply powerful methods. In his 1902 AMS address Moore embraced this entire argument. He agreed with Perry that a much larger cohort than heretofore needed to be given a genuine grasp of mathematics, and he accepted one major component of this notion, expressed bluntly by Perry: "the engineer is becoming a very important person."[37] Moore also readily adopted Perry's suggestion that his methods would ultimately benefit the training of pure mathematicians as well as engineers, providing an opening for his concept of indirection.

The evidence suggests that Moore, in the rapid run-up to the launching of his own pedagogical proposals, rather hastily appropriated the work of John Perry, a center of Anglo-American educational controversy of the day. How fully he recognized the less appealing features of Perry's thought is not clear, but he did not misinterpret Perry's basic goal of generating enthusiasm for mathematics. Harper and Dewey were of little use to Moore in this regard.

7.4. Felix Klein

Felix Klein (1849–1925) was a more fully appropriate ally for Moore's pedagogical program than Perry; nor was he a last-minute Moore enthusiasm. Indeed, in many ways Klein seems to have been a model for Moore's entire career. Both men valued and exemplified organizational as well as mathematical talent; both men put much of their energies into development of their students,

---

[36] Moore, "Foundations of Mathematics," 408.
[37] Perry, *Teaching of Mathematics*, 7–8, 16, 23.

their institutions, and their professions, at some expense to their individual accomplishments; both men saw themselves as invigorating (in Klein's case reinvigorating) mathematics in their respective universities so as to benefit their respective national mathematical communities, while facing some displeasure from a conservative old guard in the *east* (Berlin for Klein; New York and New England for Moore). As early as 1880 Klein was decrying the chasm between pure and applied mathematics that Moore would call attention to in America in 1902.[38]

We have seen that Moore's admiration for Klein extended back at least to the time of his initial appointment at Chicago; it was surely broadened and deepened by the events of 1893. Klein's Evanston lectures of that year included one entitled "On the Mathematical Character of Space-Intuition and the Relation of Pure Mathematics to the Applied Sciences." The ideas expressed here by Klein are very similar to those of Moore on intuition, on science as a sustaining source for pure mathematics, and on the conception of the historical development leading to the present period of "critical" analysis of the foundations of mathematics.[39]

Klein's philosophical views on mathematical knowledge had intimate connections with professional and pedagogical issues of long standing in Germany. Even more profoundly than in the United States, education in the German states from the beginning of the nineteenth century had been dominated by classical ideals, institutionally embodied at the secondary level in the *Gymnasium*. This institution emphasized the study of Latin, Greek, and mathematics, and came to be the primary entry point into the privileged world of the civil service and the professions. Other schools, the *Realschulen*, which came into being largely to meet growing commercial and technological needs not being met by the *Gymnasien*, often offered better opportunities for the study of natural science but were lacking in status through the nineteenth century. At the next higher educational level, the universities likewise enjoyed status privileges compared with the institutes of technology (*technische Hochschulen*). In the late nineteenth-century there were struggles on the one hand to raise the status of the *Realschulen* and the *technische Hochschulen*, and on the other hand

---

[38]Lewis Pyenson, *Neohumanism and the Persistence of Pure Mathematics in Wilhelmian Germany* (Philadelphia: American Philosophical Society, 1983), 56.

[39]Parshall and Rowe, 344–347.

to modify the curricula of the *Gymnasien* and the universities so as to better accord with the changing economic conditions.[40]

Most mathematicians remained complacent about such debates for many years, but Felix Klein was sensitive from early in his career that the reform efforts were a potential threat to the privileged position of mathematics in German education. As has been emphasized by Lewis Pyenson and David Rowe, Klein was a lifelong mathematical elitist, but his tactics for protecting his subject's exalted position changed over time. The young Klein aimed essentially at reinvigorating the neohumanist tradition, but the more mature Klein engaged in more complicated accommodations with the reformers. He was, for example, "one of the small minority of university professors who supported the technician's demands for rights commensurate with those of the universities," including the key issue of allowing the *technische Hochschulen* to confer doctoral degrees. By the 1890s, by which time Klein was securely established at the University of Göttingen, he was proposing to revamp mathematics instruction at the *Gymnasien* and the universities to emphasize spatial intuition and more practical problems, with the general aim of providing better support for the needs of students going into science and engineering. At Göttingen he enlisted the aid of private industry to build and equip new institutes for applied physics. A major inspiration for Klein's decision to seek partnership with industry was his exposure to advanced engineering education facilities at MIT, Cornell, and other American institutions he toured in 1893 after his Chicago sojourn. The University of Chicago would not have provided much of interest to Klein in this regard.[41]

Klein's ambitious program inspired opposition in Germany both from suspicious technologists and from pure mathematicians allied with the University of Berlin, long the bastion of neohumanist ideals in mathematics. But Klein achieved much, being notably successful in fostering a surge of research activity at Göttingen across the whole realm of pure and applied mathematics, overshadowing even Berlin. He brought in Hilbert and Minkowski in pure mathematics, Runge in applied mathematics, Prandtl in aerodynamics and hydrodynamics, Schwarzschild in astronomy. A number of young instructors of mathematics

---

[40]James C. Albisetti, *Secondary School Reform in Imperial Germany* (Princeton: Princeton University Press, 1983), 16–56.

[41]Schubring, 181; David E. Rowe, "Essay Review," *Historia Mathematica* 12 (1985): 282–284; Pyenson, 58.

who served in his department became illustrious as mathematicians and physicists of remarkable breadth; they included Hermann Weyl, Max Born, Arnold Sommerfeld, Richard Courant, and Theodore von Kármán.[42]

At the same time Klein sought to raise the status of mathematics at the *technische Hochschulen*, among other things ensuring that teaching positions at these institutions would be an attractive employment option for doctoral graduates of the universities. A key feature of his program was to invigorate the teaching of mathematics in the secondary schools, both technical and humanist, by stressing the function concept, with the aim of eventually giving these schools the responsibility for teaching differential and integral calculus. With the help of powerful friends such as Friedrich Althoff, Prussian minister of education, Klein scored important successes along these lines.[43]

Pyenson has interpreted Klein as cunningly "diverting" German educational reform to save mathematics as a profession. Rowe has objected that this account leaves out what he terms "obvious epistemological reasons" for Klein's achievement, namely that Klein had correctly assessed the relationship between mathematics and its applications as manifested in the West over the previous 300 years. Seeing that mathematics was in danger of cutting itself off from the sustenance of real-world problems, and conversely that the technologists were in danger of ignoring the power of mathematics, Klein had appropriately struggled to keep the lines of communication open.[44]

Both Pyenson and Rowe can be subsumed under Andrew Abbott's sociological framework, which emphasizes abstract knowledge as the key to success in conflicts over professional jurisdiction.[45] The usefulness of mathematics for science and engineering, in this view, derives precisely from its abstract nature. The reality of Rowe's epistemological reasons may be admitted, but these reasons did not simply impose themselves upon the professional negotiations in which Klein participated; they served as a resource that had to be marshaled with care. For example, Klein could expect an appreciative hearing when he proclaimed before the International Congress of Mathematicians in Zurich in 1897 that "there is a pure mathematics which after all constitutes the core of our

---

[42]David E. Rowe, "Klein, Hilbert, and the Göttingen Mathematical Tradition," *Osiris* 5 (1989): 197, 202.

[43]Schubring, 181–183, 188–189, 191–192; Pyenson, 68–69.

[44]Pyenson, 52–53; Rowe, "Essay Review," 289–290.

[45]Andrew Abbott, *The System of Professions: An Essay on the Division of Expert Labor* (Chicago: University of Chicago Press, 1988), 8–9, 102.

science, and whose prosperity forms the precondition for all other mathematical activities if they are not quickly to decline to a lower level."[46]

But this hierarchical view was not warmly received when it manifested itself in differential ratings of engineering graduates depending on their institutional origin. Despite Klein's relative liberalism regarding the *technische Hochschulen* he let slip in 1895 that he regarded its graduates as the "front officers" (*Frontoffizieren*) of industry; the graduates of the applied physics institutes he was proposing for Göttingen would then become the "general staff," (*Generalstaboffizieren*) by virtue of their greater exposure to theory. The proponents of the *technische Hochschulen* felt this slight for some time. Klein was later remembered in some circles as having given the *technische Hochschulen* graduates the even more insulting appellation of "soldiers."[47]

Moreover, as Abbott has argued convincingly, there is a strong propensity for those who claim control of valuable abstract knowledge to withdraw into the project of elaborating this knowledge rather than applying it to external problems. This is the phenomenon Abbott has labeled "professional regression,"[48] and it was something Klein had to constantly battle in holding his coalitions together. An example is provided by an anecdote involving Klein's Göttingen colleague David Hilbert, sent by Klein to a conference at the technical institute at Hannover to mend fences with the technicians. Hilbert's interpretation of this directive, as recounted by Theodore von Kármán, was to tell

> the Hannover audience in his easy-going way not to worry about Göttingen. "The university," he said, "is interested only in pure mathematics, not in engineering. The mathematician and the engineer have nothing to do with each other and never will."
>
> Klein groaned when he heard the report of what Hilbert had said. He had wanted the technical colleges to feel that they

---

[46] Felix Klein,"Zur Frage des höheren mathematischen Unterrichts," in *Verhandlungen des ersten internationalen Mathematiker-Kongresses*, ed., Ferdinand Rudio (Leipzig: B.G. Teubner, 1898), 301. My translation. The Chicago congress of 1893 has been retrospectively dubbed the "zero-th" congress.

[47] Rowe, "Göttingen Mathematical Tradition," 203; Theodore von Kármán with Lee Edson, *The Wind and Beyond: Theodore von Kármán, Pioneer in Aviation and Pathfinder in Space* (Boston: Little, Brown and Co., 1967), 53; Karl-Heinz Manegold, *Universität, technische Hochschule und Industrie: Ein Beitrag zur Emanzipation der Technik im 19. Jahrhundert unter besonder Berücksichtigung der Bestrebungen Felix Kleins* (Berlin: Duncker & Humblot, 1970), 132.

[48] Abbott, 118–119.

could get along with Göttingen, that both had a place in the teaching of engineers. He told me [Kármán] later that Hilbert's performance made it clear that a pure mathematician cannot be used for practical purposes.[49]

Indeed, it is evident that in a sense Klein had to battle the professional regression phenomenon within himself. As one of his biographers put it: "Klein was all his life striving for anything rather than Pure Mathematics, and found himself, with the shortest of interludes, doing nothing else."[50]

If one were to construct a similar capsule summary of E. H. Moore's career one would begin to see that he was not after all identical with Klein, for the summary would have to be essentially the following: Although Moore contributed almost exclusively to pure mathematics, he occasionally expressed strong interest in fostering applied mathematics. In comparison with Klein, Moore's experience was foreshortened and his efforts more superficial with regard to both applied mathematics and pedagogy.

Klein's professional advancement in the highly structured and competitive German academic world, though rapid, had necessarily to proceed by stages, taking him through several academic positions (including a stint at the *Technische Hochschule* in Munich), and allowing him to gain a thorough acquaintance with mathematical practices across Germany. In the more free-wheeling circumstances of America Moore was able to jump almost immediately to his ultimate rank. He then soon plunged into the pedagogical fray, without living with the issues for decades as Klein had done. Further, Klein had an unequaled comprehensive knowledge of the history of mathematical ideas, a knowledge whose propagation he deemed crucial to the continuing development of the subject. Moore could work up some respectable historical flourishes for a talk, but these were meager efforts compared to Klein's magisterial historical lectures. Moreover, Klein was inspired by a specific exemplar of the productive unity of

---

[49]Kármán with Edson, 54–55.

[50]W. H. Young (an English mathematician), quoted in Pyenson, 53. Capitalization in original.

mathematics and its applications, namely his great Göttingen predecessor, Carl Friedrich Gauss.[51]

It thus cannot be entirely surprising that Klein's efforts to bridge the chasm between pure and applied mathematics, although "futile in the long run," as David Rowe has noted, far outshone anything at Moore's University of Chicago. No institutes of applied physics were established at Chicago, nor even an engineering school. Nor could Chicago boast anything to match the culmination of Klein's efforts to bring mathematical enlightenment to physics: the fecund Courant-Hilbert text of 1924, *Methoden der mathematischen Physik*. Perhaps it is not entirely fair to compare this with the proclaimed pinnacle of cooperation between mathematics and physics at Chicago: Dickson solving fifteen simultaneous equations in aid of a physicist. But the fact remains that the experimental programs of Chicago physicists such as Michelson and Millikan required little sophisticated mathematics, and certainly not the new abstract products of the Chicago department of mathematics; Millikan makes only the most trivial references to the mathematicians at the University of Chicago in his autobiography.[52]

Moore was likewise less successful than Klein in bringing his pedagogical ideas to broad fruition, as will be discussed in the next chapter. One factor is clear, that the political environment encountered by Moore was fundamentally more complicated than that faced by Klein. Klein's arguments on behalf of his ideas could well have gone for naught had he not had access to the decision-makers in Berlin. But Moore operated in no such highly centralized educational system. For him there was no figure corresponding to the Prussian Minister of Education, nor any Kaiser who could effect national educational policy change with an imperial decree.[53]

---

[51]Rowe, "Göttingen Mathematical Tradition," 199; Parshall and Rowe, 312. Besides his AMS address, Moore also put historical touches into his retiring address as president of the AAAS in 1922. See G. A. Bliss, "The Scientific Work of Eliakim Hastings Moore," *BAMS* 40 (1934): 511. Many of Klein's historical lectures were collected in Felix Klein, *Vorlesungen über die Entwicklung der Mathematik im 19. Jahrhundert*, 2 vols. (Berlin Springer-Verlag, 1926–1927).

[52]The judgement of Klein's futile effort is from Rowe, "Göttingen Mathematical Tradition," 204. On the significance of the Courant-Hilbert book see Christa Jungnickel and Russell McCormach, *Intellectual Mastery of Nature: Theoretical Physics from Ohm to Einstein*, vol. 2, *The Now Mighty Theoretical Physics 1870–1925* (Chicago: University of Chicago Press, 1986), 344–345. For Millikan's bland comments on Chicago mathematicians see *Autobiography of Robert A. Millikan* (New York: Prentice-Hall, 1950), 23, 48.

[53]For example, in 1899 Kaiser Wilhelm II granted the *technische Hochschulen* the right to confer doctoral degrees. Rowe, "Göttingen Mathematical Tradition," 204.

7.5. Simon Newcomb and the Committee of Ten

There is little evidence to suggest that Moore reacted explicitly to the report of the Mathematics Conference of the Committee of Ten at the time of its publication, but the organization that he was helping to make the primary forum for research mathematicians did take note of the report. In February 1894 the *Bulletin* of the New York Mathematical Society (soon to become the American Mathematical Society) published the Committee of Ten's summary of the Mathematics Conference report, preceded by a list of all the Committee and Mathematics Conference members. The editors of the *Bulletin* (Thomas Fiske and Alexander Ziwet) did not comment on the report, but did advise their readers that they could obtain a full copy from the Bureau of Education.[54]

Less than ten years elapsed between that report and Moore's address to the American Mathematical Society, but Moore's words and actions exposed several fundamental changes in the nature of American mathematical activity and in attitudes of mathematicians towards pedagogical issues. The most visible change was that the number of individuals who consciously identified themselves as professional mathematicians greatly increased during this period. The AMS grew substantially, and what is more important, the members of the AMS markedly increased their mathematical activity, as measured by attendance at meetings, delivering of talks, and publication of papers.[55]

By the time of Moore's talk it was reasonable to speak of an American mathematical community. Moreover, the values espoused by this community were becoming clear. To have the greatest prestige and influence one should be a professor of mathematics at a college or university, one should have earned a Ph.D., and one should have a primary interest in pure mathematical research. Simon Newcomb had demerits on all these counts, while E. H. Moore's credentials were impeccable, the first AMS president of whom this could be said. Moore had been a leader in pressing for just such a national consciousness of the primacy of university-based pure mathematics. It was therefore natural that he should be much more conscious than his predecessors of speaking to his

---

[54]"The Teaching of Mathematics in the Secondary Schools," *Bulletin of the New York Mathematical Society* 3 (Feb. 1894): 127–130.

[55]The membership in the AMS grew from 239 in 1893 to 428 in 1903. See Raymond Clare Archibald, *A Semicentennial History of the American Mathematical Society, 1888–1938* (New York: American Mathematical Society, 1938; repr., New York: Arno Press, 1980), 44. For a comprehensive analysis of increasing mathematical activity during the period see Della Dumbaugh Fenster and Karen Hunger Parshall, "A Profile of the American Mathematical Research Community: 1891–1906," in Rowe and McCleary, 3:179–227.

fellow mathematicians as a distinct interest group, and in speaking on behalf of that group to those outside.

In particular Moore looked past the casual and unexamined assumption of many earlier mathematicians, Simon Newcomb and the Mathematics Conference of the Committee of Ten in particular, that mathematics had an impregnable place in the schools. Moore was sensitive to the fact that this place could indeed be challenged, and believed that defending mathematics in the schools was ultimately crucial to the health of the entire mathematical community, including the elite pure mathematical researchers. It is unlikely that this consciousness on Moore's part came entirely from personal experience of mathematics as an embattled enterprise; in 1902 the most vociferous attacks were still in the future. Moore had encountered some sharp words from president Harper, and had evidently listened to some critical remarks by scientists and engineers. He also had attended to the example of Felix Klein in Germany. But surely much of Moore's anxiety was an extrapolation resulting from precisely the professional consciousness that he was endeavoring to raise. Embracing pure mathematical research was the most effective means of distinguishing the mathematical enterprise, but this professional distinctiveness made Moore feel greater exposure to competition.

We have seen that fear of college domination had exercised secondary school educators at the time of the Committee of Ten, and that college educators had confirmed them in their fears by being oblivious, assuming that such domination was the natural state of the educational system. Moore and William Rainey Harper were far from being oblivious. They unapologetically sought college domination precisely because they did not think it inevitable; it required exercise of political power and skill. Just as Harper sought to affiliate selected secondary schools to the University of Chicago, Moore sought to bring selected secondary school teachers of mathematics into the American Mathematical Society. The sincerity of Moore's concern for the quality of secondary school mathematics is not in question, but there can be no doubt that he kept always in mind the aim of improving the prospects for mathematical research. As he wrote to a fellow mathematician in 1904,

> Parenthetically I shall be greatly surprised if the pedagogic movement which is now beginning doesn't have a marked and desirable reaction in favor of research in mathematics pure and

applied. Certainly that (feeling and) conviction has been the principal motive leading me in the matter.[56]

With regard to specific pedagogical proposals there was much continuity from Newcomb to Moore. Both championed concrete and inductive methods whose popularity had been building from earlier in the nineteenth century in parallel with the rise of natural science as a subject worthy of school study. For Newcomb and the Mathematics Conference, however, the justification of such methods was confined primarily to the realm of psychology: these methods were held to offer the most effective means of transferring mathematical knowledge to the student. Moore evidently agreed on the psychology, but here again his professional awareness introduced a new element. Moore's elaboration of concrete mathematical pedagogy in his laboratory method was in part a strategy for professional survival and growth. We have seen that the earlier generation of mathematicians did occasionally engage in the lively ongoing philosophical debates on induction versus deduction, but at the end of the century the new generation perceived that the rising tide of science might truly be a threat to mathematics both as a school subject and as a professional activity. Moore strove to accommodate to science even as he sought to distinguish mathematics from science. Absent the push to distinguish mathematics, the need to accommodate would not have been so great.

Moore sought accommodation not only with pure scientific inquiry but also with engineering. He came to the conclusion that the training of engineers had become a crucial justification for teaching mathematics in the colleges and universities, and that it was destined to become more so in the future. This marked a sharp rise in promoting mathematics as a utilitarian subject, and it was accompanied by the nearly complete disappearance of mental discipline and related concepts from Moore's pedagogical vocabulary. There was only slightly more place in Moore's pronouncements for looking on mathematics as an aspect of liberal humanism. In his AMS address he did briefly refer to mathematics as "a fundamental reality of the domain of thought,"[57] evidently as a standard around which to rally his fellow mathematicians. But for the most part he seems to have felt that in educational politics the only claims worth making for the value of mathematics were utility and research. Moreover, of these two claims, it was utility that he held must be given political priority, whatever

---

[56]E. H. Moore to Ernst Julius Wilczynski, January 23, 1904. Another idiosyncratic use of parentheses by Moore. EJWP, Box 2, Folder 2.

[57]Moore, "Foundations of Mathematics," 408.

one's intellectual allegiance to pure mathematics. This was the overarching message of his 1902 address.

Newcomb and the Committee of Ten had largely failed to recognize that American education was at the beginning of a great growth spurt. E. H. Moore, who worked within one of the major centers of this growth, did not fail to notice certain aspects. His doctoral students were readily obtaining jobs in colleges and universities, and the recipients of bachelor's and master's degrees were stepping into secondary school positions. But Moore and his mathematical colleagues in 1902 still did not grasp important implications of the coming era of mass schooling. In particular, the focus on serving science and engineering would prove to be not nearly as helpful to mathematics as Moore has envisioned, as the schools were swamped by students with more mundane needs for mathematics instruction.

CHAPTER 8

The Reception of Moore's Program

Mathematical educators at several levels responded explicitly and implicitly to the pedagogic program proposed by E. H. Moore, both to his specific proposals and to his broad vision for the future of American mathematics. Initially Moore and his University of Chicago colleagues advocated his ideas vigorously. Especially from 1903 to 1906 they taught a variety of courses using at least some features of Moore's laboratory method, they published articles related to their pedagogical experiments, and they had opportunities to make known their views through participation in regional and national meetings and committees.

From the start Moore encountered resistance from mathematicians who felt that he was lowering himself by discussing school mathematics, or that he was shirking his primary duty of promoting mathematical research. The presidents of the American Mathematical Society (AMS) who followed Moore largely ignored his call that mathematicians should pay more attention to instructional issues in the schools and the colleges. He was never able to bring a large number of school teachers into the AMS as he had hoped. Indeed, the AMS became even more focused on research, to the extent that those wishing to discuss the problems of teaching undergraduates felt compelled to form a separate organization in 1915, the Mathematical Association of America (MAA).

After 1906 Moore's pedagogical activism faded into the background of his professional life, as his time was increasingly spent on a program of research in pure mathematics he called general analysis. Even those at Chicago who remained committed to pedagogical issues toned down their rhetoric and watered down their proposals. The one specific Moore proposal that did become popular in the early decades of the twentieth century was the use of graph paper as a teaching tool in mathematics classrooms. How much this is due to Moore is hard to ascertain. Graphical techniques were on the rise in engineering education, independent of Moore, but his voice may have broadened their reach.

8.1. Promotion of Moore's Ideas at the University of Chicago

Moore did not simply announce his pedagogical proposals and passively await results. For at least three years after his 1902 address he fought hard to promote and popularize his ideas and to obtain allies. But it is also evident that his enthusiasm, and that of his Chicago colleagues, did eventually begin to wane. This was first manifested in a shift of focus from broad and ambitious pedagogical reform to more modest and specific methodological recommendations, such as graphs and the function concept. Eventually Moore's research program in pure mathematics seems to have crowded out his pedagogical interests.

Moore's address was publicized as thoroughly as could have been reasonably expected for such a work at that time. The AMS had 399 members at the close of 1902, of whom 62 attended the New York City meeting at which Moore delivered his talk on December 29. The AMS responded to his words by immediately setting up a committee to consider "the desirability of the Society undertaking to exert an effective influence on the teaching of elementary mathematics." Moore then returned home to Chicago, where he proceeded to give his talk again to the meeting of the Chicago Section of the AMS held on January 2–3, 1903. The talk was published in both the *Bulletin of the AMS* and in *Science*. In addition the full text appeared in the *School Review* and extracts appeared in the *Mathematical Supplement of School Science*, both journals published under the auspices of the University of Chicago.[1]

Meanwhile Moore expanded upon his efforts of 1902 to reform instructional methods within his own department. In January 1903 he wrote to Oswald Veblen, who was on the verge of completing his doctorate, that there would likely be employment prospects at Chicago, especially if Veblen was willing to pitch in to solve the pedagogical problems associated with "the laboratory method plans of the department," one aim of which was to "establish a unity of spirit in the staff in mathematics from secondary school through graduate school." It is evident that Veblen did indeed participate in the department's ongoing pedagogical experiments. For a trigonometry class he taught at Chicago in the fall of

---

[1]*BAMS* 9 (Mar. 1903): 281–282 includes a listing of all 62 members attending the 1902 New York City meeting. On the Chicago meeting, see *BAMS* 9 (Apr. 1903): 337. Moore's address was published as E. H. Moore, "On the Foundations of Mathematics," *School Review* 11 (1903): 521–538 and Eliakim Hastings Moore, "Pure and Applied Mathematics," *Mathematical Supplement of School Science* 1 (Apr. 1903): 25–28; "Elementary Mathematics," ibid., (June 1903): 57–63. (There is supposed to be a third Moore excerpt in the *Mathematical Supplement* which I have never been able to find.)

8.1. PROMOTION OF MOORE'S IDEAS AT THE UNIVERSITY OF CHICAGO

1903 he acknowledged trying to use some features of the laboratory method and receiving advice from Moore. Several other of Moore's colleagues and students also joined in the effort. A. C. Lunn and L. E. Dickson published reports on use of Moore-style ideas in their undergraduate teaching. N. J. Lennes recounted his experiments with using laboratory methods in teaching high-school algebra. J. W. A. Young and H. E. Slaught, who by now were confirmed pedagogical specialists, were even more active in promoting Moore's methods, as described below. In the spring quarter of 1903 Moore himself offered a course entitled "Teaching of Mathematics by the Laboratory Method." He taught the same course again in the summer and the autumn quarters. In both of these terms he also taught the introductory calculus course using the laboratory method.[2]

As noted in a previous chapter, in April 1903 Moore's Committee on Mathematics reported to the Commission on Accredited Schools of the Association of Colleges and Secondary Schools of the North Central States. This brief report contained echoes of earlier curricular initiatives as well as Moore's recent proposals. Algebra was to be taught as "generalized arithmetic." Geometry was to be taught "[i]n connection with arithmetic." "[T]he abstract form should be developed late, though foreshadowed long. ... the march should always be from the concrete to the abstract." "The pupil's attitude must be, in the main, that of an active worker, not that of a passive listener." "It is well to consider the methods of instruction in the physical laboratory. Some of these methods, suitably modified, may be of value also in the instruction in mathematics." The report also recommended that teachers familiarize themselves with "the effective movement of reform in the pedagogy of elementary mathematics in England, initiated by John Perry." A good place to find references to the Perry literature, the report noted, was in Moore's address printed in *Science*.[3]

Not all activity in mathematics education at the University of Chicago at this time can be directly traced to Moore, but his stepping forward from his prestigious position surely emboldened those who had already been thinking along similar lines. A case in point is George William Myers (1864–1931), who had done graduate work in mathematical astronomy at Chicago in the mid

---

[2]Moore to Oswald Veblen, Jan. 21, 1903, OVP, Box 8. Oswald Veblen, "Polar Coordinate Proofs of Trigonometric Formulas," *AMM* 11 (1904): 6–7. A. C. Lunn, "Outline of a Coherent Course in College Algebra," *AMM* 12 (1905): 123–129. L. E. Dickson, "Graphical Methods in Trigonometry," *AMM* 12 (1905): 129–133. N. J. Lennes, "Another Algebraic Balance," *SSM* 5 (1905): 602–605. MDLN, Box 5.

[3]"Report of the Commission on Accredited Schools," *Appendix to the Proceedings of the Eighth Annual Meeting of the North Central Association of Colleges and Secondary Schools* (1903): 182–184.

1890s and had then gone to Germany for a Ph.D. from Munich in 1898. He became a professor of mathematics at Francis Parker's Chicago Institute for teacher training, staying on as "professor of the teaching of mathematics and astronomy," when this institution was incorporated into the university to create the School of Education in 1901. Myers therefore had the opportunity to work briefly with two of the most significant pedagogues in the country, an experience that has been reported as giving him "an unusual appreciation of the trends in Colonel Parker's and Mr. Dewey's points of view."[4]

In 1903 Myers became the mathematical editor of the aforementioned *Mathematical Supplement of School Science*, a new journal devoted specifically to the problems of teaching mathematics in the secondary schools. (This journal subsequently became *School Mathematics* and later part of *School Science and Mathematics*.) From the first, Myers showed an interest in using the journal to promote ideas associated with the laboratory method. Early issues contained articles such as "The Mathematical Laboratory," "What Is the Laboratory Method?" and "The Need of a Perry Movement in Mathematical Teaching in America."[5] Myers put special emphasis on the excerpted sections of Moore's AMS address, with the following words of introduction to the first installment indicating the high aspirations he attached to Moore's pedagogical program:

> Mr. Moore... adduces reasons for the wide adoption of laboratory methods in mathematics classes, and calls upon friends of progress in mathematical methodology everywhere to join in a concerted movement for the energizing of mathematical teaching. The cause, as also its advocacy in these three papers, are so worthy and so timely, that our readers can profitably spend time for our next two issues studying them.[6]

Myers was more explicitly and proudly an educational progressive than Moore, and likewise more fond of deploring educational conservatism. Further

---

[4]On Myers, see *University of Chicago General Register of the Officers and Alumni 1892–1902* (Chicago: University of Chicago Press, 1903), 73; and Katherine C. Mayhew and Anna C. Edwards, *The Dewey School* (New York: Appleton-Century-Crofts, 1936), 366.

[5]C. E. Comstock, "The Mathematical Laboratory," *Mathematical Supplement of School Science* 1 (April 1903): 14–19; J. W. A. Young "What Is the Laboratory Method?" ibid., (June 1903): 50–56; and H. E. Cobb, "The Need of a Perry Movement in Mathematical Teaching in America," ibid., (October 1903): 121–124.

[6]Footnote to Eliakim Hastings Moore, "Pure and Applied Mathematics," *Mathematical Supplement of School Science* 1 (April 1903): 25. This footnote is signed "Ed." Myers was the mathematical editor, C. E. Linebarger the managing editor.

8.1. PROMOTION OF MOORE'S IDEAS AT THE UNIVERSITY OF CHICAGO 211

evidence of Myers' independence of Moore can be found in the added nuance the former applied to "correlation." Myers took care to distinguish it from mere "inter-relation" which "keeps the thought on subject-matter. Correlation of subjects makes prominent the reaction of the learning mind upon the subjects." Furthermore, correlation "gives to mathematics a social value." This was important because "we may be pretty sure that the school is to be raised more and more fully to the dignity of a social center as time goes on."[7] These words have a more authentic Deweyan flavor than anything found in Moore.

Myers proceeded to proselytize on behalf of the laboratory method and to work on the details of its implementation. Moore's plan to bring John Perry to Chicago, noted in a previous chapter, aroused special enthusiasm from Myers:

> All teachers who believe in the need and possibility of improvement in mathematical instruction will be pleased to learn that Prof. Perry has accepted the invitation of the University of Chicago to deliver a course of public lectures at the University during the coming summer. A real, live apostle of mathematical reform! and from the land of Euclidean idolatry!! Think of it! Sheer curiosity should make a thousand miles seem short to witness so unusual an occurrence![8]

In the very next issue it was announced that Perry was not coming after all, "for reasons not fully known to the editors."[9] It appears that Perry never did come to Chicago.

In August of 1903 Myers read a paper on the laboratory method at the Mathematical Club of the University of Chicago. He addressed some of the same themes as Moore but with harsher language, for example proclaiming that "our too early foisting of abstract mathematics upon [the pupil]" often led to the subject becoming "disgusting and stultifying to him." Myers was also considerably more explicit on the equipment needed for a mathematical laboratory. In addition to items mentioned by Moore, Myers listed such things as T-squares, logarithmic tables, a surveyor's compass, barometers, thermometers, spherical blackboards, pendulums, and a stereopticon and slides.[10]

---

[7] G. W. Myers, "Correlation of Subjects in Secondary Mathematics Teaching," *School Review* 11 (1903): 24, 27, 28.

[8] *Mathematical Supplement of School Science* 1 (April 1903): 36. This is from an editorial note signed "M."

[9] *Mathematical Supplement of School Science* 1 (April 1903): 36 and (May 1903): 84.

[10] G. W. Myers, "The Laboratory Method in the Secondary Schools," *School Review* 11 (1903): 731, 737–738.

Moore followed through on the proposal he had announced in his address to make contact with secondary school teachers. In April 1903 the Chicago Section of the AMS met jointly with the North Central Association of Science Teachers. J. W. A. Young used the occasion to report on "What is the Laboratory Method?" very much a revisiting of the themes of Moore's December address: correlation; proceeding from concrete to abstract; adopting axioms as needed and deferring the subtle analysis of foundations; teaching everything in two ways; two-hour class sessions; "evolution, not revolution."[11]

Moore also attempted to extend his reach beyond Chicago, working especially through Thomas Fiske and David Eugene Smith in New York City. As Fiske reported to Smith in April 1903:

> I recently had a letter from Professor E. H. Moore, in which he asks what has been done in the way of listing the strongest secondary-school men. He is very desirous of our getting into touch with a few of these men immediately, and suggests that it might be desirable to invite one or two of them to our festivities to be held on the evening of Saturday, April 25.[12]

In July 1903 Smith led a conference on mathematics at the annual meeting of the National Educational Association, held that year in Boston. Moore was unable to attend himself, but expressed his keen interest in the meeting program in letters to Smith. A major purpose of the NEA session was to encourage the formation of associations of secondary-school teachers of mathematics. Smith opened the session by noting the beneficent influence of these associations in Europe, and then yielded the floor to representatives of three American associations that had recently formed: in the north central states, in the middle states and Maryland, and in New England. Smith was one who would later promote the view that this organizational activity had been largely "prompted or encouraged" by Moore. Whether this was literally true in all cases (almost certainly it was for the north central states), there is no doubt that Moore was greatly interested in and approving of all such developments.[13]

---

[11] J. W. A. Young, "What is the Laboratory Method?" 50–56.

[12] Fiske to Smith, April 14, 1903, DESP. By "festivities" Fiske was referring to a dinner to be held in conjunction with the April 1903 meeting of the AMS in New York City. See "The April Meeting of the AMS," *BAMS* 9 (July 1903): 525.

[13] For Moore's communication with Smith regarding the Boston meeting see Moore to Smith, June 14, 1903 and June 24, 1903, in DESP. For Smith's conduct of the meeting see "The Mathematical Conference," *Journal of the Proceedings and Addresses of the Forty-Second Annual Meeting of the National Educational Association* (1903): 480–481. For Smith's later

8.1. PROMOTION OF MOORE'S IDEAS AT THE UNIVERSITY OF CHICAGO

The 1903 NEA mathematics session also included a report by a committee of the AMS (alluded to in Moore's address) on college entrance requirements in mathematics. Chaired by H. W. Tyler of MIT, the other committee members were Fiske, Osgood of Harvard, Ziwet of Michigan, and Young of Chicago. Among its recommendations, this committee endorsed the use of graphs in algebra instruction, in line with Moore's program. But despite the presence of Young, the report also reflected a divergence from Moore's way of thinking among AMS members that would later stand in the way of the adoption of his proposals. In particular, the report exhibited a standoffish attitude toward secondary school issues, and a great complacency regarding the place of mathematics. The committee simply assumed that all prospective entrants to colleges and technical schools should take considerable mathematics; the only question of interest was which topics ought to be studied. It also disclaimed all interest in how mathematics should be taught: "The committee understands also that the consideration of pedagogic questions is not among its duties."[14]

The Tyler report was published twice, a "preliminary" version in the *NEA Proceedings* and then a final version in the *Bulletin of the American Mathematical Society*. But there appears to be hardly any difference in the two versions, except for one footnote to be remarked on shortly. The committee gave little evidence of hard work. The section on plane geometry is less than 50 words, relying on the injunction that the course ought to include "the usual theorems and constructions of good text-books."[15] Osgood epitomized the attitude in his summary remarks on the NEA conference:

> As to the relation of these associations [of secondary school teachers] to the American Mathematical Society, there should be the greatest interest on the part of the latter in the success of the former, and without doubt members of the American Mathematical Society will be influential in such associations. Any more close relationship is not, however, advisable.[16]

Here we have the proud champion of the abstract core of mathematics condescending to take "the greatest interest" in educational matters, but wary of

---

assessment see David Eugene Smith, "Movements in Mathematical Teaching," *SSM* 5 (March 1905): 135–136.

[14] "Mathematical Conference," *NEA Proceedings* (1903): 481.

[15] Tyler et al., "Report of the Committee of the American Mathematical Society on Definitions of College Entrance Requirements in Mathematics," *BAMS* 10 (Nov. 1903): 76.

[16] "Mathematical Conference," *NEA Proceedings* (1903): 484.

8. THE RECEPTION OF MOORE'S PROGRAM

polluting the purity of the discipline through too close contact with the practitioners.

The opposite pole of the professional conflict between the research mathematicians and the mathematical pedagogues was nicely represented in another session at the same NEA meeting, with David Eugene Smith again participating. Entitled "Does the Teacher's Knowledge of Subject Differ from the Scholar's Knowledge?", this session in the "Normal Department" was an extended effort at professional uplift on the part of teachers and "normal school men." The speakers, including Smith, attempted to outdo each other in demonstrating that teachers had to know much more than "mere scholars," and that all the methodological knowledge imparted in normal schools and education departments was distinctly enhancing.[17] Here we have the resentful practitioners defending themselves against the abstract knowledge workers. It is evident that in these disputes David Eugene Smith played an intriguing mediating role, but it appears that he came to identify more strongly with the scholars as time passed.

Soon after the 1903 NEA meeting had been completed Moore started planning to make use of the 1904 meeting. He informed Smith in August 1903 that he was trying to get in touch with the appropriate NEA authorities so that at the 1904 NEA meeting "suitable provision may (if possible, surely) be made for the interests of those especially interested in math." It is instructive to look at an extended passage from this letter, illustrating as it does Moore's restless striving to properly qualify his pedagogical ideas, and the sensitivity to professional concerns which lay at the base of these qualifications:

> My understanding is (not at all that our Comm. is asked to formulate anything to be printed and published,—but) that we are to decide what our relations with the secondary field might best be →{a matter of internal policy}—and to me it is quite clear, that the men of strength and depth of insight in our ranks (even if merely for the sake of the interests of research:— to be sure, personally, I should regret to be ˆ{what seems to me} so narrow) should enter into active relation with the secondary people ˆ{especially} in the new pedagogic associations (which should always be catholic in nature, embracing teachers of mathematics in institutions of all kinds),—while, conversely,

---

[17]"Does the Teacher's Knowledge of Subject Differ from the Scholar's Knowledge?" *NEA Proceedings* (1903): 547–566.

the highest interests of mathematics as officially represented by the American M. Soc^y (even in the narrow research construction of those interests) will be furthered if the leading men of those associations join our ranks—This will come about as a matter of course in time, provided our Committee or the Council does not now do some door-shutting negative legislation—[18]

The committee to which Moore was referring was the one formed in consequence of his presidential address. The members of this committee were Moore, Smith, Osgood, Fiske, and Tyler. The "Council" was the core governing group within the AMS; Moore was automatically a member as an ex-president. Note that Moore felt he could not indubitably prevent either of these bodies from taking actions contrary to his recommendations. Indeed, the passage above suggests that Moore had already encountered resistance to his ideas. There was evidently a lack of enthusiasm for having the AMS publicly embrace the cause of secondary-school mathematics. In his address he had dramatically asked, "Do you not feel with me that the American Mathematical Society, as the organic representative of the highest interests of mathematics in this country, should be directly related with the movement of reform?" Now, however, he was retreating to allow the AMS to proceed quietly behind the scenes. Such an approach could not help but attenuate the force of Moore's proposals, especially his call to enlarge the society by bringing in secondary school teachers. Examination of the minutes of the AMS Council for 1902–1904 (which at this time listed all accessions to AMS membership), reveals no significant surge of secondary-school teachers entering the society in the wake of Moore's address.[19]

The above passage from Moore's letter to Smith also casts in a new light Moore's letter to E. J. Wilczynski quoted toward the end of the last chapter. Recall that in that letter Moore urged support of educational reform as ultimately helping the cause of research; in the letter to Smith he declared this to be an overly narrow viewpoint. Here Moore seemed to see value in educational reform quite apart from supporting the research community, but admitted that

---

[18]Moore to Smith, August 12, 1903, DESP. Parentheses and dashes as in original. The curly brackets designate passages that were inserted either above or to the side of the main text, as indicated by ˆ or →.

[19]The relevant minutes are in RAMS, Boxes 28–30, 35–37. For composition of the committee referred to, see Minutes of the AMS Council, Dec. 29, 1902, RAMS, Box 12, Folder 28. For the composition of the AMS Council, see Archibald, *History of the AMS*, 96–97. The quote from Moore is from Eliakim Hastings Moore, "On the Foundations of Mathematics," *Science* 17 (Mar. 13, 1903): 414.

the research motive might be a necessary enticement for some mathematicians. Whatever Moore's ultimate conviction, there can be no doubt that he carefully calculated his words to his audience. Wilczynski was a fast-rising research mathematician; Smith was not.

The same letter to Smith of August 1903 contains yet further revelations of Moore's attempts to fine-tune his political efforts:

> As to Tyler's Comm.—the disclaimers 1) of consideration of the pedagogic aspects, 2) of finality are quite right—**But** in my judgment the introduction ˆ{to} of that document (which is to go into the hands of thousands of teachers of mathematics) should put the ˆ{its} readers directly in touch a) with the thought that things looking towards improvement of the teaching of mathematics are happening just now in several parts of the country, and should b) exhort ˆ{all} the teachers to relate themselves to these movements. There should be a brief characterization (without prejudgment as to any special tendencies) of the situation, with precise references to officers of the various associations and to other sources of information.[20]

As already alluded to, the Tyler report's disclaimer of consideration of pedagogy was surely in violation of the spirit of Moore's presidential address, yet here he was trying to accommodate to this disclaimer, hoping to at least obtain some ringing phrases in the final report to encourage the belief that teachers and researchers were on the same side, and some specific references to the new associations of teachers. Neither the references nor the ringing phrases ever materialized. All Moore obtained for his efforts was the following footnote in the final version of the Tyler report, attached to the disclaimer of "consideration of pedagogic questions": "Reference may be made to the important work of recently formed societies for the improvement of mathematical teaching."[21]

Moore also failed to obtain any ringing published phrases from the committee on the relations of the AMS to the teaching of elementary mathematics, which reported in April 1904. The following words were the sum total devoted to this report printed in the *Bulletin of the AMS*:

> The committee appointed at the annual meeting of 1902 to consider the relation of the Society to elementary mathematics

---

[20]Moore to Smith, Aug. 12, 1903, DESP. The "but" that is bolded and underlined above was underlined three times by Moore in the original.

[21]Tyler et al., 75.

presented a final report reciting the organization, under the Society's influence, of several active associations of teachers of mathematics. The committee was discharged at its own request.

In one sense Moore did obtain a compromise; there was no "door-shutting negative legislation." The AMS retained its officially open membership policy. But it never invested the energy to attract secondary-school teachers or to engage in their concerns as advocated by Moore.[22]

Moore was aware that eastern mathematicians held suspicions of new educational ideas coming out of Chicago. At times he tried flattery to prod them along, disclaiming any final pedagogical wisdom on the part of his Chicago group: "I think you men of the East have done fine work in the pedagogic way ... Our Chicago utterances are not to be construed as other than purely optimistic and full of purpose to find out how to do." At other times he frankly expressed frustration: "But you know that our particular efforts are receiving little constructive attention from our colleagues of the Eastern states."[23]

Smith summed up the overall "eastern" reaction to Moore's 1902 address in an article published in early 1905:

> It was a fact apparent to all who heard it that the address was not favorably received by many of those present. Whatever opinions may have been publicly expressed, in private there were two adverse criticisms, (1) that a man who stood among the few recognized leaders in higher mathematics in this country should lose the opportunity offered to consider the great problems of the science *per se*, and (2) that one whose field had been so peculiarly one of research should assume to enter the realm of education and to criticize existing methods.
>
> These being the opinions of that time, it was quite apparent that the address would probably have much influence, since the ordinary paper that is pleasantly received is usually consigned to oblivion as speedily as possible. Such has proved to be the case.

---

[22]"The April Meeting of the American Mathematical Society," *BAMS* 10 (July 1904): 486. I have not discovered any written report of this committee. It is not clear whether there ever was one. The AMS Council minutes for Apr. 30, 1904 (RAMS, Box 12, Folder 30) refers to "an informal report."

[23]Moore to D. E. Smith, Feb. 8, 1904, and Moore to D. E. Smith, May 13, 1904, DESP.

Here followed Smith's laudatory survey of the associations of mathematics teachers that had sprung up in the wake of Moore's talk.[24]

It should be noted, however, that from the point of view of some AMS members, these associations of school teachers were welcome precisely because they excused the AMS from responsibility for pedagogical issues; this is certainly suggested by the lukewarm report of the AMS committee formed to respond to Moore's address. Smith also noted the disagreements that easterners had with the laboratory method: they objected to introducing physical experiments into mathematics classes; they objected to emphasizing applications of mathematics except as a means to "bring our pupils to love mathematics for its own sake"; they liked the idea of the classroom as "workshop," but felt it should be "not in applied but in pure mathematics"; in achieving this last goal Smith declared that American schools would be following the good example of the German system of education.[25]

A more positive assessment of Moore's movement was offered by G. A. Miller in two articles in 1906. Miller, who had taught briefly in Moore's department before taking positions first at Stanford and then at the University of Illinois, saw Moore and Perry and Klein as part of a great international reorientation of mathematics education. He made no claim that the organizing of associations of teachers after Moore's address was caused by Moore, only that "the time was ripe."[26] Miller wrote favorably of the aim of bringing mathematical instruction "in closer touch with applications." "The increasing use of graphic methods" also received his approval, as did Klein's emphasis on the function concept. "When such strong men as Klein, of Germany, Forsyth and Sir Oliver Lodge, of England, Moore and Fiske, of America, take active part in reform movements they are certain to be effective."[27]

Meanwhile Moore's Chicago colleagues kept up the publicity campaign. H. E. Slaught wrote a rousing piece in 1905 in which he emphasized that a teacher of mathematics should be an "*optimist*" and an "*enthusiast*," whose most important duty was to inspire the student to take an interest in the subject.

---

[24]David Eugene Smith, "Movements in Mathematical Teaching," 135. Italics in original.

[25]Ibid., 137–138.

[26]G. A. Miller, "Some Recent Tendencies in Mathematical Instruction," *Popular Science Monthly* 68 (1906): 163. Miller (1863–1951) was an instructor at Chicago during the summer of 1898. See *University of Chicago General Register 1892–1902*, 33. On Miller's career see Carolyn Eisele, "Miller, George Abram," *DSB*, 9:387–388.

[27]G. A. Miller, "Reform in Mathematical Instruction," *Science* 24 (1906): 493–495.

Slaught was also a strong advocate of professionalization at all levels of mathematics teaching, being careful to insert that the mathematics teacher should also be a *"specialist"*:

> The grammar schools now demand special preparation for all phases of their work. The high schools demand college graduates. The colleges demand men of University training. A teacher of algebra or geometry whose mathematics experience does not extend far beyond these subjects has too limited a range of knowledge and too narrow a view of related truth to be entrusted with the guidance of students in these subjects.

Slaught mentioned neither Moore nor any other educator by name, but expressed his allegiance clearly in his closing rhetorical flourish: "These constitute the triune watchword of the present mathematical renaissance,—*graphics, correlation, laboratory.*"[28]

It was J. W. A. Young's book of 1906 which gave the most comprehensive depiction of the laboratory method, as part of an ambitious attempt to capture the full scope of the reform efforts in Europe and the United States. With regard to developments on the American scene Young gave prominence to the Central Association of Science and Mathematics Teachers and to Moore's 1902 address. He devoted thirty-four pages to describing the laboratory method, expounding upon all the features previously noted. Moore's address was quoted from extensively and both Perry and Klein were cited approvingly. The laboratory method was only one of several such methods or "modes" that Young described, including "the recitation mode," and the "lecture mode":

> To characterize some of these modes in a word, it may be said that in the recitation mode the pupil works *before* the class session, in the lecture mode he works *after* it, in the laboratory mode he works *during* the session.

Young did not exclusively endorse any one mode. Each had its usefulness, although he did condemn the extreme version of the recitation mode, which he called the "examination mode," in which the teacher simply spent the class period testing the students on preassigned tasks, with no response except to declare answers right or wrong. He was happy to see this procedure in decline, and clearly suggested that modern educational progress was characterized by the increasing role of the laboratory mode and related ideas (e.g., the "heuristic

---

[28]H. E. Slaught, "Ideals in the Teaching of Mathematics," *SSM* 5 (1905): 703–705, 708. Italics in original.

mode" and the "genetic mode"). Underlying the newer methods was a larger emphasis on trying "to arouse and to hold the child's interest."[29]

The year 1906 marked perhaps a highwater mark of activism by the University of Chicago mathematics department, for it was in that year also that Moore's paper "The Cross-Section Paper as a Mathematical Instrument" appeared, his last substantial individual pedagogical effort. The paper gives additional insight into some of the obstacles to adoption of his proposals. He began expansively, touching on themes familiar from his previous writings:

> The following note is addressed to teachers and prospective teachers of mathematics in elementary schools and secondary schools and colleges, who have come to recognize as fundamental the problem of closer *correlation of arithmetic, algebra, and geometry with one another and with the various domains of application*, or the problem of *unification of elementary pure and applied mathematics* ... The problem is solved when, and only when, we put our pupils into such a physical and intellectual environment that they learn to see and to think the mathematics for and of themselves.

But as the title indicated, this was a more narrowly technical work than his 1902 address. Moore declared that the great virtue of the cross-section paper (what would later be called graph paper) was that it brought together "the three phases or dialects of pure mathematics—*number, form, formula.*" Moreover, it led directly to

> the concept of *functionality*—a concept which since the seventeenth century has dominated advanced mathematics and the sciences; a concept which in the twentieth century, according to the auspices, will play a fundamental role in the reorganization of elementary mathematical education. *Functionality is the relation or (mathematical) law of connection between two or more quantities or numbers subject to simultaneous and interdependent continuous variation.*

---

[29] J. W. A. Young, *The Teaching of Mathematics in the Elementary and the Secondary School*, 3d ed. (New York: Longmans, Green and Co., 1924), 5, 62–68, 87–121. Italics in original.

Thus this paper can be thought of as Moore's attempt to synthesize the ideas of John Perry (the utility of cross-section paper) with those of Felix Klein (the pedagogical centrality of the function concept).[30]

Moore asserted that the cross-section paper had certain naturally attractive properties for students, although in making this point he betrayed a distinctly limited view of the student population in terms of class and gender:

> Every boy loves trains, and most boys have traveled. With the cross-section paper in common-place school use, the boy of nine or ten will readily understand and create diagrams of train motion, and will enjoy making the limited express overtake the slow freight at a certain time and place.[31]

This supports the view that Moore had no special insight into the social transformation that the schools as a whole were undergoing.

Moore used a striking chemical metaphor to emphasize the effectiveness of the cross-section paper in correlating algebra and geometry:

> By maximizing the function of the cross-section paper we secure, to speak only of pure mathematics, intense reaction between geometry and algebra ... releasing, as it were, abundant stores of sub-atomic energy.

And he declared that geometry instruction was especially in need of reformation, here sounding much like John Perry:

> Indeed, for the purposes of elementary education *our current deductive geometry is of the nature of a fetish*, to be abandoned in favor of a geometry built on a richer system of geometric axioms—a system built to recognize the cross-section paper with its wealth of intuitional relations.[32]

There followed the heart of the paper, which consisted of detailed instructions for constructing graphs of certain algebraic equations and for using these graphs to facilitate computations. Moore employed "linkages," a procedure for

---

[30] Eliakim Hastings Moore, "The Cross-Section Paper as a Mathematical Instrument," *School Review* 14 (May 1906): 317–318. Italics in original. Moore's definition of functionality was not rigorous, even by the standards of the time. He almost certainly added the requirement for "continuous" variation simply for convenience; by this date it was well recognized that the utility of the function concept extended to discrete variables. In this paper Moore never specified precisely what he meant by continuous, but all the examples he used were indeed continuous in the strict technical sense.

[31] Ibid., 319.

[32] Ibid., 319, 323. Italics in original.

constraining the relationship between the vertical and horizontal coordinates which could in some cases be realized physically using "an arrangement of links or bars with slots and pins."[33] Graphical calculation, which had recently been systematized and dubbed "nomographie" (nomography in English) by the French mathematician Philbert Maurice d'Ocagne (1862–1938), essentially consists of reading off answers to selected calculations by strategically laying one graph (often a straightedge) on other graphs that have been appropriately constructed and scaled. The slide-rule can be thought of as a simple nomographic device. Moore's appeal to nomography demonstrates his broad knowledge of mathematics, as it was remote from his research interests. However, neither linkages nor nomography (with the possible exception of the slide-rule) ever became regular features of the American mathematics curriculum as Moore had hoped.[34]

In closing his paper Moore returned to a more general rhetorical stance, calling on his audience to join him in what he called "*Concluding Agreements*":

> In general, we agree to develop arithmetical, algebraic, geometric technique in a physical and intellectual environment logically and psychologically rich, full of movement, force, color, full of connotations and implications of and for real life of all kinds, including most certainly the real life of mathematics and the sciences.
>
> To secure for our young students a tolerable appreciation of the civilization of the twentieth century on the side of the theoretical and applied mathematical sciences, as teachers of mathematics we agree to shape our instruction in mathematics from the beginning from a point of view no older and no lower than that of the wonderful seventeenth century, and to this end, speaking theologically, we propose to
> *Canonize the Cross-Section Paper.*[35]

As with his 1902 address Moore did not rely on only one place of publication for his article on cross-section paper; it also appeared in *School Science and Mathematics*. Moreover, he continued to have the help of his students and colleagues in the promotion of his ideas. Directly following the Moore piece

---

[33]Ibid., 326.

[34]On nomography see Ivor Grattan-Guinness, *The Norton History of the Mathematical Sciences* (New York: W. W. Norton & Co., 1998), 514–517.

[35]Ibid., 337–338. Italics in original. Final phrase on line by itself as in original.

in the *School Review* there appeared a commentary by N. J. Lennes, based on his experiences as a high school teacher in Chicago, fully endorsing Moore's emphasis on graphs. A few months later G. W. Myers added further favorable remarks in the same periodical.[36]

Solberg Sigurdson has declared that Moore's 1906 paper represents a "surprising and abrupt change in attitude" from his 1902 address.[37] This is an exaggeration, but there can be little argument that Moore's pedagogical ambitions had narrowed. In the next several years it became clear that sharp lines of stratification had developed among the Chicago mathematical educators. Those who would carry on the agitating and organizing and broad reform efforts were those who had elected to specialize in pedagogy, such as Slaught, Young, and Myers. Those like Moore and Veblen and Dickson who had research ambitions faded into the background, pedagogically speaking, occasionally offering advice on a specific technical issues, such as graphs, or lending their names to worthy causes.

The transition is effectively symbolized by Dickson's activities in 1906. In May of that year the course offerings at the University of Chicago for the summer quarter were announced in the *American Mathematical Monthly*. Dickson was listed for two advanced courses, algebraic analysis and theory of substitutions. There then followed this interesting addendum: "Professor Dickson will offer, in addition to his advanced courses, a course in the correspondence study department in Plane Trigonometry, by the Laboratory Method."[38]

The temptation to hoot and holler at this is irresistible. A laboratory method correspondence course at the very least represented a severely restricted version of the laboratory method ideal. Such indeed was the implication of Dickson's article of the previous year on teaching trigonometry, earlier cited: graphical methods were to be employed to be sure, but there was no mention of the social aspects raised to prominence by both Moore and Young: extended class sessions to allow most work to be done during class; students working

---

[36] The second publication of the article on cross-section paper is Eliakim Hastings Moore, "The Cross-Section Paper as a Mathematical Instrument," *SSM* 6 (June 1906): 429–450. The supporting papers by Moore's colleagues are N. J. Lennes, "The Graph in High-School Mathematics," *School Review* 14 (May 1905): 339–349; and G. W. Myers, "A Class of Content Problems for High School Algebra," *School Review* 14 (Oct. 1906): 565–566.

[37] Sigurdson, 169.

[38] "Notes and News," *AMM* 13 (1906): 119. Capitalization in original.

cooperatively in groups; a head teacher aided by assistants; a classroom well fitted with appropriate apparatus.[39]

Another notice in the *Monthly* later that year further elucidates Dickson's trajectory. Since 1902 he had been assisting B. F. Finkel in editing that pedagogically oriented journal:

> We regret to announce to our readers that, owing to increased duties at The University of Chicago and to his desire to devote all his leisure moments to investigations in a most attractive field of research in which he is now interested, Editor Dickson has decided to withdraw from the active editorship of this MONTHLY ... The work Dr. Dickson has been doing so efficiently for the past four years will now be done by Dr. H. E. Slaught.[40]

Moore's own trajectory was somewhat similar, though less abrupt. His fascination with a research area that would monopolize most of his "leisure moments" for the rest of his life can already be detected by 1904. At meetings of the AMS in September and December of that year he communicated papers by the young French mathematician Maurice Fréchet (1878–1973) on "opérations linéaires." It is evident from the brief descriptions of these papers that Fréchet was already moving toward the ideas he would promulgate in his influential doctoral thesis of 1906 and other writings, ideas which have been vastly elaborated over the rest of the century, usually designated "functional analysis" or "topological analysis." Moore, greatly intrigued by such ideas and motivated especially by the attempt to unify recent work in the theory of integral equations with other areas of mathematics, dubbed his own foray into this realm as "general analysis." He first unveiled his theory at the AMS Colloquium Lecture series of 1906. Although Moore and his students revised and refined general analysis for over 30 years, the long term influence of the program was very attenuated. Saunders Mac Lane has called general analysis a "timely but failed initiative," in speculating on which he cites the difficulty of Moore's notation,

---

[39] Compare Dickson, "Graphical Methods in Trigonometry," 129–133 with Young, *Teaching of Mathematics*, 116–117.

[40] "Notes and News," *AMM* 13 (Nov. 1906): 236. Signed "F." indicating B. F. Finkel. On Dickson's association with the *Monthly* see B. F. Finkel, "The Human Aspect in the Early History of the *American Mathematical Monthly*," *AMM* 38 (1931): 313–314.

8.1. PROMOTION OF MOORE'S IDEAS AT THE UNIVERSITY OF CHICAGO

Moore's increasing reluctance to publish, and the loss of Bolza and Maschke as providers of critical feedback.[41]

The early phases of Moore's fascination with general analysis coincided with the diminishing of his pedagogical activity. As early as 1904, while he was still expressing great pedagogical enthusiasm to Smith and others, he had revealed some discontent at the demands of undergraduate teaching. Significantly this was in reply to Wilczynski sounding him out regarding a position at Chicago:

> As to lecturing in Chicago, it is quite impossible, at least on terms that would be attractive to you—for nowadays and usually we are held down very closely by the requirements of our undergraduate instruction, and we find no way to expand in the direction of graduate instruction. Needless to say this is a source of the keenest regret.[42]

By 1906 a wistful and less confident tone was beginning to creep into his correspondence with Smith:

> I am just finishing MS "The Cross-Section paper as a mathematical instrument" for May School Review, which I will send you in season—I wish much to ˆ{help} unify geometry and algebra—My guess is that before we die we shall see considerable change in that direction—And I am counting on your ˙very strong cooperation, just so far as you find me not off the track—Yours—E H Moore[43]

Moore's correspondence with Smith became much more sparse after this letter.

Traces of Moore's interest in pedagogy can still be found after 1906. For the summer of 1907 he announced a course entitled "Graphical methods in algebra especially for teachers." By 1908 the phrase "laboratory method" had disappeared from mathematics course descriptions in the University of Chicago Circular, although remnants of the rhetoric remained. The basic calculus sequence was to be taught "with much use of graphical methods"; the advanced calculus sequence was to be "developed in organic relation with the problems of

---

[41] For Moore's activities at AMS meetings during this time see "The Eleventh Summer Meeting of the AMS," BAMS 11 (Nov. 1904): 57–58, 60–61; and "The Eleventh Annual Meeting of the AMS," BAMS 11 (Feb. 1905): 233–235. On Fréchet see Angus E. Taylor, "Fréchet, René Maurice," DSB, 17:309–311.

[42] Moore to Wilczynski, Sept. 11, 1904, EJWP, Box 2, Folder 2.

[43] Moore to Smith, Apr. 4, 1906, DESP. Use of dashes for punctuation, and lack of periods, as in original. Curly brackets indicate inserted word.

Geometry, Mechanics, and Physics." The same course list contains a thirteen-line description of Moore's General Analysis sequence; no other course gets more than six lines.[44]

The demise of the laboratory method at Chicago was later explained in Bliss and Dickson's memorial notice on Moore as follows:

> In 1903–1904 and following years he modified radically the methods of undergraduate instruction in mathematics at the University of Chicago, and he himself gave courses in beginning calculus. With characteristic independence he cast aside the text books and concentrated on fundamentals and their graphical interpretations. The courses were so-called laboratory courses, meeting two hours each day, and requiring no outside work from the students. It might be added parenthetically that, as with many such new plans, the amount of work required of the instructor was exceedingly great. The two hour period was the feature which later caused the abandonment of the plan because of the very practical difficulty in finding hours on schedules which would not interfere with the offerings of other departments.[45]

This account of idealism thwarted by practical considerations is likely true in many respects. There is a strong likelihood in particular that the regime of Harry Pratt Judson, who had succeeded to the presidency of the University of Chicago when William Rainey Harper died in 1906, placed added constraints on pedagogical experiment, as alluded to in a letter by Moore in 1910:

> Will you tell me of some younger men who are emphatically making good as instructors of freshman and sophomore classes in Princeton and elsewhere?
>
> Mr. Judson is laying great emphasis on efficiency of Junior College teaching, and there is a possibility that we may be able to add a man for next year for this work.[46]

---

[44]Compare "Notes and News," *AMM* 14 (1907): 62 with *Circular of the Departments of Mathematics, Astronomy and Astrophysics, Physics, Chemistry*, University of Chicago (1908): 7–12. Capitalization as in original.

[45]G. A. Bliss and L. E. Dickson, "Eliakim Hastings Moore," *National Academy of Sciences Biographical Memoirs* 17 (1936): 88–89.

[46]Moore to O. Veblen, Jan. 12, 1910, OVP, Box 8. Underlining in original.

8.1. PROMOTION OF MOORE'S IDEAS AT THE UNIVERSITY OF CHICAGO

But the Bliss-Dickson version somewhat scants the conflict between the attractions of research and the practical requirements of teaching. Indeed, their account reflects the conflict; the less than perfect denouement of the Moore proposals being referred to only "parenthetically." In fact, pressure to stream-line the undergraduate instruction did not come only from outside the community of mathematicians.

When Moore was at last enabled to hire Ernst Wilczynski as a member of his department in 1910 (after Maschke had died and Bolza had returned to Germany) he gave him some pedagogical advice for dealing with the introductory and synoptic calculus classes:

> As to the I.C. and the Synoptic C. (but especially the former) I hope you will arrange to include considerable actual drill and quiz work in addition to the formal instruction. We find that the best results come in this way, that the very largely lecture method doesn't get to close enough quarters with the students.[47]

It would be unfair to interpret Moore's advocacy here of "drill" as representing a reversion on his part to pre-Committee-of-Ten standards, but the diminishment of his idealistic fervor is nevertheless striking. There was now no claim that his notions were part of some grand over-arching reformation, only the modest suggestion that experience had taught that mathematics instruction ought not consist of nothing but lectures. Here then was the stripped-down version of what once had been the grandly conceived laboratory method.

For the rest of his life Moore's name continued to appear in association with various pedagogical initiatives, but all indications are that his role was slender. He was listed as a joint editor on a series of secondary-school textbooks, but G. W. Myers and others did the bulk of the work.[48] He was appointed by David Eugene Smith in 1909 to an Advisory Council for the American Commissioners of the International Commission on the Teaching of Mathematics, but contributed little to this compared to his Chicago colleague J. W. A. Young.[49] He encouraged the creation in 1915 of the Mathematical Association of America (MAA), a pedagogically oriented association of college-level mathematical educators, but it was Herbert Slaught who was at the forefront of this movement.

---

[47] Moore to Wilczynski, Mar. 19, 1910, EJWP, Box 2, Folder 23. Underlining in original.

[48] Ernst R. Breslich, *First-Year Mathematics for Secondary Schools*, Fourth edition (Chicago: University of Chicago Press, 1915), vii–xvii.

[49] Moore to Smith, Apr. 5, 1909, DESP.

He was one of the original members of the National Committee on Mathematical Requirements formed by the MAA in 1916, but others were more active, especially his brother-in-law J. W. Young who acted as chairman. Moore's name was waved high during the early years of the National Council of Teachers of Mathematics, founded in 1920, but again it was Slaught who participated most significantly in this organization. These developments will be discussed further in later pages.

8.2. National Reaction to Moore's Proposals and Vision

Solberg Sigurdson has found evidence of enthusiasm for laboratory instruction in mathematics and for Moore-like correlation of mathematics and physics in the years immediately following 1902, but also finds that a reaction soon set in, especially in the east. The eastern educators displayed more attachment to the traditional mental discipline thesis, and also greater fear that the correlation idea would tend to subordinate mathematics to science instruction. Nor did correlation of physics and mathematics receive the ratification of physicists. Moore's Chicago colleagues Michelson and Millikan both objected to the over-mathematizing of elementary physics, apparently fearing to distract students from appreciating the realities of the physical world. Millikan, responding to some remarks by another Chicago colleague, philosopher George Herbert Mead, offered a warning to mathematical educators who looked to science as a means to protect their subject:

> The third problem which Professor Mead discusses, namely, the correlation of science and mathematics, is one the solution of which is much more difficult, and the source of the difficulty is found simply in the fact that *elementary science demands for its adequate presentation an extremely small amount of mathematical training* ... so that, if we regard mathematics *only* as the language of science, and postpone the development of every mathematical proposition until we actually need this proposition in our work in physics, we shall eliminate from 75 to 90 per cent of the mathematics which the high-school course now contains ... I take it therefore that we shall always wish to treat mathematics in our high-school work as something more than the mere tool or language of science—something which may profitably be studied for its own sake.

(Millikan's concluding "therefore" was a deduction whose obviousness would soon be challenged by a new generation of general educators.) By 1908 ambitious schemes to correlate mathematics and physics had been essentially rejected. Enthusiasm for unifying algebra and geometry instruction also subsided.[50]

More specific features of the Moore program received a better reception. Although nomography did not catch on as a pedagogical focus, graph paper did indeed become a staple of mathematics instruction, in secondary education and beyond. A short account in 1912 found no graphical treatment in American algebra texts prior to 1902, and attributed the topic's subsequent surge in popularity to the influence of John Perry and the 1903 AMS committee on college entrance requirements. Though unmentioned here, Moore may justly be thought of as hovering in the background. A study of algebra textbook topics in 1929 found that the use of graphs in algebra had continued to expand.[51]

Historian Larry Owens implies that Moore's agitation on behalf of graphical instruction may have had a significant impact on engineering instruction: "By 1910 the teaching of mathematics in most engineering schools thoroughly reflected the precepts of [Gardner] Anthony [dean of engineering at Tufts College], Perry, and Moore." But whether Moore can take much credit for this development is debatable; Owens offers no concrete evidence that engineers paid any attention to Moore, and engineering already had a tradition of graphical analysis, in any case. Indeed, it is more plausible that influence worked the other way, with input from engineering inspiring increased use of graphical techniques in mathematics. For example, Willard Gibbs, pioneer in vector analysis, was trained in engineering. A more realistic assessment would be that Moore was endorsing a trend that would have continued without him, rather than moving engineering education in a new direction. As Owens himself acknowledges, Moore was manifestly unsuccessful in his broader purpose of attaching pure mathematics to the ongoing expansion of engineering education. According

---

[50]Sigurdson, 143–156, 161–163, 177–183. The Millikan quote is from Robert Andrews Millikan, response to George H. Mead, "Science in the High School," *School Review* 14 (Apr. 1906): 251–252. Italics in original.

[51]The 1912 articles is Emily G. Palmer, "History of the Graph in Elementary Algebra in the United States," *SSM* 12 (1912): 692–693. The 1929 study is cited in Phillip S. Jones, ed., *A History of Mathematics Education in the United States and Canada*, NCTM Thirty-Second Yearbook (Washington, D.C.: NCTM, 1970), 158–159.

to Thornton Fry in 1941, an observer cited by Owens, the most mathematically adept industrial workers came from physics or electrical or mechanical engineering, not from pure mathematical training:

> Though the United States holds a position of outstanding leadership in pure mathematics, there is no school which provides an adequate mathematical training for the student who wishes to use the subject in the field of industrial applications rather than to cultivate it as an end in itself.[52]

On the positive side, there does seem to have been an effort in the period 1907–1912, probably owing something to Moore, to create better understanding between mathematicians and engineers. In December 1907 the Chicago Section of the AMS organized a meeting of mathematicians and engineers to discuss the role of mathematics in engineering education. Moore apparently did not participate (he began distancing himself from pedagogical issues about this time), but Slaught reported on the conference for the *American Mathematical Monthly*. One consequence of the meeting was the formation of yet another committee, this one to "make a detailed study of the teaching of mathematics to engineering students in this country."[53]

The resulting report was published in preliminary form in 1911 in several issues of the *Bulletin of the Society for the Promotion of Engineering Education*. There was no participation by anyone from the University of Chicago (recall that engineering was non-existent at that institution). The chairman was Harvard mathematician Edward V. Huntington, and W. F. Osgood of Harvard was also a member. We have already seen indications that Osgood was not in accord with Moore on pedagogical matters, and the suspicion arises that this engineering education committee had been captured by eastern conservatives. Indeed the report appears strikingly lacking in innovative approaches; it seems to be a long compilation of mathematical topics, claimed as useful to prospective engineers, and organized along traditional lines. The committee shied from prescribing any particular order of topics, and made no special effort at correlation either between different branches of mathematics or between mathematics and physical science. The committee asserted that "The defects

---

[52]Owens comments are in Larry Owens, "Vannevar Bush and the Differential Analyzer: The Text and Context of an Early Computer," *Technology and Culture* 27 (1986): 87, 90. Fry's comment is from Thornton Fry, "Industrial Mathematics," *AMM* 48 (1941): 1. Quoted in Owens, "Vannevar Bush," 89.

[53]H. E. Slaught, "Joint Meeting of Mathematicians and Engineers," *AMM* 15 (1908): 33.

in the mathematical training of the student of engineering appear to be largely in knowledge and grasp of fundamental principles."[54] This evaded the problem of how to induce the engineering student to take an interest in mathematics at all, a problem certainly in Moore's mind. Moore's influence on this effort, as on engineering education generally, must be judged mild at best.

The notion Moore derived from Klein of relying on the function concept to unify elementary mathematics instruction may also be counted as a success of the Moore program, although here the implementation was slower and more erratic than with graphs. Significant adoption of the concept does not seem to have occurred prior to the 1923 report by a committee of the Mathematical Association of America, to be described later; even then certain perplexities limited its endorsement. Moore had claimed the function as "a concept which since the seventeenth century has dominated advanced mathematics."[55]

But Moore was here leaping naively over a problem that perplexes the history of science in general and history of mathematics especially. How should we interpret the use by an individual from the past of a concept related to but not precisely the same as a modern concept? One may rightly claim to detect intimations of concept A in the work of predecessor X, but to then claim that X "really understood" concept A is another thing. In the present case it is reasonably clear that the modern concept of function as a correspondence between two arbitrary sets, in which each element of the first set is paired with one and only one element of the second set, did not come into widespread usage until the latter half of the nineteenth century. It therefore cannot be that this precise modern concept has dominated since the seventeenth century, and to claim that some less precise version has dominated, though plausible, requires elaborate historical exegesis. Moreover, the fact that it took so long to devise the modern function concept casts doubts on its pedagogical virtues. All this was succinctly expressed by Edwin Bidwell Wilson of MIT in a letter to Moore in 1918:

> Now the function idea is one of great generality, which even the best mathematicians in the world did not have until recent times, and it is out of the question to get the freshmen enthusiastic over it.[56]

---

[54] "Preliminary Report of the Committee on the Teaching of Mathematics to Students of Engineering," *Bulletin of the Society For the Promotion of Engineering Education* 1 (1911): 410.

[55] Moore, "Cross-Section Paper," 318.

[56] Edwin Bidwell Wilson to E. H. Moore, Dec. 2, 1918, EHMP, Box 2, Folder 17.

The function concept's subsequent pedagogical history has been complex.

American pure mathematicians did not rush to embrace Moore's larger vision for the future of mathematics. At the same time that Moore was declaiming against the fundamental danger of the chasm between pure and applied mathematics other mathematicians were extending the hierarchical approach to mathematics instruction advocated by Benjamin Peirce and J.J. Sylvester earlier in the century, in which the "barn yard fowl" and the "eagles" would be taught by entirely different classes of teachers. Those in this camp saw quite different dangers than those perceived by Moore. C. J. Keyser of Columbia University, writing in November 1902, was dismissive of using utility to justify mathematics. In order for the United States to achieve its full share of "mathematical productivity" it was necessary that "we learn to value the things of the mind, not merely for their utility, but for their spiritual worth." That this desirable result would come to pass was by no means a certainty, since "[w]e, as a people, have yet to learn that the value of a professor to a community can be rightly estimated, not by counting the number of hours he actually stands before his classes, but rather, ... by the fruit of quiet study and research."[57]

Four years later Keyser appealed even more forcefully to the rhetoric of efficiency to disparage utilitarian domination of education, decrying the "wickedness of the waste" created by requiring creative university professors to spend their time catering to the needs of "immense undergraduate schools thronged by young men mainly bent upon practical aims." What was needed was a division of labor among administration, teaching, and research. Keyser claimed each of these duties was equally worthy of respect ("the distinctions are not of greater and less"), although research seems to have been a little bit more equal, inasmuch as it was "the highest form of human activity."[58]

The distinctiveness of Moore's 1902 address is especially brought out by looking at the corresponding addresses of his successors as AMS president. His immediate successor was Thomas Fiske, who on December 29, 1904 spoke on "Mathematical Progress in America." Fiske began by claiming to be following the example not of Moore but of Moore's predecessor, R. S. Woodward. Woodward had spoken on the advances in applied mathematics; thus would Fiske talk

---

[57]C. J. Keyser, "Mathematical Productivity in the United States," *Educational Review* 24 (Nov. 1902): 346, 354.

[58]C. J. Keyser, "Concerning Research in American Universities," *Columbia University Quarterly* 8 (1906): 405–408.

8.2. NATIONAL REACTION TO MOORE'S PROPOSALS AND VISION

of the advances in pure mathematics. Fiske proceeded to offer a periodization of mathematics in America which has been essentially standard ever since.[59]

Prior to 1876, declared Fiske, there were only a few isolated individuals who could truly be called mathematical researchers, one being Benjamin Peirce. It is striking how firmly Fiske placed Peirce in the pure mathematical camp: "Peirce is now known chiefly for his classical memoir, Linear Associative Algebra, which was the first important research made by an American in the field of pure mathematics." We can see that the process was already well advanced whereby Peirce would become the "father of pure mathematics" in the United States, his astronomical work relegated to secondary status. With 1876 came Johns Hopkins University and Sylvester, as we have seen in Chapter 2.[60]

The next important date for Fiske was 1891, the year the New York Mathematical Society made its first drive to expand its membership nationwide. Fiske then touched on many of the events that the reader of the present work will recall: Klein's visit of 1893; the name change to American Mathematical Society in 1894; the advent of the Chicago Section in 1897. Among this list of accomplishments Fiske did pause briefly to cite Moore's address of 1902, the significance of which he interpreted thusly: "Very largely as a result of this address, the influence of the Society was exerted to bring about the organization of associations of teachers of mathematics with a view to improving the methods of mathematical teaching."[61]

Moore's sense of urgency was replaced with great complacency; from Fiske one would infer that the problems descried by Moore had been quickly dispatched. Fiske ended his talk with a complaint that professors had insufficient time for research. He proposed to improve the situation by "nothing more nor less than the relegation of the first two years of the ordinary college course to the secondary schools." He also thought that government funds ought to be solicited to support the research journals; after all, this was not uncommon in other countries.[62]

It is worth noting that Fiske was negligible as a research mathematician. He was, however, an indefatigable server on committees and was widely revered

---

[59]Thomas S. Fiske, "Mathematical Progress in America," *BAMS* 11 (1905): 238.

[60]Ibid., 239. The lauding of Benjamin Peirce as father of pure mathematics is from Carolyn Eisele, "Peirce, Benjamin," *DSB*, 10:479, quoting G. D. Birkhoff in 1938.

[61]Fiske, "Mathematical Progress," 241–244.

[62]Ibid., 245–246.

as "Founder of the Society."[63] Whereas Moore made bold to try to reorient the society, Fiske was limited to the simpler role of praising the course of the society's progress. No indirection can be found in his speech.

The next AMS president was William Fogg Osgood of Harvard. He was a more substantial mathematician than Fiske, but his presidential address shared with that of Fiske a decided complacency regarding the place of mathematics. Moreover, Osgood's talk, on the teaching of calculus in the colleges, implicitly criticized Perry and Moore. Osgood admitted that in the past there had perhaps been too much emphasis on formal drill in differentiation and integration without reference to applications, but

> Recently the pendulum has swung to the other extreme. The reformists have discovered that the engineer is well off if he can plot some simple curves from the tables and differentiate and integrate x^2 and $\sin x$, and we are told that the race has been degenerating under the old regime. Let us not lose perspective and let us not for a moment fail to recognize the fact that, whatever changes it may be desirable to make in the suggestive instruction of the course, the process by which the youth actually acquires the ideas of the calculus is to a large extent and essentially through formal work of substantial character.[64]

The following four AMS presidents devoted their talks to technical pure mathematical topics.[65] Then in 1916 came a change of pace, the first presidential address since that of Moore to suggest that American mathematics might have serious deficiencies. E. W. Brown, a mathematical astronomer in the tradition of Benjamin Peirce, Simon Newcomb, and G. W. Hill, had attained to the presidency and spoke on "The Relations of Mathematics to the Natural Sciences." Brown had been, with Moore and Fiske, one of the original editors of the *Transactions of the AMS*. This, and other active service in the society,

---

[63]These are the words under the picture of Fiske that serves as the frontispiece of Archibald's *History of the AMS*. Fiske's bibliography consists mainly of announcements of AMS or CEEB business, with a few translations and short expository notes. Ibid., 151–152.

[64]Wm. F. Osgood, "The Calculus in Our Colleges and Technical Schools," *BAMS* 13 (1907): 449–450.

[65]Henry S. White, "Bezout's Theory of Resultants and Its Influence on Geometry," *BAMS* 15 (1909): 325–338; Maxime Bôcher, "The Published and Unpublished Work of Charles Sturm on Algebraic and Differential Equations," *BAMS* 18 (1911): 1–18; H. B. Fine, "An Unpublished Theorem of Kronecker Respecting Numerical Equations," *BAMS* 20 (1914): 339–358; E. B. Van Vleck, "The Rôle of the Point-Set Theory in Geometry and Dynamics," *BAMS* 21 (1915): 321–341.

8.2. NATIONAL REACTION TO MOORE'S PROPOSALS AND VISION

helps account for his elevation to the presidency, despite his being outside the mainstream of American mathematics; he was followed in this office by a steady stream of pure mathematicians.

Brown was not so complacent about mathematics as his predecessors, in part due to anxieties associated with the Great War raging in Europe. At the end of his talk he mentioned the war explicitly, remarking that it was posing new challenges for scientific research, especially with regard to balancing research toward "a practical end" with research "on its intellectual side." This put American mathematics in an especially delicate position since, as the bulk of Brown's talk makes clear, American mathematicians had not heeded Moore's warnings on the chasm between pure and applied mathematics. Pure mathematics had undoubtedly created "a marvelous structure of thought," but Brown feared that it was becoming more isolated from the study of natural phenomena. Indeed, he noted that even the branches of pure mathematics were becoming isolated from each other. Speaking directly to Moore's point he flatly declared the situation to be getting worse:

> [I]t is noticeable that the pure science has a tendency to drive out the applied science ... or to put the matter in another form, the cleavage between pure mathematics and the experimental sciences which use mathematics has tended to increase.

Brown saw a major cause of this problem in the "almost complete disappearance" of key mediating scholars. "The temptation either to proceed to experiment or to turn to purely mathematical researches has been too strong." In order to render service to the cause of applied mathematics the pure mathematician would have to

> change to some extent the methods which he is accustomed to employ. The generality and completeness which he is accustomed to give to his researches are rarely useful in the applications. What is needed is the discovery of solutions which in a limited range can be reduced to numerical values without an excessive amount of labor.

Moreover, Brown affirmed in effect that mathematics to be useful must become commodified:

> But whatever is done in any direction, the work will be of little practical value unless it satisfies the final test, namely, that the

> form in which it is left is such that it may be applied without a full knowledge of the processes by which it is reached.[66]

This is closely related to the position of Moore and Perry on elementary education. They had held, for example, that full critical understanding of the concepts of the calculus should not be a prerequisite for applying these concepts.

In the midst of his lament on the direction of American mathematics Brown remarkably invoked a version of the American-indifference-to-basic-science thesis (a venerable theme in American intellectual history, going back at least to Alexis de Tocqueville) as cause to hope for a reorientation: "the natural bent of a nation which has been compelled by its environment to regard most of its affairs from a thoroughly practical point of view, gives promise of success."[67] He did not explain how such an innately practical people had gotten into the predicament he had just described. Brown's fundamental point was confirmed by an article of the same year by E. B. Van Vleck, who discerned "a somewhat widespread abstruse inclination in our country." He agreed with Brown that this was accompanied by "regretably [sic] less of the action and interaction of mathematics and mother nature," but was unable to hide his essential enthusiasm for the pure mathematical program: "the combination of sweeping generalization with rigor is astonishing."[68]

Brown's talk is also revealing regarding the outcome of Moore's call for the AMS to widen its membership. Brown noted that there had been internal agitation for the society to take on new activities besides the promotion of research, in particular that it engage in "efforts to improve the teaching of mathematics." Many of those so agitating were, as Brown delicately put it, "those who on account of professional duties or for other causes are unable to follow or lead far along the high-roads of research." Brown was able to report that

> The difficulty has been happily solved by the formation of the Mathematical Association of America with assumption of responsibility for the *American Mathematical Monthly* and its use as the official journal of the Association. This result has

---

[66]E. W. Brown, "The Relations of Mathematics to the Natural Sciences," *BAMS* 23 (1917): 216–217, 220–222, 227, 229. On Brown and his AMS successors see Archibald, *History of the AMS*, 59, 107, 173–183.

[67]Brown., 228.

[68]Edward B. Van Vleck, "Current Tendencies of Mathematical Research," *BAMS* 23 (1916): 13.

been brought about by the cordial cooperation of everyone concerned. They have desired not only that the interests of the American Mathematical Society should not suffer but that the new Association should be so organized and conducted as to assist in strengthening and coordinating all forms of mathematical activity throughout the country. Our best wishes go out to our young and already vigorous offspring.[69]

Much of the agitation referred to by Brown had originated with Herbert Slaught, encouraged by E. H. Moore. Slaught, as we have noted, became a mainstay of the *American Mathematical Monthly* beginning in 1906. During the next several years the financing of this journal became increasingly precarious. Slaught was able to obtain editorial and monetary assistance from Chicago and several other Midwest institutions, but by 1914 he concluded that a new foundation was needed. He secured the formation of a special committee of the Chicago Section of the AMS to consider the relationship of the AMS to the *Monthly*. This committee proposed that a special membership category of "Associate" be established for those primarily interested in the field covered by the *Monthly*. This was evidently a revival of the two-division scheme offered by Moore in 1902. But when Slaught proceeded to the national level of the AMS his plan failed. Here again a special committee was set up, to consider AMS support of the *Monthly* specifically and the relation of the AMS to pedagogical issues generally.[70]

The committee, which consisted of Fiske of Columbia, Fine of Princeton, Osgood of Harvard, Slaught of Chicago, and Earl Raymond Hedrick of the University of Missouri, in April 1915 voted three to two against AMS support of the *Monthly*. The identities of the voters were not revealed in the minutes but are easy to deduce. The two in favor of aligning the AMS with the *Monthly* were surely Slaught and Hedrick; the latter became the first president of the Mathematical Association of America. The negative votes of Fiske, Fine, and Osgood are quite in keeping with their attitudes as described in the present work. Note also that the vote was a straight east-west split. Instead, the committee adopted the following resolution, recorded in the minutes of the meeting:

---

[69]Brown, 214–215.

[70]Phillip S. Jones, "Historical Background and Founding of the Association," in Kenneth O. May, ed., *The Mathematical Association of America: The First Fifty Years* (Washington: MAA, 1972), 19-20.

It is deemed unwise for the American Mathematical Society to enter into the activities of the special field now covered by the *American Mathematical Monthly*; but the Council desires to express its realization of the importance of the work in this field and its value to mathematical science, and to say that should an organization be formed to deal specifically with this work, the Society would entertain toward such an organization only feelings of hearty good will and encouragement.[71]

Slaught took the hint and began soliciting expressions of interest in a new society to facilitate communication and publication by those interested "in the large field between the fields of secondary school mathematics and the field of pure research." The organizational meeting was held in December of 1915, resulting in the formation of the Mathematical Association of America (MAA). Although "Association" may appear to be a perfectly neutral designation it is likely that some among the founders understood it to designate them as "associated with," and indeed subservient to, the AMS. Discussions about reabsorbing the MAA into the AMS continued for several years.[72]

Some skepticism is certainly in order regarding the public expressions of "best wishes" and "good will" being dispensed by members of the AMS in conjunction with the founding of the MAA. Writing privately in 1924 Slaught recalled that in fact "[s]ome feared that it [the MAA] would work harm to the ideals and interests of the Society." He specifically placed Osgood among this group, as well as Brown (he of the "best wishes"), whom Slaught called "a violent doubter at first."[73] Yet it is clear that anxieties soon diminished (this is a main point of Slaught's 1924 letter), and that the two societies indeed established cooperative relations which have survived to the present day. The MAA immediately became larger in membership than the AMS, the MAA having more than 1,000 members, the AMS having fewer than 800 in 1916. From the first there have been many individuals who have belonged to both societies, E. H. Moore being only one. A joint national meeting is held every year.[74]

---

[71] AMS Council Minutes, Apr. 23–24, 1915, RAMS, Box 12, Folder 47.

[72] Jones, "Historical Background," 20–21.

[73] Slaught to R. G. D. Richardson, Jan. 28, 1924, OVP, Box 12.

[74] On membership figures see Thomas Fiske, "Relations Between the Association and the Society," *AMM* 23 (1916): 296. On joint meetings see Carl B. Boyer, "The First Twenty-Five Years," in May, 49.

But whatever the nature of individual perspectives on the creation of the MAA, and whatever the level of cooperation then and subsequently, the AMS was clearly attempting to rid itself of overt concern for anything other than research. It was moving in quite the opposite direction from Moore's vision of bringing in secondary-school teachers.

CHAPTER 9

School Mathematics on the Defensive

During the first two decades of the twentieth century there began to appear in the United States unprecedented hostility to the entrenched position of mathematics in school classrooms. These attacks were fueled by vast increases in the population of high school students. Scaling up the education system to march this horde of students through something approaching the classical curriculum that had been passed down from the nineteenth century proved difficult indeed. More and more students, and their parents, looked on education primarily as a means to obtaining gainful employment, and it was not at all clear how mathematics, except basic arithmetic, contributed to this goal. Many were unimpressed with the claims for mathematics as a mind trainer, or a carrier of general cultural value, and saw it more as an obstacle to advancement in life than as a help. Educators sought to placate these views and at the same time to make the system run smoothly and efficiently, and one means that occurred to many of them was to reduce requirements for school mathematics.

Moves to disparage and devalue school mathematics prompted a variety of reactions by mathematical educators. Initial unconcern turned to alarm by the 1910s, and counter attacks were formulated. E. H. Moore's attempt to link mathematics strongly with science and engineering did not prove tremendously helpful in these debates, as it was never possible to demonstrate that these fields, though growing, would offer employment to any but a small proportion of the increasing student population. Indeed, the difficulties of making a utilitarian argument for mathematics education led to a revival of the mental discipline thesis among some mathematical educators.

In part owing to the educational activism engendered by Moore, educators in the city of Chicago became especially prominent in writing and organizing to defend mathematics in the schools. The Men's Mathematics Club of Chicago, an organization of high school mathematics teachers, became the nucleus of the National Council of Teachers of Mathematics (NCTM). Chicagoans, especially

Moore's colleague H. E. Slaught, were also central to the creation of the national organization representing the interests of undergraduate teachers of mathematics, the Mathematical Association of America (MAA). It was a committee created by the MAA, the National Committee on Mathematical Requirements, which produced the culminating document of this era, *The Reorganization of Mathematics In Secondary Education* of 1923. In part this was a reaction to an earlier document deployed by the critics of mathematics, *Cardinal Principles of Secondary Education*, published in 1918. William Heard Kilpatrick, a contributor to the *Cardinal Principles*, created special consternation among mathematical educators.

The *Reorganization* report was the last major educational project in which E. H. Moore participated, though the bulk of the work was carried by his brother-in-law, J. W. Young. The report was a venture of mathematical educators operating on their own, apart from general educators and specialists in other subjects, distinguishing it from the Mathematics Conference of the Committee of Ten. This illustrates that the population of mathematical educators had grown and now commanded more resources (they were able to obtain a grant from John D. Rockefeller's General Education Board), but also that the place of mathematics in the curriculum was no longer so automatically accepted by non-mathematical educators in this new era. The *Reorganization* report has often been lauded, but its actual influence on American education must be judged minor, especially in relation to the large hopes for change expressed by E. H. Moore at the beginning of the century.

9.1. Questioning the Place of Mathematics in the School Curriculum

The great increase in secondary-school enrollments, alluded to several times earlier in this book, was underway by 1890. From that year to 1900 the number of students going to public high schools increased more than two and half times, and the numbers of these schools more than doubled. Another doubling of enrollment occurred between 1900 and 1912, and yet another between 1912 and 1920. During the course of the thirty years from 1890 to 1920 the percentage of American youth between 14 and 17 enrolled in public high school grew from less than 7 per cent to more than 28 per cent. We will not dwell on the fascinating question of why this surge in students occurred when it did; the popular historiographical triad of urbanization, industrialization, and immigration is surely implicated, but tracing these influences in detail is a difficult task, beyond the

9.1. QUESTIONING THE PLACE OF MATHEMATICS IN THE CURRICULUM

scope of the present work. There can be little doubt that the student population growth ultimately had a major effect on how educators viewed the role of the secondary schools. By about 1906 the moderate revisionist consensus that had motivated the Committee of Ten was breaking down. The moderate revisionists had succeeded in gaining respectability for the modern academic subjects, and had brought more uniformity to college entrance requirements, but other issues were emerging rapidly.[1]

One of these emerging issues, evidently influenced in part by the huge influx of students, was the demand for industrial or vocational education. Pressure had been building for some time to tie education of students more closely to their future employment, and now this pressure became impossible to ignore. Krug cites the 1906 report of the Massachusetts Commission on Industrial and Technical Education as a key event. In the same year the National Society for the Promotion of Industrial Education was formed. A crisis in the schools was discerned, namely that they were training students "away from the farm and the workshop." These last words were issued in 1907 by a well-placed spokesman for the movement: President Theodore Roosevelt. Even Charles Eliot felt obliged to raise his voice in support of trade schools, in the course of which he seemed to blatantly contradict his long-held views against differentiating college-bound students from other students at an early age.[2]

With the rise of the vocational education movement the key distinction was no longer between classical and modern educational subjects but between academic and practical, and the academic subjects were being put on the defensive.[3] Mathematics had successfully crossed from classical to modern, but had a much more difficult time shaking the disparaging label of "academic." Our earlier chapters suggest that the mathematicians would have provided little help in this regard. The Mathematics Conference of the Committee of Ten had openly sought to distance itself from commercial mathematics. E. H. Moore had talked of the importance of applications of mathematics, but his enthusiasm was mainly confined to substantial scientific and engineering problems; the

---

[1] For secondary-school enrollment see Joel Spring, *The American School 1642–1990* (New York: Longman, 1990), 197; and Herbert M. Kliebard, *The Struggle for the American Curriculum 1893–1958* (Boston: Routledge & Kegan Paul, 1986), 8. Edward Krug indicates that adding in private school pupils would not have increased the enrollment figures by more than a couple of percentage points. Edward A. Krug, *The Shaping of the American High School 1880–1920* (New York: Harper & Row, 1964), 173.

[2] Ibid., 217–226.

[3] Ibid., 236, 244.

mundane tasks of the farm and the workshop did not greatly engage him or other mathematicians.

Also emerging at full strength, after building through the last third of the nineteenth century, was a preoccupation with efficiency, a major theme of the Progressive Era generally. In education, efficiency concerns were notably manifested in increased attention to devising ways to compare and evaluate students, schools, and programs. Instances that have been mentioned include the advent of a national system of college entrance examinations, and the adoption of some variant of credit-hour units as a standard measurement of student progress. Clearly such developments were greatly invigorated by the simple need to deal with more students.

But there was more to the efficiency movement in education than wrestling with the administrative operation of overburdened educational institutions. There was also a widespread desire to better regulate the output of these institutions in a socially beneficial manner. This interest in education for social control or social efficiency contained the vocational education movement as a special case; all students should be fitted to their proper role. Moreover, school subjects were now asked to justify themselves as supporting socially efficient goals.[4]

The rise of social efficiency as an educational aim was accompanied by an increasingly direct skepticism about mental discipline. We have seen that the unadulterated mental discipline thesis had already lost much of its appeal by the time of the Committee of Ten. Even before 1900 socially minded educators were looking upon mental discipline as an unduly individualistic doctrine. Meanwhile psychologists were dismissing the faculty psychology upon which mental discipline had originally rested. Experiments in 1890 by William James had thrown doubt on the existence of an isolated faculty of memory, and after the turn of the century James's pupil Edward Lee Thorndike proceeded to publish apparently devastating refutations of the "transfer of training." In 1917 one educator writing on mathematics expressed the view that "We can not any longer make a philosophy of education out of the doctrine of formal discipline."[5]

---

[4]Ibid., 307.

[5]Ernest C. Moore, "Does the Study of Mathematics Train the Mind Specifically or Universally?" *School and Society* 6 (1917): 491. For pre-1900 skepticism about mental discipline see Krug, *High School*, 250–251, 307. On the James and Thorndike experiments see Kliebard, *Struggle for the Curriculum*, 105–108.

By the second decade of the twentieth century mathematics in the schools was thus under sharp attack from several related directions. Many of the criticisms were clear descendants of those we have encountered in our survey of the nineteenth century, but the new educational environment appeared to give them greater vitality. Considered purely administratively, mathematics was found to produce undesirable obstacles to the passage of students through the system. Charles Judd, who took charge of the School of Education at the University of Chicago in 1909, opined:

> Algebra is one of the greatest sources of retardation and elimination in our schools. A student who is confronted in his first year with Latin and ancient history, and the usual type of algebra, has a grievance against our civilization. What we need, and what we shall get shortly, is some reorganization of our algebra and geometry.[6]

By "retardation" and "elimination" Judd meant that students were being slowed in their transit from grade to grade by failure to pass algebra, and that some among them were becoming so discouraged that they were dropping out of school.

In addition mathematics was often judged to be a failure as a promoter of social efficiency, now that it was in competition for time in the curriculum not only with English, natural science, and the modern languages, but also with physical education, home economics, industrial arts, and hygiene. According to one educator, social efficiency was defined by five "sub-aims" for "individual and social happiness and growth": physical health, gainful employment, recreational enjoyment, good citizenship, and ethical behavior. This educator asked a question whose negative answer was to him so obvious as to require no articulation: "Do algebra and geometry promote these five aims better for all pupils than anything that can be substituted?"[7]

Individual mathematical educators recognized the changing environment without at first being greatly alarmed for their subject. William Betz, a high school mathematics teacher from Rochester, New York, just emerging as an activist in pedagogical matters, sketched the new terrain in 1908. Betz observed that it was indeed proper to talk of a "reform movement in the teaching

---

[6]Charles H. Judd, "Meaning of Science in Secondary Schools," *SSM* 12 (1912): 97. On Judd see Thomas Wakefield Goodspeed, *A History of the University of Chicago* (Chicago: University of Chicago Press, 1916), 329.

[7]Louis W. Rapeer, "A Core Curriculum for High Schools," *School and Society* 5 (1917): 542–543. See also Krug, 278–280.

of mathematics," and that "Prof. E. H. Moore's memorable address" was the proximate cause. As evidence he cited, like others we have encountered, the forming of associations of mathematics teachers subsequent to Moore's talk. Betz then proceeded to list five "transforming influences" that made it imperative to reform mathematics teaching, especially for geometry.[8]

First Betz evoked "modern industrialism, with its demand for tangible success," which was leading to the unhappy prospect of "the direct bread-winning power of a subject" becoming "the sole criterion of its usefulness." This was hard on geometry. Second, he noted that the concentration of people in cities was leading to a great diversification of the high school population. He was pleased that education was becoming "truly democratic," but saw the schools faced with a great problem of "assimilation of so much raw material from homes giving no cultural impulses, and of so many students having no intention of entering higher institutions of learning." But third, and according to Betz "more far reaching" than the first two influences, was the rise to "great prominence" of natural science, characterized by the "inductive method of investigation." Education in this manner, also referred to by Betz as the "laboratory method," was a fine thing, but it required increased resources of time and material to be truly effective. Fourth, Betz acknowledged the rise of the "new education," a collection of psychological and sociological ideas among which "the doctrine of interest" figured prominently. He did not dismiss these ideas entirely, but found them subject to "exaggerations" which often rendered them harmful. Students were too often led to avoid "all real difficulties" and thus the opportunity "to develop a strong character" was missed. "Last, not least," Betz wrote,

> I must refer to the molding power of recent research in the domain of pure mathematics. The labors of men like Pasch, Peano, Veronese, Hilbert, Klein, Russell, and others are making it clear that the subject of rigor in geometry is one of extreme delicacy. It appears that our text-books are full of hidden assumptions and that their usual boast of rigorous presentation is ludicrous.[9]

One thing missing from Betz's admirably comprehensive and balanced list of factors is the professional dimension of the last factor above. The "recent research" referred to presented a conundrum. On the one hand it could be seen as

---

[8]William Betz, "The Teaching of Geometry in its Relation to the Present Educational Trend," SSM 8 (1908): 625.

[9]Ibid., 625–627.

a justification for claims of jurisdiction by research mathematicians over school mathematics; the mathematicians were needed to cleanse school instruction of falsehood and lack of rigor. But on the other hand it could easily be seen as an excuse for mathematicians to withdraw from school mathematics altogether; the domains were simply too far removed from each other. The Moore program offered a compromise attempt to keep the mathematicians interested, without overburdening the schools: the schools would not be forced to assimilate the latest research, but they would be encouraged at least to incorporate the mathematics of the seventeenth century, as interpreted by twentieth-century mathematicians.

Betz in his 1908 article went on to craft a careful middle-of-the-road position. He welcomed the demise of "the old inflexible schedule of studies," but feared that the further advance of the elective system would mean more students avoiding mathematics. He praised the efforts of Moore, Perry, Myers and others to modify the "compartment" system in mathematics instruction, but noted that their ideas had proved difficult to implement. Betz also opined that "over-insistence on discipline for its own sake has done incalculable harm to the interests of mathematics." He praised John Dewey for seeking to break down the sharp distinction between intuitive and demonstrative geometry. He admitted that the crowding of the curriculum necessitated some omission of mathematical topics, and looked to Felix Klein, "one of the greatest of living mathematicians," as a guide to determining the topics of most value. Most specifically, Betz recommended that a "new committee of ten" ought to be set up to "work out a national geometry syllabus." He concluded on positive note:

> Never in the history of the world have the opportunities for sound training been greater, never has the unique importance of mathematics been more generally recognized, nor has there been more mutual sympathy and cooperation.[10]

But Betz's faith in cooperative relations among mathematicians, schoolteachers, and pedagogical theorists would not long endure. By 1911, as we shall shortly note, he was already more alarmed at the menace to mathematics represented by the social efficiency educators.[11]

Earle R. Hedrick, professor of mathematics at the University of Missouri, was still mainly optimistic in 1909. Speaking on geometry instruction at a

---

[10]Ibid., 628–633.

[11]On Betz's evolution see George M. A. Stanic, "The Growing Crisis in Mathematics Education in the Early Twentieth Century," *Journal for Research in Mathematics Education* 17 (1986): 197.

meeting of the NEA he promoted a view much like that of E. H. Moore: student intuition should be nurtured; logical hair-splitting should be avoided. Hedrick thought geometry especially valuable as a means of unifying mathematics and tying it to science. "Should we not," he asked, "take our stand, as a unified subject, rather in the ranks of general science than in those of the moribund metaphysical studies?" His conclusion expressed the view that cooperation by mathematical educators with general trends in educational reform would be largely beneficial, and that most of the responsibility for change belonged to the mathematical educators:

> In conclusion, let me say that it seems to me that my remarks are in tune with the spirit of the age. We must come to a saner and more practically useful attitude in all of our mathematics, or we shall be pushed to the wall; already is to be heard a not uncertain demand that mathematics be not required in high schools, and this will inevitably come if the mathematical topics be allowed to fall into the disrepute of being merely useless playgrounds of the imagination.

Hedrick thus shared with Moore professional anxiety regarding the future of mathematics education, and indeed foresaw the potential for even more far-reaching difficulties than Moore had discerned only seven years earlier.[12]

James F. Millis, head of the mathematics department at a private secondary school in Chicago, spoke to the same NEA meeting that Hedrick had addressed, and likewise agreed that mathematics should cooperate with the spirit of the age. The particular form of cooperation advocated by Millis was what he called the "real-problem movement," which he defined as "an attempt to reform the teaching of secondary mathematics by teaching the different subjects in relation to their uses in solving the real problems that are actually encountered in life." Millis declared himself against

> the old idea of education that is now obsolete, namely the idea of education as a *discipline* ... As opposed to the disciplinary idea education is seen to be the growth or development of the individual from within—it is the process of living itself.

The Central Association of Science and Mathematics Teachers had organized an effort to collect such problems in a wide variety of practical fields, and Millis and others were experimenting to determine which problems were best adapted

---

[12]Earle R. Hedrick, "The Treatment of Geometry For Secondary Instruction," *Journal of Proceedings and Addresses of the NEA* (1909): 515–516, 519.

to instructional purposes. Millis touted this scheme as an excellent means to winnow the curriculum:

> Certain topics of elementary algebra, for example, which find no applications in practical work, should be eliminated, or at least given less prominence in the secondary schools, where the great body of students will never have occasion to pursue more advanced courses in mathematics, and will leave the secondary school directly to enter upon some practical vocation.[13]

Note Millis's assumption that the curriculum should be slanted toward the needs of the "great body of students" rather than to those few who would go on to take advanced mathematics. This put mathematics instruction in a delicate position. Moreover, the elimination criterion proposed by Millis could be readily used, in the hands of more radical proponents of social utility and efficiency, to support severe reductions in mathematics instruction. In one notable study, some 4,000 sixth, seventh, and eighth grade students were asked to follow their parents around for two weeks, collecting mathematical problems solved by the adults in the course of business and household tasks. Most adults in this survey, conducted in the middle west during 1916–1917, were found to use little except the most elementary arithmetic processes. Square root, for example, was among those "processes appearing in this study so few times as to suggest their omission from the arithmetic work of the elementary grades," and no one ever had occasion to extract a cube root at all.[14] That anyone would consider making curricular decisions on the basis of such a survey illustrates how the environment had changed since the Mathematics Conference of the Committee of Ten proposed to eliminate curricular items for failing to provide "any really valuable mental discipline."

The new committee that Betz called for in 1908 to propose a national geometry syllabus became a reality in 1909, under the auspices of the NEA. It was not a committee of ten but of fifteen. It included eight representatives of secondary schools, including Betz himself, as well as Mabel Sykes, a Chicago high

---

[13]James F. Millis, "The Real-Problem Movement in Its Relation to the Teaching of Geometry and Algebra in Secondary Schools," *Journal of Proceedings and Addresses of the NEA* (1909): 519, 520, 522. Italics in original.

[14]G. M. Wilson, *A Survey of the Social and Business Usage of Arithmetic* (New York: Bureau of Publications, Teachers College, 1919), 5, 15, 52–53. See also George M. A. Stanic, "Mathematics Education in the United States at the Beginning of the Twentieth Century," in Thomas S. Popkewitz, ed., *The Formation of School Subjects* (New York: Falmer, 1987), 162–163.

school teacher who had done some graduate work with E. H. Moore. There were seven university teachers on the committee, four of whom we have met: Earle Hedrick, David Eugene Smith, Florian Cajori, and Herbert Slaught. Slaught was the chairman. The committee produced a provisional report in 1911 and a final report in 1912.[15]

The committee demonstrated its awareness that it was functioning in the midst of uncertainty: "[O]wing to the condition of unrest in the entire field of secondary education it is at present impossible to give any final advice along any of these lines of change." But it still felt confident about requiring that all high school students take a rigorous geometry course. Overall it interpreted most reform efforts rather benignly:

> The most noteworthy tendency in secondary education is the desire for more organic teaching and hence the desire for more time. This tendency finds its most significant expression in the movement toward a *six-year curriculum*. It is undoubtedly true that in a six-year curriculum many of the problems of correlation would be brought nearer to a solution, that many difficulties arising from the present tandem system would disappear, and that mathematics would be given a place in the curriculum more nearly commensurate with its importance.[16]

"Organic teaching" would appear, in context, to be the committee's conception of an intellectually respectable reference to the calls for practicality and social efficiency. The "tandem system" referred to what the committee saw as the unduly loose linkage between the grammar school and the secondary school, usually configured as eight years of the former followed by four years of the latter. The committee, and others, thought that if the high schools took control of grades seven and eight then various desirable preparatory work could be accomplished in mathematics, thus solidifying its position in the schools. The desirability of six-year secondary schools had been alluded to by the Committee

---

[15]"Provisional Report of the National Committee of Fifteen on Geometry Syllabus," *SSM* 11 (1911): 329–355, 434–460, 509–531. *Final Report of the National Committee of Fifteen on Geometry Syllabus* (NEA, 1912). There appear to be only two significant differences between the two versions. First, Cajori's survey of the history of geometry instruction which appeared in the provisional report (pages 330–355) was not reprinted in the final report, although readers were encouraged to seek it out. *Final Report*, 3. Second, the final report included several pages (47–54) of comment and discussion not included in the provisional report. On Mabel Sykes see Moore to D. E. Smith, Apr. 4, 1906, DESP.

[16]*Final Report of the Committee of Fifteen*, 10–11. Italics in original.

of Ten and put forward more explicitly by the Committee on College Entrance Requirements.

The six-year curriculum did indeed gain in popularity, but not quite as the committee would have hoped. Only rarely did high schools expand to six years, creating a six-six system. More common was the emergence of "junior high schools" and the six-three-three system. Indeed, the junior high school became in many parts of the country a venue for establishing early differentiation of students into academic and vocational, with social efficiency educators arguing that advanced mathematics should be reserved for the high school. This was hardly what the Committee of Fifteen had in mind.[17]

It is clear from the Committee of Fifteen report that the prestige of England's John Perry had subsided among American mathematical educators since the boost provided by Moore's speech of 1902. Cajori's historical introduction in the provisional report noted that "the Perry laboratory method has led to some severely practical works" in England, but that "a middle ground has met with greater favor." It was also acknowledged that the "Perry movement" had had some influence in America, but Moore's talk was not mentioned.[18] Perhaps the extent of Moore's support of Perry was already considered a bit embarrassing. In the final report the history of enthusiasm for less formal geometry was summed up in the following fashion, with Perry presumably classed with the "extreme school in England":

> But this formal side has been attacked time after time, by the astrologers and mystics, by the cathedral builders of the Middle Ages, strongly by the French writers of the seventeenth and eighteenth centuries, recently by an extreme school in England, and at present in a less formidable fashion in our own country. The results of these attacks in so far as they have meant the abandoning of formal proofs have been futile.[19]

The Committee of Fifteen, retainers of the moderate revisionist tone of the Committee of Ten, claimed to recommend a "judicious fusion of theoretical and applied work, ... free from radicalism in either direction." Apparently the committee did not even feel the need to be defensive at referring approvingly to the "disciplinary value" of geometry. But individual members of the committee

---

[17] See Krug, *High School*, 142, 239–240, 327–335. See also David Eugene Smith, "Certain Problems in the Teaching of Secondary Mathematics," *MT* 5 (1913): 178.

[18] "Provisional Report of the Committee of Fifteen," 352, 355.

[19] *Final Report of the Committee of Fifteen*, 11.

were certainly already aware of the attack on mental discipline. Writing in 1913 E. R. Hedrick acknowledged that "the modern theory of education renders uncertain the possibility of disciplinary training." In the same year David Eugene Smith advocated emphasis on the "potential utility" of mathematics: "Call it a claim for mental discipline if you please,—this is a mere question of fashionable or unfashionable phraseology; it is a claim for serious attention to a vital issue in education."[20]

Smith was one of the mathematical educators most alert to the growing threat to school mathematics, both by virtue of his wide national and international contacts and because some of the loudest anti-mathematical agitation came from his own institution, namely Teachers College, Columbia University. At the International Congress of Mathematicians at Rome in 1908 Smith had set in motion an ambitious effort to study mathematical education around the world. Felix Klein was secured as the president of the resulting International Commission on the Teaching of Mathematics. The participating countries included almost all the nations of Europe, as well as Japan, Brazil, Australia, and the United States. The three American Commissioners were Smith himself, Osgood of Harvard, and J. W. A. Young of Chicago. Numerous other mathematical educators at all levels were recruited to staff various specialized committees, including Millis and Betz, as well as Slaught, Van Vleck, Wilczynski, Lunn, and G. D. Birkhoff, university professors who have appeared earlier in this work. For the next several years reports appeared under the auspices of the commission. Most reports had appeared by 1912, but some appeared as late as 1918.[21]

No general assessment of this vast undertaking will be attempted here. George Stanic notes that very disparate views can be found in these reports, depending on which particular report one selects. Educators can be observed struggling to choose the most politic position on mental discipline and related issues. Betz, for example, served on the committee on Public General Secondary Schools, and was chairman of the subcommittee on "Failures in the Technique of Secondary Teaching of Mathematics: Their Causes and Remedies." In the

---

[20]Ibid. E. R. Hedrick, "Foreword on Behalf of the Editors," *AMM* 20 (1913): 3. Smith, "Problems in Teaching Secondary Mathematics," 164.

[21]Krug, *High School*, 347–348. J. W. A. Young, "The Fifth International Congress of Mathematicians," *SSM* 12 (1912): 705–707. International Commission on the Teaching of Mathematics, *Graduate Work in Mathematics in Universities and in Other Institutions of Like Grade in the United States* (Washington: Government Printing Office, 1911), 2. Phillip S. Jones, ed., *A History of Mathematics Education in the United States and Canada*, NCTM Thirty-Second Yearbook (Washington, D.C.: NCTM, 1970), 182–183, 311–312.

report of this latter subcommittee he took the liberty of reprinting his 1908 discussion of the "transforming influences in education," but now he added some language suggesting additional sensitivity on his part to the dangers of the "new gospel of social efficiency" which taught that "mental discipline as ordinarily conceived is a myth." Betz felt compelled to draw a distinction: "It is not so much this new theory itself, as the hasty inferences drawn from it by superficial minds that we must regard as dangerous." Commissioners Smith, Osgood, and Young, as Stanic points out, tried to straddle the mental discipline issue by claiming that the new precision in experimental psychology should be viewed as a triumph of mathematics, and yet that judicious mathematical educators were still entitled to doubt that the "doctrine of formal discipline" had been "exploded" by these experiments.[22]

Smith's experience on the Commission led him to observe that European countries were not so subject to "the blind attack upon mathematics, which seems to be a phase of degeneracy in some of our educational circles today."[23] Smith, although not a research mathematician himself, had come increasingly to identify with the interests of such mathematicians, and indeed outdid most of the true researchers in public expressions of enthusiasm. He seemed to see his own work in history and education as continuous with the research endeavors, but at the same time he felt more exposed to anti-mathematics agitation than the researchers did. In trying to rally secondary-school teachers of mathematics to his views his rhetorical flights rose high:

> And above all, it seems to me to be our duty to stand for the interest of mathematics for its own sake, for setting forth its beauty of symmetry, for voicing its poetry, for living its religion, and for exalting it for the truth that it sets forth so clearly and for the invariant properties that characterize it in every branch. It is only by being imbued with such feelings and ambitions that we can bring our pupils to love the subject and to feel the great mental uplift that comes from its study.

---

[22]Stanic, "Mathematics Education in the United States," 159–161. Betz's words can be found in *Mathematics in the Public and Private Secondary Schools of the United States* (Washington: Government Printing Office, 1911), extracts reprinted in James K. Bidwell and Robert G. Clason, *Readings in the History of Mathematics Education* (Washington, D.C.: NCTM, 1970), 329, 351.

[23]David Eugene Smith, "The International Commission on the Teaching of Mathematics," *Educational Review* (1913): 6; quoted in Krug, *High School*, 348.

This sort of sentiment was present only peripherally in E. H. Moore, whose educational proposals might be summarized as *higher vocationalism*. Smith, however, had concluded that vocationalism of any sort was a dangerous argument to make in support of school mathematics. His position might be termed *higher mental discipline*. By supporting mathematics "for its own sake" he was trying to avoid the traps of vocationalism and of crude versions of the "transfer of training," in both of which mathematics stood or fell by virtue of its service to other endeavors. Yet the "mental uplift" of mathematics at the core of his defense was surely a descendant of mental discipline.[24]

Two notable critics of mathematics unmoved by Smith's rhetoric, and with whom he had to deal at Teachers College, were David Snedden and William Heard Kilpatrick. Snedden, who served on the committee on General Elementary Schools of the International Commission, had taught some high school mathematics in California early in his career. He did not deny that some mathematics was socially useful, mainly elementary arithmetic. He also allowed that "No one disputes that a young man expecting to study engineering should equip himself, as a preliminary thereto, with a good knowledge of algebra." But most students would never use algebra in their lives, and therefore it was nonsensical to require it of all students. The defenders of algebra were often, in Snedden's view, merely expressing their unexamined faith in the subject for purposes of "mental training" and for "general culture." He did not believe these purposes could withstand scrutiny, and held that in truth algebra remained in the curriculum mainly as the beneficiary of the continued deference to outdated notions from the nineteenth century and earlier.[25] Writing to Smith in 1917 on the question of "how far specified levels of mathematical studies should be prescribed, advised, or left optional in our various public schools," Snedden summed up his position:

> From personal experience, I have become painfully impressed with the injustice often resulting from prescriptions of certain mathematical studies, either for admission to higher institutions of learning, for graduation from secondary school, or for earning those forms of recognition and approval which are often so dear to parents and vital to children.[26]

---

[24]See Smith, "Problems in Teaching Secondary Mathematics," 164, 179.

[25]David Snedden, *Problems of Secondary Education* (Boston: Houghton Mifflin, 1917), 222–227. On Snedden see Walter H. Drost, *David Snedden and Education for Social Efficiency* (Madison: University of Wisconsin Press, 1967).

[26]Snedden to Smith, Mar. 21, 1917, DESP.

9.1. QUESTIONING THE PLACE OF MATHEMATICS IN THE CURRICULUM

Smith and Snedden sparred for many years. The nature of this interchange was reviewed by Smith in 1919:

> I don't suppose I will ever understand your position in this matter, and it seems fairly certain that you will never understand mine. It is therefore very good that we should interchange thoughts as well as we can, and do so with such pleasant relations.[27]

William Heard Kilpatrick had much more first hand acquaintance with advanced mathematics and mathematicians than Snedden; indeed, his mathematics background was comparable to that of Smith. But Kilpatrick never became similarly socialized into the mathematical community, and ended by becoming nearly as adamant as Snedden in arguing that the role of mathematics in the schools be diminished. After graduating from Mercer College in his native Georgia in 1891, Kilpatrick spent a year of graduate study in mathematics and physics at Johns Hopkins. Finding himself somewhat ill-prepared for the level of instruction at Hopkins he was advised to improve himself by reading Simon Newcomb's calculus textbook. Among the Hopkins faculty, Thomas Craig was especially impressive to Kilpatrick as a teacher: "Professor Craig really wanted to help us. I never knew that such wonderful things went on in the world."[28]

Kilpatrick's continued desire for learning led him to the University of Chicago summer school of 1898, where he took classes in both education and mathematics. He was unimpressed with the education professor, John Dewey, which is ironic in view of Kilpatrick's later admiration for Dewey. The mathematics professor (Heinrich Maschke) incited Kilpatrick's interest more successfully, but ultimately caused him to lose confidence in his ability to do research, by assigning him a tough problem to take home to Georgia. In the course of teaching mathematics at his alma mater, Mercer College, Kilpatrick became more and more estranged from the subject:

> Professor Kilpatrick loved to teach, but he was not happy teaching his subject. Mathematics, he pointed out, is a closed system. The deeper the instructor probes the subject the further behind he leaves his students; the less close he is to his students and their needs. One hour of original math thinking on the instructor's part might equal for students enough

---
[27]Smith to Snedden, Jan. 14, 1919, DESP.
[28]Samuel Tenenbaum, *William Heard Kilpatrick: Trail Blazer in Education* (New York: Harper & Brothers, 1951), 17–18.

work for a semester. Further, he had begun to doubt the theory of formal discipline, namely, that the study of mathematics trained the mind, making it a sharper and better instrument in solving practical life situations. But most of all, he wanted to be close to his students—in their emotional, social, and attitudinal thinking and living—and mathematics, he felt, provided a weak and tottering bridge for such communication.

Thus began Kilpatrick's turn toward the theory and practice of education, culminating in his becoming first a student and then a professor of education at Teachers College. Here Kilpatrick's estrangement from mathematics was amplified by Edward Thorndike's work, which Kilpatrick interpreted as effectively disposing of the theory of mental discipline and therefore of the traditional justification for studying mathematics.[29]

In 1915 Kilpatrick was selected to chair the mathematics subcommittee of the NEA's Commission on the Reorganization of Secondary Education (CRSE). This commission, an important landmark in the social efficiency education movement, produced its main report in 1918 under the title *Cardinal Principles of Secondary Education*, a document often compared and contrasted with the Committee of Ten Report. Those seeing the Committee of Ten as a barrier to progress have frequently enthused over the *Cardinal Principles*, while others have seen the latter as a sad falling away from virtue. One striking difference was the standing of the individuals staffing the two endeavors. While the Committee of Ten and its conferences was dominated by college and university educators, including presidents like Charles Eliot and elite subject-matter specialists such as Simon Newcomb, the CRSE and its subcommittees contained many more representatives of secondary schools and schools of education. University mathematicians in particular were entirely unrepresented. The chairman was Clarence D. Kingsley, a graduate of Columbia's Teachers College, who had been a teacher of mathematics at the Brooklyn Manual Training High School, and later Inspector of High Schools in Massachusetts.[30]

The *Cardinal Principles* explicitly acknowledged the greater number and diversity of students attending secondary schools and that these students would

---

[29]Ibid., xi, 37, 42–43, 102–103. On Kilpatrick see also Cremin, *Transformation of the School*), 215–220. Kilpatrick took both the Theory of Invariants and the Theory of Functions from Maschke in the summer of 1898. MDLN, Box 5.

[30]On the Cardinal Principles, see Krug, *High School,* 378–406.

be graduating into "a more complex economic order." Industrialization and urbanization were cited. The CRSE analyzed the needs of education in a democratic society and concluded that there were seven main objectives: "1. Health. 2. Command of fundamental processes. 3. Worthy home-membership. 4. Vocation. 5. Citizenship. 6. Worthy use of leisure. 7. Ethical character." Without attempting to analyze the full implications of these words and phrases, it is notable that much mathematics instruction in particular, and much else in traditional education as understood by the Committee of Ten and others, was relegated to only one of these objectives, namely command of fundamental processes; the only reference to mathematics in the report was under that heading. The CRSE also unapologetically endorsed vocationally based "differentiated curriculums." But the commission was firm in believing that the appropriately democratic solution was not a set of different schools for the needs of different students but rather "the comprehensive high school." The fact that such high schools did indeed come to predominate is often seen as a measure of the success of the *Cardinal Principles*.[31]

Like the Committee of Ten the CRSE created a number of subcommittees to consider specialized aspects of the school curriculum. As one might expect from the foregoing sketch of the CRSE's philosophy, its subcommittees examined not only the academic subjects but also such topics as agricultural education, art education, business education, household arts, industrial arts, and physical education.[32] The subcommittee reports were never gathered together with the main report but were published separately over several years.

The very name of the mathematics subcommittee suggested a lack of sympathy with the subject: "Committee on the Problem of Mathematics in Secondary Education." Chairman Kilpatrick circulated a list of possible committee members for comment. One to whom he sent this list was his colleague David Eugene Smith, who was disturbed by the names he found, many of whom he claimed had only a "local following" and some of whom he implied were "cranks." But the absence of any "first-class man" was even more disturbing to Smith. Here Smith clearly exposed his allegiance:

[31] CRSE, *Cardinal Principles of Secondary Education* (Washington: Government Printing Office, 1918), 7–8, 10–11, 22, 24–27. See also Cremin, "Revolution in American Secondary Education," 306–307.

[32] *Cardinal Principles*, 6. William Duren, in a tendentious survey, incorrectly claims that the CRSE had no subcommittee on mathematics. See William L. Duren, Jr., "Mathematics in American Society 1888–1988, A Historical Commentary" in Peter Duren, ed., *A Century of Mathematics in America* (Providence: American Mathematical Society, 1989), 2:405.

It seems to me, in the first place, you must have on a first-class mathematician who will command the respect of the mathematical world, and who is an earnest advocate of the disciplinary value of mathematics. Such a man would be Professor W. F. Osgood, of Harvard University. Without some such man your report would command no attention on the part of the scholars in mathematics in this country.

Smith also recommended several others, including H. W. Tyler of MIT, L. C. Karpinski of the University of Michigan, J. W. A. Young of the University of Chicago, and J. F. Millis of the Francis Parker School, Chicago. Smith's characterizations of the latter two show his attempt to appear broadminded. "[Young] is a progressive man, but is not a crank upon the destruction of mathematics ... Though I believe [Millis] to be an extremist in many ways, he is a gentleman and a scholar, and he commands the respect of all teachers, even those who do not believe in his particular policies."[33]

Kilpatrick ignored Smith's advice. Besides himself his committee consisted of two school superintendents, two high school principals, an education professor, a normal school mathematics teacher, and a high school mathematics teacher. In February 1917 Kilpatrick sent Smith a draft of his committee's report, asking for comment and criticism. Smith's reply was courteous, but reflected great displeasure. He had two primary objections. First, he asserted that the report "tells us nothing new," and that the committee had not "kept pace with recent progress in the teaching of mathematics." Second, Smith was offended by

> the language used in the report, and which I assume to be due to the efforts of your fellow members. I feel confident that no report expressed in such language or written in such style will receive any serious consideration on the part of the thinking members of our profession.

It is amusing to observe Smith and Kilpatrick trying to retain a courteous relationship, as in Smith's diplomatic assumption above that Kilpatrick was not to blame for the worst aspects of the report. Later on, writing to a third party, Smith would suggest that the offensive tone, indeed the entire report, was the work of Kilpatrick alone. For his part Kilpatrick affected to interpret Smith's criticism in the following astounding manner: "It is a matter of gratification to us to have you say that you approve so generally the principles annunciated."

---

[33]Smith to Kilpatrick, Jan. 13, 1916, DESP.

9.1. QUESTIONING THE PLACE OF MATHEMATICS IN THE CURRICULUM

Perhaps this was his way of saying that he had no intention of acceding to Smith's point of view.[34]

Smith and others proceeded to work behind the scenes to prevent publication of the report. It finally appeared in 1920, but with the concession that it was to be taken as no more than a "preliminary report." CRSE chairman Kingsley admitted in the preface that "problems of great difficulty" had been encountered in its writing, and that therefore it was "submitted primarily for the purpose of stimulating discussion."[35]

Krug notes that it was probably the tone rather than its specific recommendations that caused Smith and his allies to label the Kilpatrick report as anti-mathematics. In some cases the Kilpatrick committee seemed to be careful to take moderate or familiar positions. It commended the ongoing study being conducted by the Mathematical Association of America, which we will discuss shortly. It noted that by speaking of orienting education toward "felt need" it was "psychologic and not economic need which acts as the factor in learning." Therefore the utility of "theoretic interest" could not be ruled out on this basis. It was also judged highly desirable to "make the study of mathematics more nearly approximate a laboratory course," although neither E. H. Moore nor any other advocate of the laboratory method was referred to. On mental discipline the committee found that "transfer of training" had been exaggerated in the past, but that the popular idea that "all transfer was denied" had no "serious support." But because of continuing disagreement among psychologists on the issue the committee elected not to use "formal discipline" as a factor in its recommendations.[36]

The Kilpatrick committee was firm on the subject of differentiation. It claimed this was "now generally accepted," because "[t]he fact of marked individual differences has been scientifically established." Students were classified into four groups, "not sharply marked off from each other": "general readers";

---

[34]The Kilpatrick-Smith exchange is from Kilpatrick to Smith, Feb. 6, 1917, Smith to Kilpatrick, Feb., 15, 1917, and Kilpatrick to Smith, Feb. 23, 1917, all in DESP. See also CRSE, *The Problem of Mathematics in Secondary Education* (Washington: Government Printing Office, 1920), 3 and Sigurdson, 417.

[35]Sigurdson, 417–419; Krug, *High School*, 349–350, *Problem of Mathematics*, 8. Tenenbaum indicates that Kilpatrick was aware of Smith's activity against the report. However, this account seems a bit garbled, asserting that the report was published by the U. S. Bureau of Education after Smith dissuaded the CRSE from publishing it on its own. Tenenbaum, 107. In fact the Bureau of Education published all the CRSE reports. I have not examined the draft report. Sigurdson asserts that it does not differ greatly from the final version.

[36]Krug, *High School*, 352. *Problem of Mathematics*, 9, 12, 16–17.

those preparing for practical trades such as machinists and plumbers; those preparing for science and engineering; and the "specializers." Smith and many mathematicians would have undoubtedly viewed this last group as the standard to which all should aspire, but the Kilpatrick committee seemed to consider the specializers as frankly eccentric: "This group will include those pupils, both boys and girls, who 'like' mathematics."[37]

The Kilpatrick committee recommended that all four groups take a common introduction to mathematics through the ninth grade, concentrating on essentials in arithmetic, algebra, and geometry while avoiding strict compartmentalization. Later mathematics would be entirely elective; many students would take no mathematics after the ninth grade. This was alarming to many mathematical educators. Moreover, the committee seemed to take pleasure in eliminating items from the curriculum. Rather than displaying any of Smith's reverence for mathematics the committee wrote of wielding "a grim pruning hook to the dead limbs of tradition." In discussing the "criteria for exclusion" the committee demonstrated again a certain disdain for those peculiar few who enjoyed mathematics for its own sake. Note the consistently invidious comparisons made between theory and practice in the following:

> It is probably correct to say that these exclusions relate to material introduced from considerations of theory rather than of intelligent practical mastery; from considerations of the pleasures that theorizers (teachers mostly) get from the study of mathematics rather than from a conscious purpose to give that familiarity and grasp which the future practical man will need.[38]

9.2. Organized Responses By Mathematical Educators

Chicagoans at varying levels of the hierarchy of mathematical educators were conspicuously active in devising new organizational structures to cope with the more hostile environment, and although E. H. Moore's personal participation was not always evident, his general influence seems undeniable. His 1902 address was often used as reference or inspiration.

One instance of such Chicago activism is provided by the formation of the Men's Mathematics Club of Chicago (MMC). This originated about 1914 in some informal gatherings of Chicago area teachers of high school mathematics,

---
[37]Ibid., 10, 15, 20.
[38]Ibid., 15, 18–19, 21–23.

united in their concern about threats to their subject. The fact that this was the Men's Mathematics Club is itself fraught with political significance. The official history of the club regales the reader with the claim that women refused to join the club because of excessive tobacco use by the men. This claim is rendered dubious by correspondence in the club records, and by the contentious history of educational labor in Chicago in the early twentieth century, often involving female teachers championing progressive causes. The women teachers in Chicago formed their own mathematics club. There was cooperation between the clubs, but they did not finally merge until the 1970s.[39]

Among the mathematical educators who early joined the MMC were Herbert Slaught of the University of Chicago mathematics department and Ernst Breslich of the University High School. The latter had taken a variety of undergraduate and graduate mathematics courses at Chicago from 1898 to 1902, including courses from both Moore and Slaught.[40] In 1960 one of the founders of the MMC, Charles M. Austin of Oak Park High School, recollected the atmosphere of the early meetings:

> The members of this group were all quite conservative in their thinking about educational matters. We considered Mathematics to be an important body of knowledge, and that its elementary principles should be a required part of the high school curriculum ... Even at that time the progressive ideas of education were creeping into the schools. Principals and Superintendents were adopting the so-called progressive notions. They wanted to substitute easier subjects such as Social Science and Civics for Algebra and Geometry. Of course, we opposed this idea. These other subjects had no inherent value and they did not furnish a basis for future education. It is only in recent days that we see what dire results came from this Progressive Education.[41]

---

[39] Glenn Hewitt, "History of the Men's Mathematics Club of Chicago," in David Rappaport, ed., *A Half Century of Mathematics Progress* (Chicago: privately printed, 1965), 192–210; RMMC, Box: "MMC Data on MMC Meetings"; David B. Tyack, *The One Best System: A History of American Urban Education* (Cambridge, MA: Harvard University Press, 1974), 255–268.

[40] MDLN, Box 5.

[41] "Club History," by C. M. Austin, May 1960, RMMC, Box: "MMC Early History Pre 1930–1950s"; Folder: "Early History (prior to 1930)." Capitalization as in original.

George Stanic has observed that some forty years earlier Austin and his allies had displayed a different attitude toward the word "progressive." At that time they felt it was a word unwise to denigrate; indeed Stanic astutely notes that the mathematical educators sought to appropriate the word for their own initiatives.[42] The younger Austin distinguished between "[s]o-called educational reformers" and "friends of mathematics." He wrote of aiming to "help the progressive teacher to be more progressive," and to "arouse the conservative teacher from his satisfaction."[43]

One of the club's early initiatives was to defend mathematics against the aspersions of "would-be reformers" by conducting a survey of "prominent doctors, lawyers, merchants, bankers, etc., in the city of Chicago." These pillars of the community were asked how they valued their high school instruction in mathematics, and whether algebra and geometry should be retained in high school.[44] The published responses were very positive for mathematics. This incited a response by David Snedden, who questioned the representative character of the individuals surveyed by the club:

> I suggest, therefore, as a next step, that the committee assemble the opinions of the classmates of the prominent men. Possibly in the same college class with Mr. Bank President was a future member of the I.W.W. Why not get the latter's opinion of algebra also? Very likely, Mr. Prominent Broker went to high school with a number of girls, several of whom got better marks than he. Most of these girls are now housewives—why not get their testimony as to the values of algebra and geometry?[45]

Snedden was confident that the housewives would testify that school mathematics had proved of little use to them, helping his contention that they should not have been required to take the subject.

---

[42]See Stanic, "Growing Crisis in Mathematics Education," 198–199.

[43]C. M. Austin, "The National Council of Teachers of Mathematics," *MT* 14 (1921): 1, 3. Austin died in 1967 at the age of 93. "In Memoriam," *MT* 60 (1967): 870. This obituary reveals nothing about Austin's early background and education, which remains unknown to me.

[44]Alfred Davis, "The Status of Mathematics in Secondary Schools," *SSM* 18 (1918): 25. Another version of this same study was published in *School and Society* 6 (Nov. 17, 1917): 576–582.

[45]See David Snedden, "Mathematics in Secondary Schools," *School and Society* 6 (Dec. 1, 1917): 652.

The MMC proved to be the nucleus of a much larger organization of mathematical educators, the National Council of Teachers of Mathematics (NCTM). There had been discussion of such a national teachers association soon after the formation of the regional associations in the wake of Moore's 1902 address. What resulted at that time, however, was merely a federation of the regional associations of science and mathematics teachers; membership in the federation was not conferred upon individual teachers but upon the associations, who sent representatives to national meetings. This federation seems to have declined into inactivity sometime prior to 1919. In February of that year the NEA met in Chicago. Austin and other MMC members, shocked by the anti-mathematical sentiments they heard at this meeting, vowed to organize nationally to protect their profession. By this time they had before them the successful examples of the National Council of Teachers of English, primarily for secondary-school teachers of that subject, and of the Mathematical Association of America, primarily for undergraduate teachers of mathematics.[46]

The resulting NCTM was designed primarily to support the interests of secondary-school teachers of mathematics. It was inaugurated at the February 1920 meeting of the NEA, and Austin was named the first president. One of its first acts was to acquire an existing publication, the *Mathematics Teacher*, as the official journal of the NCTM; this journal had been published by the Association of Teachers of Mathematics of the Middle States and Maryland since 1908. Herbert Slaught became an active participant in the NCTM just as he had in the MMC: he worked on internal committees of the NCTM, served as associate editor of the *Mathematics Teacher* for a time, and shortly before his death in 1937 was named honorary president.[47]

Another early act of the NCTM was to enthusiastically support the MAA's National Committee on Mathematical Requirements (NCMR), to which we now turn. The existence of this committee was used by Austin to help make the

---

[46]On the early agitation for a national association for mathematics teachers, see "Notes," *AMM* 12 (1905): 166. On the federation mentioned, see "Second Annual Meeting of the American Federation of Teachers of the Mathematical and Natural Sciences," *SSM* 8 (1908): 78. On the genesis of the NCTM see "Notes and News," *MT* 12 (1919): 77–78; Austin, "The NCTM," 1–4; and C. M. Austin, "Historical Account of Origin and Growth of the National Council of Teachers of Mathematics," *MT* 21 (1928): 204–213.

[47]Austin, "Historical Account," 209; Edwin W. Schreiber, "A Brief History of the First Twenty-Five Years of the National Council of Teachers of Mathematics (Inc.)," *MT* 38 (1945): 372.

case for forming the NCTM. He was grateful to the MAA for its organizational example, but in acknowledging the MAA's attention to secondary-school education he hinted at jurisdictional disputes which would indeed emerge:

> The pity of it is that this work [the NCMR], wholly in the realm of the secondary schools, should have to be done by an organization of college teachers. True they have generously called in high school teachers to help, but the fact is that it remained for the college people to initiate the work. They could do it because they possessed a live, vigorous organization.[48]

The formation of the National Committee on Mathematical Requirements was one of the first acts of the MAA, and like the formation of the association itself, it illustrated both the pedagogical and professional sensitivities of a core of college-level mathematical educators, and the penchant of many of their colleagues to consider themselves above the battle. In 1915, simultaneously with the AMS discussions of whether or not to take responsibility for the *American Mathematical Monthly*, the Society was also considering "whether any action should be taken by the Society in regard to the movement to displace mathematics in the schools."[49]

In April 1914 an AMS committee had been established, composed of H. W. Tyler of MIT, W. F. Osgood of Harvard, and E. B. Van Vleck of Wisconsin. Their report was submitted in January 1915 and considered by the Council of the AMS at the April meeting. The one and one-half page report reflected concern that attacks on mathematics in the schools could ultimately affect the professional interests of the mathematicians, although the notion that jobs might be at stake was relegated to a parenthetical statement in preference to more exalted concerns:

> [I]f mathematics is not generally studied in the schools, the supply of students of higher mathematics (and of teaching positions) will decline, to the detriment of scientific progress.

Therefore it was recommended that the AMS appoint a special committee on the "Status of Preparatory Mathematics." This committee would keep abreast of developments in the secondary schools and would "so far as may seem wise, cooperate with similar committees representing mathematical teaching associations." We see here the extreme caution with which some mathematicians approached any contact with the lower orders of their realm. But, the report

---

[48] Austin, "The NCTM," 1.
[49] AMS Council Minutes, Jan. 2, 1915, RAMS, Box 12, Folder 47.

9.2. ORGANIZED RESPONSES BY MATHEMATICAL EDUCATORS

avowed, the Society was now "large enough, and strong enough" to risk such contact "without any interference whatever with its primary interests." Such were the apparently cautious and moderate views of this report, but they seem to have been too radical to even command a majority of the three-man committee; Tyler alone subscribed to the entire report. Osgood and Van Vleck added the following proviso at the end:

> We concur to the extent of bring [sic] the above recommendations before the Council, expecting however to present certain objections to the proposed action.[50]

The Council seems to have found the objections persuasive, judging by the result:

> After further discussion it was decided that action on the part of the Society in the matter of the movement against mathematics in the schools is inadvisable at this time.[51]

An amusing typographical error in the Tyler report provides us with spurious, but irresistible, interpretative guidance. In the very first sentence it was proclaimed that the question at issue was "whether any action is desirable on the part of the Society in the matter of the movement against mathematicians [sic] in the schools."[52] Mathematicians did not want to be in the schools; ergo, there was no problem requiring action. More justly, many mathematicians, whatever their concern with education, wanted to keep the AMS purely concerned with the promotion of research; it was at the same meeting of the AMS Council that the resolution was adopted encouraging the formation of the Mathematical Association of America, an organization to which educational issues could be directed in future.

The MAA almost immediately proceeded to engage with these issues. At the first summer meeting of the Association in early September 1916 a Committee on Mathematical Requirements was formed, eventually to be known as the NCMR. E. H. Moore was one of the five initial members. The others

---

[50] "Report to the American Mathematical Society," Jan. 2, 1915, RAMS, Box 12, Folder 47.

[51] AMS Council Minutes, Apr. 23–24, 1915, RAMS, Box 12, Folder 47. The Council at this time consisted of the president, two vice-presidents, the secretary, the treasurer, the living ex-presidents, the three editors of the *Bulletin*, the three editors of the *Transactions*, and twelve additional specially elected members. Archibald, *History of the AMS*, 96–97. The Council Minutes reveal no details about who was present at meetings or how they voted on issues.

[52] "Report to the AMS," Jan. 2, 1915, RAMS, Box 12, Folder 47.

were Moore's brother-in-law, J. W. Young of Dartmouth (named as chairman), Moore's former student Oswald Veblen of Princeton, D. E. Smith of Columbia, and A. R. Crathorne of the University of Illinois. Tyler of MIT was added at the first meeting. The committee set itself to inquire about which topics and teaching methods were appropriate for mathematics in secondary schools and colleges and how teachers should be prepared. In particular it asked: "What general values (utilitarian, disciplinary, cultural) can actually be secured by the study of mathematics?"[53]

The NCMR did not hesitate to take an action that had made the AMS queasy, namely to contact organizations of secondary school teachers. It soon added three members representing these organizations: Vevia Blair, Horace Mann School, New York City, representing the Middle States and Maryland Association; G. W. Evans, Charlestown High School, Boston, representing the New England Association; and J. A. Foberg, Crane Technical High School, Chicago, representing the Central Association. The members of the committee clearly wished to respond to attacks on mathematics, but also felt that prudence was required. In October 1917 Smith wrote to Moore that he and Young had "thought somewhat of starting some publicity work. It is desirable, however, to go rather slowly lest we make the lower type of educator think that we take any stock in his style of attacks." He also asked Moore whether Harris Hancock and C. N. Moore of the University of Cincinnati mathematics department would be worth enlisting in the coming struggle. Moore replied as follows:

> Our men consider that both Hancock and Moore would be altogether likely to make successful propagandists for mathematics as speakers as well as writers—I hope the thought of organizing opposition to these misguided or at least misguiding pedagogy-people may fructify.

In addition to showing that Moore too had become sensitive to the existence of undesirable critics of mathematics, this passage also indicates that he was no longer so enthusiastic about organizing the response of the mathematicians himself as he had been in the previous decade.[54]

---

[53]"Preliminary Report of the Committee on Mathematical Requirements," *AMM* 23 (1916): 283. Although originally the committee seems to have wanted to consider both college and secondary school education, the focus gradually shifted over time to be almost entirely on the secondary schools. I cannot find any explicit acknowledgment of this shift. Some attention was still given to college entrance examinations.

[54]The correspondence referenced is D. E. Smith to E. H. Moore, Oct. 24, 1917 and Moore to Smith, Nov. 4, 1917, both in DESP. Information on the NCMR is from J. W.

Slow progress initially by the NCMR was due not only to judicious avoidance of controversy, but was also partly the effect of the world war; Veblen resigned from the committee in 1918 to become a major in the Army, involved with ballistics studies.[55] The pace began to pick up in 1919 when the committee successfully solicited a grant from the General Education Board. This was an action of some political delicacy, but J. W. Young proved able in this regard. The General Education Board (GEB) had been founded in 1902 as the main channel for John D. Rockefeller's philanthropy in education. A 1916 paper written by a member of the GEB had caused considerable stir: "A Modern School," by Abraham Flexner (1866–1959). Flexner had become a leading educational commentator, best known for his 1910 critique of American medical education. "A Modern School" proposed a utilitarian education having some resemblance to the *Cardinal Principles* and other instances of progressive and socially efficient educational thought. The capability of Latin and algebra to "train the mind" was judged unproven. Four main fields were held to be especially appropriate for the "modern curriculum": science, industry, aesthetics, and civics. Flexner, a disillusioned former teacher of high school Greek and Latin, was especially hard on these subjects, bringing down upon his head much abuse by classicists. He also proposed, citing Snedden in support, that mathematics needed a "radical reorganization," quite likely involving severe reduction of its role in the curriculum. As we will see, "reorganization," sans "radical," made its way into the title of the report of the NCMR.[56]

Some mathematical educators long carried a grudge against Flexner, as can be seen in Harris Hancock's anti-Semitic comments of 1925, in a letter to E. H. Moore:

> Abraham Flexner, an ordinary little jew school teacher of Louisville, Ky. was put on the Rockefeller Education Board by

---

Young, "The Work of the National Committee on Mathematical Requirements," *AMM* 24 (1917): 463.

[55]*The Reorganization of Mathematics in Secondary Education* (n.p.: MAA, 1923), viii; Loren Butler Feffer, "Mathematical Physics and the Planning of American Mathematics: Ideology and Institutions," *Historia Mathematica* 24 (1997): 72–73; David Alan Grier, "Dr. Veblen Takes a Uniform: Mathematics in the First World War," *AMM* 108 (Dec. 2001): 922–931.

[56]On the GEB see Cremin, *Transformation of the School*, 81–82; Krug, *High School*, 342–343. On Abraham Flexner and his ideas see Abraham Flexner, *A Modern College and a Modern School*, (Garden City: Doubleday, Page & Co., 1923), x–xi, 96–97, 102–103, 115–116; Michael R. Harris, "Flexner, Abraham," *DAB*, (supp. 6):207–209 and Steven C. Wheatley, "Flexner, Abraham," *ANB*, 8:120–121.

his brother Simon, another little jew of great ability in certain phases of medicine. He wrote a monograph "On the Modern School" in which he minimized the study of mathematics. This monograph was sent by the Education Board by the car load all over the country. And every school teacher considered it as gospel truth![57]

At the time of the NCMR it is likely that many besides Hancock would have seen courting the GEB as dealing with the enemy; and conversely, it is not likely that the GEB, with Flexner sitting on it, would have looked favorably upon a project designed simply to defend the status quo in mathematics education. J. W. Young nevertheless perceived an opening for his committee. His surviving correspondence with E. H. Moore makes clear that it was Young who was the aggressive actor in this instance, with Moore merely one among those whose support Young sought. Sometime in early 1919 Young circulated his initial GEB proposal to the members of the NCMR. In late April he wrote to try to rouse Moore's interest in his plan:

> I have received replies from all the members of the Committee except yourself. All of the replies were heartily in favor of the proposed plan with the exception of Tyler, who is in some doubt as to the wisdom of the proposed alliance.

In response to Tyler's criticism Young had revised the proposal so that "the signing of it does not commit any member of the Committee to any of the more or less radical views expressed by 'certain friendly critics', etc."; and at the same time to make it "perfectly clear that we are by no means pre-judging the case, and that the Committee is still entirely open-minded regarding suggestions and criticisms, etc."[58]

In early May Young circulated another letter and a final draft proposal. He informed the committee that "negotiations are proceeding favorably with the General Education Board" and that he would be meeting personally with Flexner in a few days to "lay the matter officially before him on behalf of the Committee." This circular letter ended as follows:

> You will note a number of changes from the original draft, which I have no doubt you will approve. They have all been made in the interest of greater clarity and the avoidance of

---

[57]Harris Hancock to E. H. Moore, Dec. 26, 1925, EHMP, Box 2, Folder 1. Lack of capitalization as in original.

[58]J. W. Young to E. H. Moore, Apr. 29, 1919, EHMP, Box 2, Folder 14.

possible misconception as to the attitude of the Committee on possibly contentious points.

I would state also, for your information, that according to the vote of the Committee, Mr. Raleigh Schorling has been elected a member.[59]

Young's last point of information seems likely to have been some sort of bargaining ploy. Schorling was well known in both New York and Chicago, and seems to have been able to maintain good relations with all competing educational camps. At the time of his appointment to the Young committee Schorling was a mathematics teacher at the Lincoln School, a secondary school recently created by Teachers College, Columbia University, financially underwritten by the GEB, and expressly designed to realize Flexner's proposals in "A Modern School." On the other hand, in the early 1910s Schorling had been a teacher at the University High School of the University of Chicago, associated with Ernst Breslich. He was also a founding member of the Men's Mathematics Club of Chicago. When Schorling moved to New York in 1917 he not only joined the faculty of the Lincoln School, but also became a member of the Kilpatrick committee on the "Problem of Mathematics in Secondary Education."[60]

The copy in the Moore papers of the NCMR's proposal for the GEB appears to be a late draft, since Raleigh Schorling's name is included among the members of the NCMR. This proposal began by acknowledging that "the mathematical curriculum is in need of reformation," but was then careful to declare that "this reformation can be successfully undertaken only by those who are familiar with the science and who are sympathetic with scholarship and with the interests and needs of the student." The GEB's support for the Lincoln School was specifically hailed:

> Without entering upon the details of the experiment, we may say that the general plan of adapting the curriculum to the needs of the world of today and tomorrow instead of the world of yesterday is precisely what the Committee has in mind.

---

[59] J. W. Young to the members of the NCMR, Circular Letter No. 15, May 6, 1919, EHMP, Box 2, Folder 14.

[60] On the Lincoln School, see Cremin, *Transformation of the School*, 281–291. On Schorling see Sigurdson, 271; "Club History," by C. M. Austin, May 1960, RMMC, Box: "MMC Early History Pre 1930–1950s"; Folder: "Early History (prior to 1930)"; and Krug, *High School*, 352.

This was a fine example of Young's ability to appear sympathetic to progressive reform without committing to anything in particular. The NCMR then admitted that "[t]he work in mathematics in our secondary schools is and always has been dominated unduly by the college entrance requirements." Here we see recognition of the changed role of the secondary schools since the days of the Committee of Ten and of E. H. Moore's address. The AMS committee on college entrance requirements of 1903, chaired by Tyler, was specifically cited as being out of date. Recall the essential complacency of that committee. The NCMR, in contrast, saw a need for much greater vigilance. It retained the faith of the Committee of Ten that a curriculum could be designed that could accommodate both the prospective college students and the terminal high school students, but with this latter contingent now much larger the NCMR acknowledged that the problem had become more difficult.[61]

The essence of the solution proposed by the NCMR was largely in consonance with Moore's program of seventeen years before, but with added sensitivity to the recent attacks on mathematics. Time previously spent on drill and abstract work should be devoted instead to

> real uses of algebra, the applications of intuitive geometry, modern methods of computation, and the significance of trigonometry. Among specific topics which have of recent years been suggested and whose availability in this connection should receive careful and open-minded consideration may be mentioned the mathematics of business and finance, the graphic representation of facts, mechanical drawing and perspective, elementary notions of statistics, the use of logarithms (and of tables in general), the slide rule, elementary principles of surveying with field work, etc.

Note that the committee felt it prudent to be "open-minded" regarding commercial mathematics, a clear distinction from Simon Newcomb and the Mathematics Conference of the Committee of Ten.[62]

The NCMR proposal then proceeded to welcome the complaints of the "friendly critics" of mathematics, whose views were especially commended as being "not the opinions of extremists in the schools of education," and "not the opinions of destructionists." We see here the careful distinctions the NCMR

---

[61]NCMR to Wallace Buttrick, President of the GEB, dated April 1919, EHMP, Box 2, Folder 14. I have not located copies of earlier drafts.

[62]Ibid.

was attempting to make; the desired classification of Abraham Flexner is clear. The recommendations of the friendly critics were left vague, but it was affirmed that "constructive suggestions and criticisms should receive careful study on the part of secondary school and college teachers." From such study the NCMR hoped to be able to "place the subject of secondary school mathematics in an entirely new light and make its position secure without in any way jeopardizing the scholarship for which it has always stood."[63]

Finally the NCMR claimed that mathematics had special claims for attention among all other school subjects: the work of the International Commission on the Teaching of Mathematics made possible useful comparisons with educational practices in other nations; mathematics was better organized and more "definite" than any other subject, "with the possible exception of Latin"; "the war has greatly increased the interest in mathematics"; and the committee had "already collected much material." Thus the NCMR was emboldened to ask for \$16,000. This money would support two committee members full time for a year (a full time "college professor" at \$400 per month, and a full time "secondary school man" at \$350 per month), and would also defray clerical and traveling expenses incurred by the committee.[64]

The \$16,000 request was duly approved by the GEB; more than 25,000 additional dollars were later bestowed by the GEB to help finish the work of the committee and to publish its report. The member designated to work full time, in addition to J. W. Young, was J. A. Foberg of Chicago's Crane Technical High School. The latter had at one time been a graduate student at the University of Chicago, and had taken the same summer school classes with Maschke as W. H. Kilpatrick in 1898. Four more representatives of the secondary schools were also added, and progress was soon reported on various sections of the final report, most being written by individual members of the committee. The committee moved briskly to publicize its work.[65]

---

[63]Ibid.

[64]Ibid.

[65]J. W. A. Young, *The Teaching of Mathematics*, 406. "The National Committee on Mathematical Requirements," *AMM* 26 (1919): 279. "The Work of the National Committee on Mathematical Requirements," *AMM* 26 (1919): 439. On Foberg's University of Chicago education see MDLN, Box 5.

The NCMR completed most of its work by the end of 1921. Some portions of the committee findings were published by the United States Bureau of Education, but the full report was not finally published until 1923 by the MAA itself, underwritten by the GEB.[66]

9.3. The *Reorganization* Report of 1923

The NCMR was a hard-working group, and its final document far surpasses in bulk, and in evident research effort, all the other reports that have been discussed earlier in this book. *The Reorganization of Mathematics in Secondary Education* runs to 652 pages, including index. A close and comprehensive reading of this report will not be attempted here; we will confine ourselves to a brief survey of those features that illuminate the evolving response of mathematicians to educational issues, from the Committee of Ten of 1893, through E. H. Moore's 1902 address, to the 1920s. The exertion of the NCMR is itself indicative of the changed environment for mathematics education. The committee had no doubts that its subject was under attack and that a massive response was required, along a broad front. A long weekend of brainstorming to produce a few well chosen words would not suffice for this committee.

In the preface the NCMR disclaimed all intention of providing a history of the developments that had led up to itself, save for mention of one key event which it declared to be a "convenient starting point for the history of the modern movement in this country." This event was none other than the 1902 address of E. H. Moore, himself a member of the committee.[67]

The report was divided into two parts. Part I, "General Principles and Recommendations," consisted of eight chapters, totaling 85 pages. The second part, "Investigations," consisted of eight more chapters covering more than 500 pages, including an extensive bibliography. An appendix (listing cooperating organizations) and an index rounded out the report. Chapter II, entitled "Aims of Mathematical Instruction—General Principles," provides a succinct summary of most of the important ideas presented in the report, and thus is appropriate as the focus of the present description. Other chapters will be referred to as needed.

The aims of education promoted in the *Cardinal Principles* were referred to briefly in a footnote at the beginning of Chapter II. There was no attempt here or elsewhere to directly engage with that report, but many elements of the

---

[66] "The Sixth Summer Meeting of the Association," *AMM* 28 (1921): 357–358. *Reorganization of Mathematics*, viii–ix.

[67] Ibid., ix.

NCMR report can be read as implicit criticisms or modifications of the *Cardinal Principles* and the Kilpatrick report, although one particular proposal of the *Cardinal Principles* was endorsed: secondary education should consist of six years, divided equally into junior and senior high school.[68]

The NCMR made clear that it was concerned with "general" education. It sought to offer recommendations suitable for "large sections of the student population." Here again one can observe an increased acknowledgement that the secondary schools had now become sites through which many students would pass. But the committee was by no means ready to admit that only a few specializers would benefit from a good bit of mathematics. Three overall "aims" (corresponding to the original charge the committee had given itself in 1916) were touted in support of mathematical education: practical, disciplinary, and cultural. The practical usefulness of mathematics was judged to derive not only from mastering the *"fundamental processes of arithmetic"* (echoing the language of the *Cardinal Principles*), but also from *"an understanding of the language of algebra,"* including "study of the *fundamental laws.*" The committee was thus here championing the abstract core of mathematics, as the Mathematics Conference of the Committee of Ten had done with its reference to the "commutative law." The NCMR's recommendations that students become familiar with *"geometric forms"* and exercise their *"space-perception"* were also not greatly different from the Mathematics Conference. But the NCMR went beyond the Mathematics Conference in affirming a notion that Moore had emphasized: the importance of understanding *"graphic representations."*[69]

In turning to "disciplinary aims" the NCMR admitted that it was venturing into an area rife with controversy, but it was a controversy that the committee judged should not be avoided.[70] Kilpatrick had claimed to sidestep formal discipline as a factor, but it is clear that the mathematical educators on the committee had concluded that they must address the issue directly. Attacks on mathematics so frequently involved attacks on mental discipline that it was imperative that the latter be defended in key respects. This illustrates again the lack of prescience in Moore's original program. In 1902, with the anti-mathematical fervor still faint, he had thought to justify mathematics primarily as a tool of science and engineering, without resorting to mental discipline. His successors (and apparently he himself) had concluded that this would not work.

---

[68]Ibid., 5, 15, 19.
[69]Ibid., 5–7. Italics in original.
[70]Ibid., 8.

The strategy adopted by the NCMR with regard to mental discipline was to directly engage their adversaries on the field of psychology. An entire chapter (Chapter IX) was devoted to "the present status of disciplinary values in education," written by one of the original secondary school members of the committee, Vevia Blair of the Horace Mann School of New York City, an institution affiliated with Teachers College, Columbia University. Blair admitted that the disciplinary view had been founded on the old-fashioned "faculty psychology," which had now been superseded by "the new psychology," relying much more on experimental tests. She summarized the findings of some of these new psychologists, beginning with William James and E. L. Thorndike. She also reported on the results of a survey of forty prominent educational psychologists, one of whom was Thorndike. From the responses she found that the "two extreme views for and against disciplinary values practically no longer exist ... the psychologists quoted here almost unanimously agree that transfer does exist." The experiments that had been conducted could not accurately quantify transfer in particular educational settings, but Blair was gratified to be able to claim that many of the psychologists thought transfer was "very largely dependent on methods of teaching." From such support came the NCMR's claim to hold a moderate, psychologically informed position on mental discipline: that "general mental discipline is a valid aim in education."[71]

The particular disciplinary aims of mathematics education should include several elements, according to the NCMR. First, students should acquire, "in precise form, ... those *ideas or concepts in terms of which the quantitative thinking of the world is done.*" Second, students should develop the *"ability to think clearly in terms of such ideas and concepts."* Third, students should acquire the appropriate *"mental habits and attitudes."* All of these aims are readily seen to have deep roots in nineteenth-century educational thought; they would not be much out of place in the Yale Report of 1828. But the fourth and last disciplinary aim of the NCMR was decidedly more novel:

> Many of these disciplinary aims are included in the broad sense of *the idea of relationship or dependence*—in what the mathematician in his technical vocabulary refers to as a "function" of one or more variables. Training in "functional thinking," that is thinking in terms of and about relationships, is one

---

[71]*Reorganization of Mathematics*, 8, 90–104. On the Horace Mann School see Cremin, *Transformation of the School*, 287.

9.3. THE *REORGANIZATION* REPORT OF 1923

of the most fundamental disciplinary aims of the teaching of mathematics.[72]

We have seen that the emphasis on the function concept was a post-Committee-of-Ten development in the United States, and that E. H. Moore was a pioneer. Here we see the NCMR following Moore's lead on functions, but in addition making a remarkable attempt to insert the "technical vocabulary" of the mathematician into a "disciplinary" context. Recall that Moore referred to functions in a more utilitarian context of nomograms, linkages, and graph paper.

This wielding of the function concept by the NCMR was yet another claim for the primacy of the abstract core of mathematics, but at the same time there was sensitivity about going too far in this direction. Chapter VII of the report, by E. R. Hedrick, was entitled "The Function Concept in Secondary School Mathematics." This chapter was mainly a collection of examples from algebra, geometry, and trigonometry that were judged to be especially suitable for encouraging functional thinking on the part of the student. Hedrick also included more general comments, some alluding to the complications attendant upon the effort to introduce such an abstract concept into the curriculum. Foreshadowing similar debates about the set concept during the New Math era of the 1950s and 1960s, he was emphatic that the intent should not be to teach "any sort of function *theory*." The teacher should not be trying to force upon the student "any definition to be recited by the pupil," and in fact "the word 'function' had best not be used at all in the early courses."[73]

Indeed, it is notable that nowhere in this chapter, nor anywhere else in the report, is the concept of function given a definition. This illustrates a profound problem for the Moore program and its descendants. Mathematicians like Moore and Hedrick sought to bring the unifying power of abstraction into the curriculum, but they sensed that this had to be done in a concrete manner, so as not to scare the clients with "theory," and to avoid reminding anyone of the bad old ways of rules and rote learning. Thus the pedagogically paradoxical handling of the function concept: it was claimed as pervasive and "fundamental," yet to define it precisely, or even to name it, had to be reserved for specially prepared audiences.

Hedrick's further comments in his chapter provide evidence that, whatever the merits of a precisely defined function concept for research purposes, it was useful to keep the concept vague for the purposes of educational rhetoric. For

---

[72]Ibid., 9. Italics in original.
[73]Ibid., 64–65. Italics in original.

example, the function concept was commended to "[m]echanics, farmers, merchants, housewives, as well as scientists and engineers," as the key to "clear thinking for maximum efficiency." Further, when it came to "public questions" such as tariffs, regulation of insurance rates, and income taxes, "functional thinking" was "a vital element toward the creation of good citizenship." And finally, since the function concept united a wide variety of examples originally thought of as disparate, it was the very best candidate to promote transfer of training; "whereas the transfer of the training given by courses in mathematics that do not emphasize functional relationships might be questionable."[74] The function concept seems here to have taken on rather magical properties, allowing mathematicians to simultaneously claim allegiance with the social efficiency educators and with old-fashioned mental discipline, while overcoming their longstanding distaste for commercial applications.

With regard to the "cultural aims" of mathematics education the NCMR held that the student should be inculcated with appreciation of the beauty of geometrical forms and taught to value the perfection of logical reasoning. The benefits could be "intellectual, ethical, esthetic, or spiritual." Religious effects, "in the broad sense," were alleged to derive from "the study of the infinite and of the permanence of laws in mathematics." The greatest influence here (cited in two footnotes) was the thinking of committee member David Eugene Smith, examples of whose exaltation of mathematics we have earlier seen.[75]

Having described the practical, disciplinary, and cultural aims of mathematics teaching, the NCMR then synthesized them into a "general point of view governing instruction":

> The primary purposes of the teaching of mathematics should be to develop those powers of understanding and of analyzing relations of quantity and of space which are necessary to an insight into and control over our environment and to an appreciation of the progress of civilization in its various aspects, and to develop those habits of thought and of action which will make these powers effective in the life of the individual.

In explicating this point of view the committee built bridges to the wider world of educational thought. John Dewey's insistence on not imposing strict logical structure on early instruction was mentioned favorably. There was no talk of any "grim pruning hook," but it was acknowledged that topics that did not

[74]Ibid., 72.
[75]Ibid., 9–10.

serve the "powers" mentioned above ought to be eliminated from the curriculum. Useless drill was decried; drill should be limited to that which supported "*probable applications either in common life or in subsequent courses.*"[76]

Echoes of the Mathematics Conference of the Committee of Ten and of E. H. Moore's earlier pronouncements are readily discernible. "Formal demonstrative" geometry ought to be preceded by instruction in "intuitive, experimental, and constructive" geometry. The "water-tight compartment" method of teaching mathematics was disparaged. The committee was pleased to find that "progressive teachers" were significantly departing from the old "rigid division" of mathematics into arithmetic, algebra, and geometry. "Correlated" courses were praised. Later in the report the committee wrote of the need "to correlate the work in mathematics with the other courses of the curriculum, especially in the sciences." This was entirely in line with Moore's earlier program. There was no direct acknowledgment that many educators had already found it difficult of realization. The chapter on "Experimental Courses in Mathematics," written by Raleigh Schorling, described the University of Chicago High School as maintaining the faith in correlated mathematics inspired by Moore's address.[77]

There was also a separate chapter on correlation (Chapter X, by A. R. Crathorne of the University of Illinois), but here the word was used in a technical sense, quite alien to the Herbartians and others who had originally championed this term 30 years before. Titled "The Theory of Correlation Applied to School Grades," this chapter proved to be a short introduction to mathematical statistics. The Pearson correlation coefficient, a number between zero and one measuring the relation between two paired sets of numbers, was defined. High values suggested a strong relationship between the pairs; low values suggested a weaker relationship. This notion was then applied to data on student grades from a number of high schools. For instance, one might ask whether there was a strong correlation between a student's grades in algebra and Latin, or geometry and manual training, etc. This then was a more modest, but more quantifiable, approach to issues that had been the subject of the mental discipline controversy. In particular, Crathorne found that success in algebra seemed to be moderately correlated with success in other school subjects, but no claim for a causal relationship in either direction was made. The NCMR showed itself here and elsewhere to be interested in the then recent manifestation of the social efficiency movement that emphasized precisely measuring the knowledge of

---

[76]Ibid., 10–11. Italics in original.
[77]Ibid., 12–13, 28, 202–209.

students by means of standardized tests. Nearly one quarter of the entire report was devoted to a chapter entitled "Standardized Tests in Mathematics for Secondary Schools," written by Clifford Brewster Upton of Teachers College, Columbia University.[78]

The NCMR provided recommendations for mathematics instruction in a six-year secondary education program. Like the Kilpatrick committee, it recommended that all students should take mathematics through ninth grade. The courses should be so designed that all students attained a "broad outlook over the various fields of mathematics." The two committees also seem to have largely agreed that mathematics for students in grades ten through twelve should be essentially elective, but the tone and emphasis with which the NCMR expressed this was very different from that of the Kilpatrick committee:

> The committee believes nevertheless that every standard high school should not merely offer courses in mathematics for the tenth, eleventh, and twelfth years, but should encourage a large proportion of pupils to take them.

Mathematics, for the NCMR, was not an obstacle to other more worthy educational pursuits. The committee affirmed that applications of mathematics were proliferating throughout "the activities of the world." The importance of mathematics in the recently concluded world war was cited as prime evidence, but it was also confidently asserted that such applications were just as important in "those fields of human endeavor which are of a constructive nature."[79]

The NCMR strongly recommended improved training for secondary school teachers of mathematics, evidently with some cognizance of the professional support this would provide to college and university mathematics. It maintained that the United States lagged behind most European nations, where graduate training was commonly required for secondary school teachers. Mathematics teaching in the United States was still too often subject to the notion that "anybody can teach mathematics." The committee realized, however, that the situation would have to be changed gradually; immediate installation of high standards would unduly limit the supply of teachers. Chapter XIV, by R. C. Archibald of Brown University, was devoted to describing current practices for

---

[78]Crathorne's finding on algebra is on page 122. The standardized testing chapter covers pages 279–428.

[79]Ibid., 14, 20, 32, 33.

9.3. THE *REORGANIZATION* REPORT OF 1923

training teachers throughout the world, largely based on the reports of the International Commission on the Teaching of Mathematics.[80]

The NCMR report also included chapters on college entrance requirements, on key propositions in geometry, on terms and symbols in elementary mathematics, on mathematical curricula in foreign countries, and on "questionnaire investigations." This last included a revised version of the earlier cited survey conducted by the Men's Mathematics Club of Chicago, in which prominent individuals were asked to rate the value of their mathematics education.[81]

Indications are that E. H. Moore was not among the most active committee members, but we have seen that the report of the NCMR certainly reflected his influence, such as in promoting the pedagogical benefits of graphic representation and the function concept. The committee was aware, however, of the emergence since 1902 of more hostile elements in its environment, and seems to have concluded that some of the more ambitious elements of Moore's original program were now counter-productive or irrelevant. For instance, allowing students greater latitude to assume various results without proof was still encouraged, but there was no more talk of having high school pupils develop competing axiom systems. There was favorable mention of Moore's vision of breaking down compartments and of correlation, but there was no discussion of the chasm between pure and applied mathematics. Although the large attention to graphs may be seen as a sort of monument to John Perry, there was only the most fleeting personal reference to him: in the chapter on "Experimental Courses in Mathematics," The Horace Mann School in New York City was described as using a graphical procedure referred to as "Perry's 'black thread' method." The only reference to "laboratory" was a brief appendix to that same chapter, entitled "A Mathematics Laboratory Equipment." It would appear that the high hopes for the laboratory method had largely dissipated, nor was there any mention of the desirability of aligning mathematics education with engineering. The most potent threats to mathematics had come from a different direction, and in consequence the committee departed from Moore's original approach, instead proposing renewed buttressing of the mental discipline thesis.[82]

Much effort was expended to publicize the NCMR report. Even before its full publication, portions of it were printed and discussed in the pages of the *Mathematics Teacher*, now the organ of the National Council of Teachers of

---

[80]Ibid., 16–17, 429–508.
[81]Ibid., 523–538.
[82]Ibid., 25–26, 230, 277–278.

Mathematics. The support of the General Education Board made it possible to distribute copies of the final report to a large number of school superintendents, normal schools, and libraries. By 1926 the original printing of 25,000 copies had been exhausted. As with the Report of the Committee of Ten in 1893, the report of 1923 did not lack for extravagant praise greeting its publication. J. W. A. Young, in the third edition of his *Teaching of Mathematics*, published in 1924, added a chapter of more than forty pages summarizing the report, which he introduced as follows (recall that this was the other Young, so that although he was hardly a disinterested observer he was not lauding his own report):

> By far the most important event of the decade 1913–1923 is the work and *Report of the National Committee on Mathematical Requirements*. It not only dwarfs every other development in the domain of secondary mathematics in this decade, but it is unique in the history of the country, and in certain respects probably unparalleled in the annals of all nations.

Most commentators have been somewhat more restrained, but it is possible to find attributions of great influence for the *Reorganization of Mathematics in Secondary Education* into more recent years.[83]

If one looks at the legacy of the NCMR in relation to the ambitions of its members, and more generally in relation to the ambitions of the segment of the mathematical community that had sought to follow the vision of E. H. Moore's 1902 address, one must come to a more modest assessment. Sigurdson sees the report having "little immediate effect." William Duren has come to an even more negative conclusion: "It is fair to say that these authoritative and thoughtful recommendations came to nothing." Moreover, he has characterized the period 1915–1940 as a "twenty-five year depression" in mathematics. Duren has deficiencies as a historian, but he deserves respect as a witness; he

---

[83]On the printing history of the NCMR report see Jones, *History of Mathematics Education*, 202, 205 and the *Reorganization of Mathematics in Secondary Education (Part I)* (Boston: Houghton Mifflin, 1927), iii. This latter was a reprinting, including all of Part I of the report, with extracts from Part II. The Young quote is from J. W. A. Young, *The Teaching of Mathematics in the Elementary and the Secondary School*, 3d ed. (New York: Longmans, Green and Co., 1924), 405. For later commentary see Charles H. Butler, "The Reorganization Report of 1923," *MT* 44 (1951): 90–92, 96; Jones, *History of Mathematics Education*, 208–209; John Servos, "Mathematics and the Physical Sciences in America, 1880–1930," *Isis* 77 (1986): 624–625; and Richard Askey, "Good Intentions Are Not Enough," in Tom Loveless, ed., *The Great Curriculum Debate: How Should We Teach Reading and Math?* (Washington, DC: Brookings Institution, 2001), 163–175.

came to maturity just at the time that the report was being released. Important elements of Duren's gloomy view are confirmed by others. Carl Boyer indicates that the attacks on mathematics in the secondary school curriculum became even more strident after the appearance of the report. High schools were increasingly dropping mathematics as a graduation requirement. Even required ninth-grade mathematics, the point at which the NCMR had chosen to make a stand, was under challenge. During the period 1910 to 1934 percentages of students taking algebra in grades nine through twelve of the public schools declined from 57% to 30%, with students in geometry going from 31% to 17%. This surely justifies George Stanic's characterization of the period as a "crisis."[84]

It is clear in particular that the NCMR did not succeed in silencing or rendering impotent those it saw as "extremists" and "destructionists." One social efficiency reformer, Franklin Bobbitt, complained that

> As one reads the pages [of the NCMR report] one feels one's self wholly within an academic atmosphere, and never at any time does he get a real whiff of the world's actual life, and of the mathematics that actually functions in the real lives of men and women.

Another critic, Harold Rugg, a professor of education at Teachers College, Columbia, specially affiliated with the Lincoln School as its "educational psychologist," scored the NCMR for exclusively employing the "method of expert teaching opinion." This was a "subjective" procedure, engaged in by well-intentioned specialists biased by the "reverence in which teachers hold the subject-matter that they teach." Rugg blithely waived off as irrelevant the NCMR's psychological defense of mental discipline. He complimented Vevia Blair on her survey of psychologists, but found the conclusion drawn a mere truism: "Does training transfer? Of course it transfers." A more essential question was whether high school mathematics, either as presently taught or in a reorganized version, increased a student's ability to think. And even more

---

[84]Sigurdson, 427. William L. Duren, "Mathematics in American Society," 408. W. L. Duren, Jr., "CUPM, The History of an Idea," *AMM* 74-II (1967): 24. On Duren's career see W. L. Duren, Jr., "Graduate Student at Chicago in the Twenties," in Douglas M. Campbell and John C. Higgins, eds., *Mathematics: People. Problems, Results* (Belmont, CA: Wadsworth International, 1984), 1:180. Carl B. Boyer, "The First Twenty-Five Years," in Kenneth O. May, ed., *The Mathematical Association of America: The First Fifty Years* (Washington: MAA, 1972), 32–33. Stanic, "The Growing Crisis in Mathematics Education," 195.

important, might not other subjects serve this purpose as well or better? Such questions could only be answered by rigorous experiment, not by gathering the opinions of psychologists.[85]

Textbook writers did pay their respects to the report by mentioning it in their prefaces, as they had done earlier with the Committee of Ten. It is also reasonable to conclude that the report further solidified the place of elementary graphing in the curriculum, and that the endorsement of the function concept was of some long-term consequence, as earlier indicated. The rise of graphs and graph paper can be looked on as a partial realization of the project of correlating algebra and geometry, but the grander aspects of this concept proved to be largely chimerical. Overall, the *Reorganization* report of 1923, like E. H. Moore's more radical proposals of twenty years before, left a few traces, but did not substantially reform mathematics education in the United States.[86]

---

[85]Bobbitt is quoted in Stanic, "Mathematics Education in the United States," 170–171. The Rugg quotes are from Harold Rugg, "Curriculum-Making: What Shall Constitute the Procedure of the National Committees?" *Journal of Educational Psychology* 15 (1924): 23, 32–34.

[86]Edward A. Krug, *The Shaping of the American High School 1920–1941* (Madison: University of Wisconsin Press, 1972), 90–91. Phillip S. Jones, "The History of Mathematical Education," *AMM* 74-II (1967): 53.

CHAPTER 10

Conclusion

During the period 1893–1923 the position of mathematics in the schools of the United States became progressively less secure. Challenges to the place of mathematics in the curriculum had already arisen earlier in the nineteenth century, especially from advocates of the natural sciences. This is reflected in the 1893 Mathematics Conference of the Committee of Ten, chaired by Simon Newcomb, which emphasized experiment, concrete methods, and inductive procedures, while liberalizing the venerable mental discipline thesis to account for student interest. But overall the Conference, like most mathematical educators of the time, remained complacent, confidently prescribing that all secondary-school students should take a full array of mathematics, and receiving only muted complaints in reply. In contrast, the report of the National Committee on Mathematical Requirements (NCMR) of 1923 was explicitly defensive in nature, striving to publicize the case for mathematics in the schools against increasingly vocal critics who explicitly advocated displacing mathematics from the school curriculum.

This book has traced the evolution of American attitudes towards school mathematics from the early nineteenth century, with emphasis on the views of American college and university mathematicians from the time of the Committee of Ten to that of the NCMR. The mathematicians were not, without extra effort, at the center of debates about school mathematics. They were not administering the schools or teaching classes in the schools, though most of them were teaching the products of the schools, and some of them were writing school textbooks. But peripheral as they often were, a few mathematicians came to have special anxieties about school mathematics, coming to believe that the success of their own endeavors depended on what transpired in the schools. Some mathematicians also held the view that they deserved jurisdictional rights over school mathematics, by virtue of their greater knowledge of the subject and their superior position in the educational hierarchy, although the more politically astute among them realized that it could be counterproductive to insist too overtly on this jurisdiction. On the other hand, there were many mathematicians who were entirely oblivious, or disdainful, of all school matters, creating political problems of a different sort.

A mathematician who was especially attuned to these issues was E. H. Moore of the University of Chicago, a central figure in this book. In the decade after the Committee of Ten Report, as Moore was promoting the growth of the American mathematical research community and orienting it toward pure mathematics, he became anxious about the place of mathematics in the schools. Unlike his predecessors he did not take for granted the high place of mathematics in the curriculum. Moore's pride in distinguishing mathematics from other related academic endeavors was accompanied by a sensitivity to the jurisdictional competition between mathematics and these other fields. If mathematicians were not vigilant, Moore warned, they might lose control of the mathematical education of scientists and engineers, which Moore saw as an increasingly important segment of the educational market.

Moore's 1902 address to the American Mathematical Society was a full-fledged response to the rise of science in nineteenth-century education, and a scheme for positioning mathematics to survive and prosper from this development. Building upon ideas of Felix Klein in Germany and John Perry in England, Moore concluded that mathematics ought to advertise itself as an indispensable tool of science and engineering, and that it should model itself pedagogically after the scientific laboratory. By this means he hoped to maintain production of a small cadre of pure mathematicians similar to himself, while keeping a wider clientele convinced of the utility of mathematics. In making these proposals Moore came close to jettisoning the mental discipline thesis as a justification for teaching mathematics.

Moore enjoyed a number of advantages as an educational spokesman: he was widely considered to be one of the preeminent mathematicians of the country; he was associated with an institution, the University of Chicago, of considerable educational prestige; and he was a skilled academic politician. Nevertheless, mathematics education in the United States did not develop as he had envisioned. There were experiments along lines he recommended or approved, but the long-term success of these experiments was not great. Moore's "laboratory method" did not become the dominant method of instruction; the "thoroughly coherent four years' course" of algebra, geometry, and physics " he called for did not become the norm in the secondary schools; mathematical instruction in the colleges and universities did not undergo a major reorientation to better serve the needs of engineering and physical applications as he had wished. This book suggests several reasons why Moore was a mediocre educational prophet or promoter.

One factor was surely the decentralized structure of American education: there was a distinct lack of political levers whereby an individual, a committee, a university, or a professional association, however aggressive, could effect change. Consequently, Moore was largely limited to exhortation and example. Within these constraints many teachers of secondary school mathematics did respond enthusiastically to Moore, especially to his suggestion that their status ought to be improved. But although these teachers used his words to rally their forces and raise their self-esteem in reaction to external attack, they were in no position to implement the core of his proposals. In addition Moore had to deal with regional prejudice; easterners from well-established educational institutions often distrusted ideas originating from the upstart institutions of the middle west. Further, those outside of mathematics with whom Moore sought to ally did not provide much sustenance. Physicists failed to wholeheartedly embrace his proposals. Engineers may have been more positive, but the case is unproven.

But what is most substantially revealed by this book is that Moore's educational program was in conflict both with the environment enveloping the secondary schools and with the professionalization project of American college and university mathematicians. With regard to the school environment Moore, like many other educators, largely failed to foresee the consequences of the changing demographics. The early twentieth century saw a huge increase in students attending secondary schools. These students were not all training to be scientists and engineers, far from it. Why then should the great majority of students be required to take mathematics? This was an awkward question for a educational program that emphasized application of mathematics to science and de-emphasized studying mathematics as general mind training.

Moreover, in the face of this surge in students the calls for efficiency that had emerged during the last half of the nineteenth century became yet more insistent. The efficiency advocates were claiming to offer means to control the flood of students by carefully circumscribing requirements in terms of time and effort. In contrast, Moore's mathematical laboratory called for such extravagances as performing all demonstrations in two different ways and for blurring of subject-matter boundaries. Doing away with "water-tight compartments" within the mathematics curriculum could well be seen as a prescription for waste and confusion. In addition, although the mathematical laboratory was doubtless far less costly than a chemistry laboratory in terms of materials, it was labor-intensive, requiring extra class periods and more individualized instruction.

As for the mathematicians, they had embarked on a long-term commitment to pure mathematical research, a movement Moore did much to strengthen throughout his entire career. Many proponents of this professionalization project, following the lead of nineteenth-century forebears such as Benjamin Peirce, sought to disclaim all responsibility for matters pedagogical, except for training graduate students or advanced undergraduates. Even those like Moore who thought this attitude unwise had great difficulty keeping their attention consistently focused on lower levels of the educational system.

Furthermore, there was an unavoidable incongruity in the spectacle of a mathematician such as Moore, the purest of pure mathematicians, arguing on behalf of intuition and applications. Even as he was preaching the value of developing geometric intuition for instructional purposes, a fellow American mathematician was characterizing the exploration of the axiomatic foundations of geometry as follows: "Geometric intuition has no place in this order of ideas which regards geometry as a mere division of pure logic." This observer named the international leaders of this anti-intuitional approach; in America it was E. H. Moore.[1] Another of Moore's American contemporaries declared that "one of the facts most vividly brought home to pure mathematicians during the last half century is the fatal weakness of intuition when taken as the logical source of our knowledge of number and quantity."[2]

Thus for Moore and other pure mathematicians of his ilk to espouse intuition and concrete methods, while they themselves ascended to higher and higher levels of abstraction, involved a profound tension. Would the students invited into mathematics via visual and tactile experience eventually gain full entry to mathematical knowledge, or would they remain second-class citizens, with the deeper abstract mysteries reserved for the mathematical priesthood?

This dilemma is at the core of the phenomenon labeled professional regression by sociologist Andrew Abbott. Once a profession has differentiated the fundamental from the peripheral in its knowledge base, the prestige of the fundamental becomes hard to resist, and attempts to raise the status of *applications*

---

[1] James Pierpont, "The History of Mathematics in the Nineteenth Century," *BAMS* 11 (1904): 158.

[2] Maxime Bôcher, "The Fundamental Conceptions and Methods of Mathematics," *BAMS* 11 (1904): 116.

may merely serve to reinforce the distinction.[3] When Ernest Brown called for increased production of mathematical tools that could be used without full understanding of their derivation he was in part disparaging the other-worldliness of pure mathematicians, but simultaneously affirming their high position, since it was they who possessed the full understanding.

We can observe one of E. H. Moore's best students tending more and more to privilege abstraction in his career, though occasionally avowing the importance of physical intuition. In 1918 Oswald Veblen, by then well established at Princeton University, completed a comprehensive two-volume treatise on projective geometry in collaboration with Moore's brother-in-law J. W. Young. The memory of the laboratory method seems faint indeed in Veblen's preface:

> I shall pass by the opportunity to discuss any of the pedagogical questions which have been raised in connection with the first volume and which may easily be foreseen for the second. It is to be expected that there will continue to be a general agreement among those who have not made the experiment, that an abstract method of treatment of geometry is unsuited to beginning students.
>
> In this book, however, we are committed to the abstract point of view.[4]

This last phrase was one that Veblen continually employed as the motto of his approach to mathematics. In 1922 he explained that "the abstract point of view in geometry" had become familiar to mathematicians twenty years before. Veblen was excited by the invigoration of geometric study caused by the advent of Einstein's general theory of relativity, and was even briefly inspired to comment on "the problems of teaching." He admitted some pedagogical shortcomings arising from abstraction:

> The branch of physics which is called Elementary Geometry was long ago delivered into the hands of mathematicians for the purposes of instruction. But, while mathematicians are often quite competent in their knowledge of the abstract structure of the subject, they are rarely so in their grasp of its physical

---

[3]Andrew Abbott, *The System of Professions: An Essay on the Division of Expert Labor* (Chicago: University of Chicago Press, 1988), 118–119.

[4]Oswald Veblen, "Preface," in Oswald Veblen and J. W. Young, *Projective Geometry*, vol. 2 (Boston: Athenaeum Press, 1918), iii. The second volume was largely Veblen's work. The first volume had appeared in 1910. See Saunders Mac Lane, "Veblen, Oswald," *DSB*, 13:599.

meaning. In recent years this defect has become glaringly apparent and the teachers of elementary geometry are beginning to cultivate the experimental technique of the subject.

But Veblen quickly added a qualification:

At the same time it will not be forgotten that the physical reality of geometry cannot be put in evidence with full clarity unless there is an abstract theory also.[5]

Similarly in 1934 he announced that "In arriving at this clear cut view of geometry it was necessary to regard it as an abstract science." He specifically recalled that it was his studies at the University of Chicago under Moore that had introduced him to the power of "the abstract point of view."[6]

In the National Committee on Mathematical Requirements we see the concluding phase of the Moore educational program. The greatest challenge to American mathematics education in the twentieth century had come not from scientists and engineers who insisted on applying Fourier series without rigorously proving all the supporting theorems, but rather from those who thought that most students could do without much mathematics at all.

It appears likely that Moore's profound respect for Felix Klein caused him to misperceive the extent to which the future of American mathematics depended on accommodation with engineering and physics. Attempting to correct Moore's mistake, the NCMR considerably reduced his original emphasis on such an accommodation. Laboratory methods became peripheral while mental discipline was revived as a major justification for studying mathematics. Moore's specific recommendations to give more attention to graphs was given further impetus, although there was nothing so elaborate as Moore's proposal to use linkage diagrams and nomograms. The NCMR also endorsed Moore's promotion of the function concept, but with a new feature: it was found to be central to the value of mathematics as a mental discipline. Although the NCMR touted the unifying power of the function concept, it was skittish about explaining this in terms of the unadorned abstraction defined by mathematicians.

The NCMR manifested a substantial degree of cooperation between certain university mathematicians and secondary school educators, but the very

---

[5] Oswald Veblen, "Geometry and Physics," *Science* 57 (Feb. 2, 1923): 129–131. This is the text of a talk Veblen gave in December 1922 as retiring vice-president of the American Association for the Advancement of Science. Capitalization as in original.

[6] Oswald Veblen, "The Modern Approach to Elementary Geometry," *MT* 60 (1967): 99. The article originally appeared in 1934 in the *Rice Institute Pamphlet*.

10. CONCLUSION

circumstances of the creation of the NCMR illustrate the powerful trend toward the stratification of mathematical activity which the NCMR did little to modify. The mathematicians carefully circumscribed their participation in the NCMR, creating a separate organization, the Mathematical Association of America, to keep the research oriented activities of the American Mathematical Society untainted by educational issues.

J. W. Young, chairman of the NCMR, some years later proclaimed the lack of class feeling (in an intellectual sense) within the American mathematical community, almost surely revealing in the process that there must have been enough to distress him:

> We can't all be research men. Some of us do not even want to be. I have sought to show that there exist wide fields of enquiry and activity other than research that are important, interesting and worthy of intensive cultivation, and that are being sadly neglected. I have incidentally attempted to combat the attitude, if and wherever it exists, that would make of research a fetish, that proclaims that the only worthy function of a mathematician is research and that other activities are to be looked on with contempt. There is fortunately very little of this sort of self-righteous snobbery in our two organizations.[7]

In this same talk Young provided further evidence of stratification, revealing that all was not well in the relations between the mathematicians and the National Council of Teachers of Mathematics (NCTM). This latter organization had been founded in 1920 largely by school teachers concerned about attacks on mathematics from "educationists" and others. Moreover the NCTM supported Young's NCMR and claimed descent from Moore's program. In 1926 the NCTM published its first annual "Yearbook," subtitled "A General Survey of Progress in the Last Twenty-Five Years." This yearbook included essays by David Eugene Smith and Herbert Slaught, the first a long time friend of Moore who had written widely on mathematics education and its history, and the other a student and colleague of Moore who had taken the lead on many mathematics education projects at all levels of instruction. The yearbook also reprinted Moore's 1902 address, since it was by this address, according to the Yearbook Committee (chaired by Charles Austin, a Chicago school teacher),

---

[7]J. W. Young, "Functions of the Mathematical Association of America," *AMM* 39 (1932): 15. This paper was delivered by Young as retiring president of the MAA on Sept. 8, 1931.

that "this era of progress seems to have been inspired."[8] Nevertheless, by 1931 Young felt compelled to make the following statement:

> Unfortunately, from some points of view at any rate, the National Council reflects the point of view of our schools of education rather than that of our departments of mathematics.[9]

This apprehension of Young about the growing power of schools and professors of education echoed similar complaints outside of mathematics. An incisive commentator on this theme was William C. Bagley, who wrote in 1930 that

> the professor of education, in the last analysis, is the controlling agent. With the growth of his influence, the subject-matter specialists—the scholars in the differentiated fields of knowledge—have a constantly diminishing influence in determining where our educational system will go and how it will get there.[10]

Similar complaints would be voiced periodically by other champions of academic subject matter through the remainder of the twentieth century. Among the notable expressions of this tradition of what might be called academic idealism is Arthur Bestor's *Educational Wastelands: The Retreat from Learning in Our Public Schools* (1953). More recent representatives include E. D. Hirsch, Jr.'s *The Schools We Need and Why We Don't Have Them* (1996), and Diane Ravitch's *Left Back: A Century of Failed School Reform* (2000).[11] Hirsch and Ravitch both make a point of extolling Bagley, who they see as a prophet without honor, but neither seems to have taken sufficient note of a crucial feature of Bagley's analysis:

> That the professor of education now has a virtual monopoly of this strategic influence has been through no greed of his own. It is a monopoly that has been almost literally handed over to him by his colleagues in the subject-matter fields. The latter

---

[8]Raleigh Schorling, ed., *The First Yearbook of the National Council of Teachers of Mathematics* (n.p.: NCTM, 1926), "Foreword," no page number.

[9]J. W. Young, "Functions of the MAA," 7.

[10]William C. Bagley, "Professors of Education and Their Academic Colleagues," *Mathematics Teacher* 23 (May 1930): 277.

[11]Arthur Bestor, *Educational Wastelands: The Retreat from Learning in Our Public Schools* (Urbana: University of Illinois Press, 1953; repr., 1985); E. D. Hirsch, Jr., *The Schools We Need and Why We Don't Have Them* (New York: Doubleday, 1996; Anchor Books edition with new introduction, 1999); and Diane Ravitch, *Left Back: A Century of Failed School Reform* (New York: Simon and Schuster, 2000).

have failed to grasp the tremendous significance of the transformation of the high school from an institution of selective class-education to an institution of non-selective mass-education.[12]

Bestor likewise suggested that the subject matter specialists were in part responsible for their own lack of influence on school education:

Most faculties of liberal arts and sciences failed to take seriously the problem of devising sound and appropriate curricula for the education of teachers, and thus left a vacuum into which the professional educationists moved.[13]

Such notions are much less evident in Hirsch and Ravitch, who in the above books place all causal emphasis on the fiendish skill of the anti-academic educationists in promoting their misguided aims. The evidence provided in the present work demonstrates the inadequacy of the Hirsch-Ravitch view in the case of mathematics and lends support to the qualifying comments of Bagley and Bestor. The mathematicians of the late nineteenth and early twentieth centuries surely did, in general, fail to grasp the transformation of the schools. They displayed a fitful interest in school mathematics and an ambivalence about claiming responsibility for teacher training. They surely did, as Bestor asserted, leave a jurisdictional vacuum into which other actors moved. For all the efforts of E. H. Moore and like-minded mathematicians, the gaps between school mathematics and research mathematics continued to grow. In accordance with the ideas of sociologist Andrew Abbott, mathematicians in achieving professional self-awareness had come to value devotion to the abstract core of mathematics over serving clients of mathematics, the largest class of clients being students in the schools.

The estrangement of mathematicians from school mathematics did not result in obvious disaster for pure mathematics at the university level, which continued to prosper, although much of the research produced became increasingly obscure to many users and potential users of mathematics. This obscurity did not prevent Moore and other pure mathematicians from receiving some recognition beyond the community of mathematicians, although none approached the visibility of Simon Newcomb as a public intellectual in the nineteenth century. In 1921 Moore served a term as president of the American Association for the Advancement of Science (AAAS). Three years later that same organization gave a $1,000 award for notable contributions to science to Moore's former student

[12]Bagley, 282.
[13]Bestor, 137.

L. E. Dickson, now himself a senior professor at the University of Chicago. Nor did Oswald Veblen's espousal of the "abstract point of view" do any apparent harm to his career; in the early 1930s he proceeded to transcend even the airy precincts of the Princeton University department of mathematics by being appointed as one of the initial denizens of the Institute for Advanced Study (IAS). It was the first director of the IAS, Abraham Flexner, who chose mathematics as the foundation subject for this institution, contrasting interestingly with his skeptical attitude toward school mathematics requirements noted in Chapter 9.[14] In doing this Flexner was implicitly denying a close, organic connection between the upper and lower reaches of the mathematics hierarchy, an issue that has continued to perplex observers over the years.

Despite such evidence of prestige, mathematicians such as E. H. Moore were not thereby enabled to develop much contact between their research activities and the school curricula. This reality would eventually give rise to laments such as that of research mathematician Saunders Mac Lane, one of the last mathematicians to receive instruction directly from Moore. Mac Lane began a 1954 essay entitled "The Impact of Modern Mathematics on Secondary Schools" with the following words:

> My subject is vacuous; the lively modern development of mathematics has had no impact on the content or on the presentation of secondary-school mathematics.[15]

This comment, suggesting a pent-up desire to bring the wisdom of mathematicians to bear on school mathematics, came at the beginning of the period of educational reform often called the "new math," which among other features constituted another cycle of educational enthusiasm and disillusionment such as has been described in the present book. The pattern set during 1893–1923 has proved resilient.

---

[14]R. C. Archibald, ed., *A Semicentennial History of the American Mathematical Society 1888–1938* (New York: American Mathematical Society, 1938), 144, 183. Abraham Flexner, *An Autobiography* (New York: Simon and Schuster, 1960), 235.

[15]Saunders Mac Lane, "The Impact of Modern Mathematics on Secondary Schools," *MT* 49 (1956): 66. Reprinted from *The Bulletin of the National Association of Secondary-School Principals* 38 (May 1954). Mac Lane was a graduate student at the University of Chicago during 1930–1931, taking courses with Moore, Dickson, and Bliss. Moore died the following year. See Donald J. Albers, Gerald L. Alexanderson, and Constance Reid, *More Mathematical People* (Boston: Harcourt Brace Jovanovich, 1990), 202–206.

APPENDIX A

Acronyms

AMS	American Mathematical Society Oriented toward mathematical research. Founded as New York Mathematical Society in 1888, became AMS in 1894.
CCER	Committee on College Entrance Requirements Produced report 1899. Mathematics subcommittee chaired by Jacob William Albert Young of the AMS.
CEEB	College Entrance Examination Board Founded 1900.
CRSE	Commission on the Reorganization of Secondary Education. Sponsored by NEA. Produced the *Cardinal Principles of Secondary Education*, 1918.
GEB	General Education Board Rockefeller philanthropic agency. Founded 1902.
MAA	Mathematical Association of America Oriented toward college-level teaching. Founded 1915.
MMC	Men's Mathematics Club of Chicago Local association of secondary school teachers, founded ca. 1914.
NEA	National Educational Association (National Education Association, from 1908).
NCMR	National Committee on Mathematical Requirements. Sponsored by the MAA. Produced *Reorganization of Mathematics In Secondary Education* in 1923, under chairmanship of John Wesley Young.
NCTM	National Council of Teachers of Mathematics National association of secondary school teachers. Founded 1920.

TABLE A.1. Organizations and Committees

A. ACRONYMS 295

AMM	*American Mathematical Monthly* Founded 1893. Official publication of the MAA from 1915.
BAMS	*Bulletin of the American Mathematical Society* Founded as *Bulletin of the New York Mathematical Society* in 1891, became *BAMS* in 1894.
MT	*The Mathematics Teacher* Founded 1908. Official publication of the National Council of Teachers of Mathematics from 1920.
SSM	*School Science and Mathematics* Founded as *School Science* in 1901, became *SSM* in 1905.

TABLE A.2. Serials

ANB	*American National Biography* New York: Oxford University Press, 1999
DAB	*Dictionary of American Biography* New York: Scribner, 1973
DSB	*Dictionary of Scientific Biography* New York: Scribner, 1970–1990
NCAB	*National Cyclopedia of American Biography* Clifton, N.J.: James T. White, 1975

TABLE A.3. Encyclopedias

CBP	Carl Barus Papers Brown University Archives
DESP	David Eugene Smith Papers Columbia University, Rare Book and Manuscript Library
EHMP	Eliakim Hastings Moore Papers University of Chicago Special Collections.
EJWP	Ernst Julius Wilczynski Papers University of Chicago Special Collections
JHUPP	Johns Hopkins University Presidents' Papers Ferdinand Hamburger, Jr. Archives, The Johns Hopkins University
MCR	Mathematical Club Records 1893–1921, University of Chicago Special Collections
MDLN	Mathematics Department Lecture Notes 1894–1913, University of Chicago Special Collections
OVP	Oswald Veblen Papers Library of Congress, Manuscripts Division
RAMS	Records of the American Mathematical Society Brown University Manuscripts Division
RMMC	Records of the Men's Mathematics Club of Chicago Northeastern Illinois University Archives
SNP	Simon Newcomb Papers Library of Congress, Manuscripts Division
UCPP	University of Chicago Presidents' Papers 1889–1925, University of Chicago Special Collections
WRHP	William Rainey Harper Papers University of Chicago Special Collections

TABLE A.4. Archival Sources

APPENDIX B

Bibliographical Note

This note aims to guide the interested reader into the secondary literature utilized in the writing of this book, but not always explicitly cited in the footnotes. I will begin by highlighting some particular predecessors that I have enlisted throughout or reacted against, and then I will proceed chapter by chapter, with less critical commentary.

On the history of higher education generally, I have often relied on Burton J. Bledstein, *The Culture of Professionalism* (New York: Norton, 1976); Frederick Rudolph, *The American College and University: A History* (New York: Knopf, 1962); and Julie A. Reuben, *The Making of the Modern University: Intellectual Transformation and the Marginalization of Morality* (Chicago: University of Chicago Press, 1996). But no source proved more consistently stimulating than Laurence R. Veysey's classic account, *The Emergence of the American University* (Chicago: University of Chicago Press, 1965). Although later scholars have offered revisions to Veysey's views, his categorization of rival conceptions of higher learning in nineteenth-century America (discipline, utility, research, and liberal culture), and his description of the rising appeal of science as a rhetorical resource in colleges and universities of that era, remain eminently useful. But neither Veysey nor other general works on higher education has had much to say about mathematics or mathematicians, especially after the heyday of the classical curriculum, thus leaving ample room for the present volume.

For critiques of Veysey see Reuben, 12; James McLachlan, "The American College in the Nineteenth Century: Toward a Reappraisal," Teachers College Record 80 (Dec. 1978): 287–306; John R. Thelin, "Laurence Veysey's The Emergence of the American University," *History of Education Quarterly* 27 (Winter 1987): 517–523; and Caroline Winterer, *The Culture of Classicism: Ancient Greece and Rome in American Intellectual Life, 1780–1910* (Baltimore: Johns Hopkins University Press, 2002), 77.

This book's treatment of secondary school education is informed by the following well-known works: Lawrence A. Cremin, *American Education: The Metropolitan Experience 1876–1980* (New York: Harper and Row, 1988); Lawrence A. Cremin, "The Revolution in American Secondary Education, 1893–1918," *Teachers College Record* 56 (Mar. 1955): 295–308; Herbert M. Kliebard, *The Struggle for the American Curriculum 1893–1958* (Boston: Routledge & Kegan Paul, 1986); Herbert M. Kliebard, "Constructing a History of the American Curriculum," in *Handbook on Research on Curriculum* (American Educational Research Association, 1992), 157–184; Theodore R. Sizer, *Secondary Schools at the Turn of the Century* (New Haven: Yale University Press, 1964); and Edward A. Krug, *The Shaping of the American High School 1880–1920* (New York: Harper & Row, 1964).

The last of these has been especially valuable. Krug offers a wealth of information and thoughtful interpretations, and is also a pleasure to read. The present book follows Krug's lead in the treatment of numerous topics, including the elective system, differentiation, correlation, uniformity, articulation of school and college, vocationalism, social efficiency, and the "moderate revisionism" of the Committee of Ten. Krug makes some useful passing observations on mathematics education, but neither he nor the other scholars of school education mentioned have directed full attention to it.

Of works explicitly devoted to mathematics education, there is helpful material in Phillip S. Jones, ed., *A History of Mathematics Education in the United States and Canada, NCTM Thirty-Second Yearbook* (Washington, D.C.: NCTM, 1970); as well as the more recent two-volume anthology, George M. A. Stanic and Jeremy Kilpatrick, eds., *A History of School Mathematics* (Reston, VA: NCTM, 2003), although this latter work scandalously lacks an index. An anthology published by the American Mathematical Society (AMS), Peter Duren, ed., with the assistance of Richard A. Askey, Harold M. Edwards, and Uta Merzbach, *A Century of Mathematics in America*, 3 vols. (Providence: AMS, 1989), has likewise proved helpful, although the one article most directly relevant, William Duren's "Mathematics in American Society 1888–1988, A Historical Commentary," (vol. 2, pp. 399–448), is unreliable. An earlier work, Solberg Einar Sigurdson, "The Development of the Idea of Unified Mathematics in the Secondary School Curriculum 1890–1930" (Ph.D. diss., University of Wisconsin, 1962), treats mathematics education in the United States in the period covered by the present book, but Sigurdson's references to Simon Newcomb, E. H. Moore, and other mathematicians interested in pedagogy are superficial.

George Stanic's dissertation and subsequent publications are a considerable improvement on Sigurdson, but still do not treat the role of the mathematicians in educational reform with the precision sought for in the present work. See G. M. A. Stanic, "Why Teach Mathematics? A Historical Study of the Justification Question" (Ph.D. diss., University of Wisconsin, 1983); George M. A. Stanic, "Mathematics Education in the United States at the Beginning of the Twentieth Century," in Thomas S. Popkewitz, ed., *The Formation of School Subjects*, (New York: Falmer, 1987), 145–175; and George M. A. Stanic, "The Growing Crisis in Mathematics Education in the Early Twentieth Century," *Journal for Research in Mathematics Education* 17 (1986): 190–205).

The shortcoming of Stanic's interpretation for the purposes of the present book arises from his application of a framework proposed by Herbert Kliebard for the study of American education at the turn of the century period. (Kliebard, *Struggle for the Curriculum*, 27–29) distinguishes four different "interest groups" operating to influence American education in the 1890s and beyond. The status quo of 1893 he sees as being maintained by the "humanists": "guardians of an ancient tradition tied to the power of reason and the finest elements of the Western cultural heritage." Kliebard counts Charles Eliot and most academic intellectuals as humanists in his sense, and thus sees the Committee of Ten Report as preeminently a humanist document. The educational reformers who challenged the humanists were of three kinds, according to Kliebard: "developmentalists," such as G. Stanley Hall, who emphasized student psychology; "social efficiency educators," of whom journalist Joseph Mayer Rice was a forerunner; and "social meliorists," such as sociologist Lester Frank Ward. Although this is a useful scheme for Kliebard's purposes, it is not adequate for dealing with mathematics education, especially for understanding the university mathematicians, almost all of whom have to be lumped together as humanists by default, as is done by Stanic ("Mathematics Education in the United States," 159–161).

The present study seeks a finer grained picture, able to accommodate sharp differences of opinion among university mathematicians as well as between such mathematicians and others with a stake in education. How did university mathematicians come to play any role at all in discussions of school pedagogy? How did some mathematicians come to dispute their fellows on such issues? How did a mathematician's place in the profession affect that mathematician's influence on education? What role was played by the major organizational structures harboring mathematicians, primarily universities and professional associations? These are the sorts of questions that especially concern this book and with which the previous work on the history of mathematics education has not come to grips.

As should be obvious to any reader, this book relies on excellent studies of two of its central figures, Simon Newcomb and E. H. Moore. Albert E. Moyer, *A Scientist's Voice in American Culture: Simon Newcomb and the Rhetoric of Scientific Method* (Berkeley: University of California Press, 1992) provides a rich portrait of that multifarious individual, in a thoroughly developed context of nineteenth-century intellectual life in America. Moyer treats Newcomb's pedagogic excursions ably but concisely, leaving ample room for the fuller treatment offered here. Similarly, the superb book of Karen Parshall and David Rowe, *The Emergence of the American Mathematical Research Community, 1876–1900: J. J. Sylvester, Felix Klein, and E. H. Moore* (American Mathematical Society, 1994) has been invaluable on the roles of Sylvester, Klein, and Moore in fostering an American mathematical research community. Parshall and Rowe have provided indispensable guidance on the significance of such key phenomena as the following: Johns Hopkins University and the University of Chicago as path breaking institutions in American mathematics, and the differences between them; Felix Klein and the German model for American higher education in mathematics; and the swift emergence of pure mathematics as the dominant mode of mathematics in the universities of the United States. In Parshall and Rowe's book Moore as a pedagogue naturally takes third place to Moore as a researcher and as an academic politician. It is one purpose of the present work to shed new light on Moore, and on American mathematics generally, by reversing the order; pedagogy first, research last. It is pleasant to report that the middle term, academic politics, gains additional clarity from this procedure.

Moore's pedagogic work has not gone unnoticed. Since his own time Moore's 1902 address before the American Mathematical Society has intrigued and startled observers. Some historians, I believe, have been too credulous in proclaiming Moore's success as an educational thinker. See John Servos, "Mathematics and the Physical Sciences in America, 1880–1930," *Isis* 77 (1986): 627; Larry Owens, "Vannevar Bush and the Differential Analyzer: The Text and Context of an Early Computer," *Technology and Culture* 27 (1986): 90; and Sidney Ratner, "John Dewey, E. H. Moore, and the Philosophy of Mathematics Education in the Twentieth Century," *Journal of Mathematical Behavior* 11 (1992): 105–116. My own view is more in accord with Loren Butler Feffer, "Mathematical Physics and the Planning of American Mathematics: Ideology and Institutions," *Historia Mathematica* 24 (1997): 70–71. Although I find claims for Moore's great pedagogic influence to be mistaken, a more dispassionate look at the realities surrounding his educational efforts makes him for me an even more fascinating figure.

Scholarship on professionalization has had a special place in the development of this book. Works in this field which made an imprint on the analysis without being explicitly cited include the following: JoAnne Brown, "Professional Language: Words That Succeed," *Radical History Review* 34 (1986): 33–51; Magali Sarfatti Larson, "The Production of Expertise and the Constitution of Expert Power," in T. L. Haskell, ed., *The Authority of Experts*, (Bloomington: Indiana University Press, 1984, 28–80); Eliot Freidson, *Professionalism Reborn: Theory, Prophecy and Policy* (Chicago: University of Chicago Press, 1994); Elliott A. Kraus, *Death of the Guilds: Professions, States, and the Advance of Capitalism, 1930 to the Present* (New Haven: Yale University Press, 1996); and Rue Bucher and Anselm Strauss, "Professions in Process," *American Journal of Sociology* 66 (Jan. 1961): 325–334. The last of these confirmed me in the conviction that mathematicians should not be viewed as an undifferentiated profession but as "segmented." But most influential of all has been Andrew Abbott, *The System of Professions: An Essay on the Division of Expert Labor* (Chicago: University of Chicago Press, 1988). Abbott does not address mathematics directly, but his comments on the related areas of operations research and computer programming are so perceptive as to give high confidence in the applicability of his work to mathematics and mathematics education.

Abbott's concepts of abstraction and jurisdiction provide a useful framework in which to view both mathematical teaching and research. For Abbott "professions are exclusive occupational groups applying somewhat abstract knowledge to particular cases," and "the characteristic of abstraction is the one that best identifies the professions" (Abbott, 8). On its face this naturally tends to give mathematics a high professional standing, and is thus very agreeable to partisans of the subject, but Abbott does not stop at this. He proceeds to a careful analysis of the advantages and the liabilities of abstraction, going a long way toward explaining the strengths and weaknesses of mathematics and related activities as professions.

As an example of the advantages conferred by abstraction Abbott cites the ability of physics graduates to enter engineering:

> Relentlessly abstract training enabled physics graduates to assimilate new techniques and methods to a core of fundamental and largely unchanging abstractions (Abbott, 181).

This applies with even greater force to mathematics graduates, who have been enabled thereby to make incursions into engineering, physics, and many other fields. Being the abstract base of numerous activities appears to be a great source of social power for mathematics, and may in some way help explain the deference paid to it even by those to whom its inner workings are entirely opaque.

However, there are liabilities to reliance on abstraction. As Abbott points out, there have been cases where "graduates actually had training too abstract for application and proved unemployable." But the major disadvantages of abstraction are more subtle than this. First, merely because a profession is the proprietor of abstract models of wide applicability does not mean that this profession takes actual jurisdiction over these other areas. As Abbott notes in the case of medicine, the more likely outcome is for the jurisdiction to be metaphorical. But metaphors that spread widely "become less and less the property of the originating profession," and they become vulnerable to "specialization within and to diffusion into the common culture without" (Abbott, 87, 88, 364). An example of this latter phenomenon is the escape of the mathematical notion of "chaos" into the common culture, resulting in notable unhappiness by mathematicians unable to control the use of the term. See Paul R. Gross and Norman Levitt, *Higher Superstition: The Academic Left and Its Quarrels With Science* (Baltimore: Johns Hopkins University Press, 1994), 92–106.

An even more fundamental danger posed by reliance on abstraction is that which Abbott terms "professional regression": the tendency of professions "to withdraw into themselves, away from the task for which they claim public jurisdiction." This follows from the centrality of a profession's knowledge system for its status rankings; the more one is involved directly with the knowledge system the higher one's status. There thus arises a situation where "professionals admire academics and consultants who work with knowledge alone; the public admires practitioners who work with clients." This is very suggestive for many phenomena in the history of mathematics education in the United States: the high status of the research mathematicians within the mathematical community; the reluctance of these mathematicians to become involved with the client-intensive arena of secondary school teaching; but the persistent pressures by the public which have prevented the mathematicians from withdrawing entirely, and which sporadically have resulted in forays by researchers into the lower levels of the curriculum (Abbott, 118).

Abbott's work enables a striking interpretation of the early rise of pure mathematics in this country. For his scheme supports the view that the most immediate road to status for mathematicians was not to service physics and engineering but rather to cultivate the abstract roots of the subject. It was even possible to obtain public approval of this endeavor (including that of university administrators and philanthropic benefactors), owing to "the public's mistaken belief that abstract knowledge is continuous with practical professional knowledge." And once started on the path, withdrawal to cultivate the abstract core of the subject became self-reinforcing, since "the mechanism of professional regression is irreversible" (Abbott, 54,119). This last statement is too unconditional and ahistoric, but it is nevertheless illuminating for the history of mathematics and mathematics education. This book has attempted to utilize such insights in studying mathematics education in the late nineteenth and early twentieth centuries.

B.1. Introduction

For an example of a recent debate over the need for algebra in school, see Nel Noddings, "Algebra for All? Why?," *Mathematics Education Dialogues* 3 (Apr. 2000): 2 and Dorothy S. Strong and Nell B. Cobb, "Algebra for All: It's a Matter of Equity, Expectations, and Effectiveness," *Mathematics Education Dialogues* 3 (Apr. 2000): 3. On the "math wars" see Suzanne M. Wilson, *California Dreaming: Reforming Mathematics Education* (New Haven: Yale University Press, 2003) and David Klein, "A Quarter Century of US 'Math Wars' and Political Partisanship," *Journal of the British Society for the History of Mathematics* 22 (2007): 22–33. A concise overview of the "new math" of the 1950s and 1960s can be found in David L. Roberts and Angela L. E. Walmsley, "The Original New Math: Storytelling Versus History," *MT* 96 (Oct. 2003): 468–473.

B.2. Mathematics Education in Nineteenth Century America

For examples of the ingenious arithmetical methods used in the nineteenth century in the absence of algebra, see Joel Silverberg, "The Teaching and Study of Mercantile Mathematics in New England during the Colonial and Early Federal Periods: Sources, Content, and Evolution," *Proceedings of the Canadian Society for History and Philosophy of Mathematics* 17 (2004): 219–240.

For discussion of using local tax money to support schools (the Kalamazoo decision of 1874 is often cited as setting a precedent), see Edward Krug, *Salient Dates in American Education, 1635–1964* (New York: Harper & Row, 1966), 91–94; and Lawrence A. Cremin, *American Education: The National*

Experience (New York: Harper and Row, 1980), 162–63. On the private dispensations for higher education establishing Johns Hopkins University (1876), Clark University (1888) and the University of Chicago (1890), see W. Carson Ryan, *Studies in Early Graduate Education: The Johns Hopkins University, Clark University, and the University of Chicago* (New York: The Carnegie Foundation, 1939). For a more general view of the financing of the new universities, see Rudolph, 424–427. The federal government entered higher education in a crucial way with the Morrill Act of 1862. See Rudolph, 188, and 249–254.

A survey of the history of the mental discipline concept is provided by Walter B. Kolesnik, *Mental Discipline in Modern Education* (Madison: University of Wisconsin Press, 1958). See especially 94–95 and 114–116. As Laurence Veysey has noted, there was no agreed upon canonical list of mental faculties in the nineteenth century. See Veysey, 22–23. For an elaborate list of faculties from 1883, see Edward Brooks, *Mental Science and Methods of Mental Culture*, excerpted in James K. Bidwell and Robert G. Clason, eds., *Readings in the History of Mathematics Education* (Washington, D.C.: NCTM, 1970), 78–85.

On the Yale Report, see Rudolph, 131–32; Reuben, 25–26; George E. DeBoer, *A History of Ideas in Science Education* (New York: Teachers College Press, 1991), 3–4; Kliebard, 5–6; Cremin, *National Experience*, 272, 404–5, 566–67; and Bledstein, 239–240. On the Yale curriculum at the time of the Report, see Cremin, *National Experience*, 404–05. Relevant details of the career of Yale president Jeremiah Day can be found in Helena M. Pycior, "British Synthetic vs. French Analytic Styles of Algebra in the Early American Republic," in David E. Rowe and John McCleary, eds., *The History of Modern Mathematics* (Boston: Academic Press, 1989), 1:126; John S. Whitehead, "Day, Jeremiah," *ANB*, 6:272–73; Harris Elwood Starr, "Day, Jeremiah," *DAB*, 3:161–62; and "Day, Jeremiah," *NCAB*, 1:169–170.

Issues of social class in higher education are discussed in Stanley M. Guralnick, "The American Scientist in Higher Education," in Nathan Reingold, ed., *The Sciences in the American Context: New Perspectives* (Washington, DC: Smithsonian Institution Press, 1979), 99–14. Guralnick notes that it was not until the Gilded Age that the eastern colleges "finally became associated with the upper classes" (Guralnick, 130). He also vigorously attacks the view that science had a consistent lowly status in American higher education until the rise of the research university. From Guralnick's point of view Eliot is one of the "myth makers." I contend that Eliot was entirely sincere in his perception that the status of science in education needed enhancement.

Late nineteenth-century concerns about college enrollments are expressed in William Graham Sumner, "Our Colleges Before the Country," in *Collected Essays in Political and Social Science* (New York: Holt, 1885), 160–173; and noted in Bledstein, 240–242, and David Tyack and Elizabeth Hansot, *Managers of Virtue: Public School Leadership in America, 1820–1980* (New York: Basic Books, 1982), 137. The reality of any true decline is doubtful. See James McLachlan, "The American College in the Nineteenth Century: Toward a Reappraisal," *Teachers College Record* 80 (Dec. 1978): 296. On efficiency as a general concern of the progressive era, see Daniel Rodgers, "In Search of Progressivism," *Reviews in American History* (Dec. 1982): 126–127.

The most prominent liberal culture champions of the late nineteenth and early twentieth centuries seem largely to ignore mathematics. For example, see Irving Babbitt, *Literature and the American College*, (Los Angeles: Gateway Editions, 1956; originally published in 1908); and Andrew Fleming West, Short Papers on American Liberal Education (New York: Charles Scribner's Sons, 1907). Babbitt was aware of the craze for laboratory teaching methods; he scoffed at "a recently published Laboratory Method for the study of poetry." See Babbitt, 60.

Julie Reuben has observed that in the latter part of the nineteenth century many American thinkers dropped their allegiance to purely Baconian inductive science in reaction to the perceived success of Darwin's theory of evolution. See Reuben, 36–50. I would argue, however, that many Baconians and Darwinians were virtually indistinguishable regarding their attitude toward mathematics, which from either point of view was considered deficient in observable facts, although mathematicians then and now would consider this an absurdly blinkered view of their subject. On the general American attraction to fact-driven science in the nineteenth century, see Ronald Kline, "Constructing 'Technology' as 'Applied Science': Public Rhetoric of Scientists and Engineers in the United States, 1880–1945," *Isis* 86 (1995): 201–202. On T. H. Huxley as an educational propagandist see DeBoer, 8–12 and Herbert M. Kliebard, *Forging the Curriculum* (New York: Routledge, 1992), 34–38.

On the influence of Pestalozzi and Froebel in American education, see DeBoer, 21–24; Daniel Calhoun, *The Intelligence of a People* (Princeton: Princeton University Press, 1973), 66, 105–111; and Sally Gregory Kohlstedt, "Parlors, Primers, and Public School Schooling: Education for Science in Nineteenth-Century America," *Isis* 81 (1990): 441. Warren Colburn's inductive methods derived largely from Pestalozzi. See Patricia Cline Cohen, *A Calculating People: The Spread of Numeracy in Early America* (Chicago: University of

Chicago Press, 1982), 137. On industrial and manual education see Lawrence Cremin, *The Transformation of the School* (New York: Vintage, 1964), 23–34; and Kliebard, *Struggle for the Curriculum,* 128–135.

For more on Herbert Spencer's educational thought, see DeBoer, 12–16 and Kliebard, *Forging the Curriculum,* 29–34. Spencer's father, a school teacher, had written a geometry book along inductive lines, reprinted in the United States after the son had become famous. See William George Spencer, *Inventional Geometry,* with a prefatory note by Herbert Spencer (New York: D. Appleton and Co., 1877).

Terms used to designate educational institutions were in flux through much of the nineteenth century. Common school was a generic term that could apply to virtually any school open to all comers in a region. In the early nineteenth century primary school usually referred to a school serving very small children who were too young to assist on the farm. Often the instruction included nothing more advanced than reading and spelling. A grammar school (sometimes also called a writing school) was a step up, for somewhat older children. By the late nineteenth century the distinction between primary and grammar school had become less clear-cut. See Willard S. Elsbree, *The American Teacher* (New York: American Book Co., 1939; repr., Westport, CT: Greenwood Press, 1970), 193–94; and Cremin, *National Experience,* 388–390. On academies see Krug, *Salient Dates,* 17–20; and Joel Silverberg, "Higher Mathematics Education in the United States: The Role of the Academy in the Years following the War for Independence," *Proceedings of the Canadian Society for History and Philosophy of Mathematics* 16 (2003): 234–249.

David Eugene Smith (1860–1944) taught at the State Normal School, Cortland, New York from 1884 to 1891 and at the Michigan State Normal School in Ypsilanti from 1891 to 1898. The latter had higher standards than many normal schools. In the 1890s, under Smith's guidance, it became one of the first such schools to offer courses in the teaching of high school mathematics. On Smith see Eileen F. Donoghue, "The Emergence of a Profession: Mathematics Education in the United States, 1890–1920," in George M. A. Stanic and Jeremy Kilpatrick, eds., *A History of School Mathematics* (Reston, VA: *NCTM,* 2003), 160–61; Albert C. Lewis, "Smith, David Eugene," *ANB,* 20:159–61 and Carolyn Eisele and Lyle G. Boyd, "Smith, David Eugene," *DAB,* (supp. 3):721–22.

The distinction between analysis and synthesis in mathematics is discussed in C. B. Boyer, "Analysis: Notes on the Evolution of a Subject and a Name," *MT* 47 (1954): 452, 458–60; Felix Klein, *Elementary Mathematics from an Advanced Standpoint: Geometry,* trans. E. R. Hedrick and C. A. Noble (New York:

Dover, 1939), 55–56; Michael Otte and Marco Panza, eds., *Analysis and Synthesis in Mathematics: History and Philosophy* (Dordrecht: Kluwer Academic Publishers, 1997); Amy Ackerberg-Hastings, "Analysis and Synthesis in John Playfair's Elements of Geometry," *British Journal for the History of Science* 35 (2002): 64; and Helena M. Pycior, "British Synthetic vs. French Analytic Styles of Algebra in the Early American Republic," in David E. Rowe and John McCleary, eds., *The History of Modern Mathematics* (Boston: Academic Press, 1989), 132. Pycior notes that the "synthetic style" was especially prominent in British mathematical textbooks of the early nineteenth century, while the "analytic style" was found most readily in French textbooks (Pycior, 146). Ivor Grattan-Guinness has cited Vieta in the late sixteenth century and Lagrange in the mid eighteenth century as contributing to the confusion over these terms. Cauchy's early nineteenth-century use of "analysis" ("analyse" in French) to refer to his theory of limits seems to have fixed the habit of using this term to describe the entire huge field of inquiry stemming from the calculus of Newton and Leibniz. See Ivor Grattan-Guinness, *The Norton History of the Mathematical Sciences* (New York: W. W. Norton, 1997), 190, 326–327, 373–374. The late nineteenth-century decline in debating analysis versus synthesis in mathematics is noted in Moritz Epple, "Styles of Argumentation in Late 19th Century Geometry and the Structure of Mathematical Modernity," in Otte and Panza, 180–81.

B.3. Simon Newcomb: One Mathematician's Educational Theory and Practice

In the United States it is estimated that among the white population aged five to nineteen the proportion who at least occasionally attended school was 35% in 1830, 38% in 1840, 50% in 1850, and 58% in 1860. See Cremin, *National Experience*, 179. In Nova Scotia enrollments jumped from 15,000 in 1835 to 34,000 in 1847. Yet by 1863 out of the 83,000 children between five and fifteen only 31,000 attended school. Patrick Wilfred Thibeau, "Education in Nova Scotia before 1811" (Ph.D. diss., Catholic University of America, 1922), 115–6. The rise of a stable system of popular education in Nova Scotia was long hampered by religious disputes. See W. S. Macnutt, *The Atlantic Provinces: The Emergence of a Colonial Society 1712–1857* (Toronto: McClelland & Stewart Ltd., 1965), 265.

On George Biddell Airy's "culture of calculated precision," see William J. Ashworth, "The Calculating Eye: Baily, Herschel, Babbage, and the Business of

Astronomy," *British Journal of the History of Science* 27 (1994): 427–28. Airy and Babbage were both highly involved with the problem of "menial mental labor," (Ashworth, 432). Babbage, as is well known, aimed to entirely mechanize such labor. See Harry Braverman, *Labor and Monopoly Capital* (New York: Monthly Review Press, 1974), 82, 316–318.

The financial and administrative struggles at the early Johns Hopkins University are described in John C. French, *A History of the University Founded by Johns Hopkins* (Baltimore: Johns Hopkins University Press, 1946), 64–71, and Hugh Hawkins, *Pioneer: A History of the Johns Hopkins University, 1874–1899* (Ithaca: Cornell University Press, 1960), 242–244. Both emphasize that Gilman was intent on developing an undergraduate program from the beginning. The number of matriculates in residence rose slowly but steadily, from 12 in 1876 to 129 in 1889 (Hawkins, Pioneer, 243). For a jaundiced view of the growth of undergraduate instruction at Johns Hopkins, see Clifford Truesdell, "Genius and the Establishment at a Polite Standstill in the Modern University: Bateman," in *An Idiot's Fugitive Essays on Science: Methods, Criticisms, Training, Circumstances* (New York: Springer-Verlag, 1984), 410–414.

So far as I am aware, the full story of the Newcomb-Morrison episode has never been published, but related matters are discussed in Howard Plotkin's "Astronomers Versus the Navy: The Revolt of the American Astronomers over the Management of the United States Naval Observatory, 1877–1902," *Proceedings of the American Philosophical Society* 122 (1978): 385–399.

On Henry Holt and his company, see Ellen D. Gilbert, *The House of Holt, 1866–1946: An Editorial History* (Metuchen, N. J., Scarecrow Press, 1993); and Henry Holt, *Garrulities of an Octogenarian Editor* (Boston & New York: Houghton Mifflin Co., 1923). Holt was educated at Yale, where librarian Daniel Coit Gilman first attracted his attention to publishing as a profession. Holt, 74. Holt had been involved with textbooks from his first days in publishing in the late 1860s, but markedly stepped up his interest about the time he began to publish Newcomb. In 1882 he established a separate Textbook Division of his firm, headed by E. N. Bristol. See Gilbert, 231.

My claim that Newcomb's textbooks were not very remunerative is based on the following. Holt's usual practice was to offer authors 10% of the retail price of all copies sold after the first 1,000 or 1,200. John Tebbel, *A History of Book Publishing in the United States*, vol. II, *The Expansion of an Industry 1865–1919* (New York & London, R. R. Bowker, 1975), 134. Noting from the price list in the front of an 1887 edition of Newcomb's A School Algebra (New York: Holt, 1887), that most of his books sold for $1.60 or less, and generously assuming that figures on numbers of copies printed in, Raymond Clare Archibald, "Simon Newcomb 1835–1909, Bibliography of his Life and Work," *National Academy of Sciences Memoirs* 17 (1924): 55–56, can be equated to numbers of copies sold, it is doubtful that Newcomb earned as much as $6,000 total from his mathematics textbooks over the entire period from 1881 until his death in 1909. In comparison, Newcomb's government salary was $3,500 per year in 1881 (Moyer, 161), and he was paid $2,500 per year for his work at Johns Hopkins starting in 1884 (Hawkins, Pioneer, 137). George Wentworth, in contrast, grew genuinely rich from his textbooks, which were "used in over two-thirds of the schools in the country." See George Middleton, "The Text-book Game and Its Quarry," *The Bookman* 33 (Apr. 1911): 146–47. Wentworth was able to retire from teaching in 1891 to devote full time to his textbook empire. David Eugene Smith, "Wentworth, George Albert," *DAB*, 10:655.

Newcomb's daughter became Dr. Anita Newcomb McGee (1864–1940). Always encouraged to pursue her intellectual interests by her father, she obtained a medical degree at Columbian University (now George Washington University) in Washington, D.C., and engaged in further medical studies at Johns Hopkins. She played a major role in organizing the Army Nurse Corps during the Spanish-American War. See Mary R. Dearing, "McGee, Anita Newcomb," in Edward T. James, ed., *Notable American Women 1607–1950* (Cambridge: Harvard University Press, 1971), 2:464–466.

On Cauchy's approach to limits, see Carl B. Boyer, *A History of Mathematics* (Princeton: Princeton University Press, 1968), 563. Presumably Newcomb would have absorbed some of this by way of Benjamin Peirce, who introduced Cauchy's work at Harvard. See Parshall and Rowe, 18.

B.4. The Committee of Ten and Its Mathematics Conference

For examples of using the Committee of Ten to bolster educational arguments more than 100 years later see Paul Gagnon, "What Should Children Learn?" *Atlantic Monthly* 276 (December 1995): 65–78, and Diane Ravitch, *Left Back: A Century of Failed School Reform* (New York: Simon & Schuster, 2000), 41–51, 378.

On the history of the National Education Association, see Krug, *High School*, 8, and Ralph Dickerson Schmid, "A Study of the Organizational Structure of the National Education Association 1884–1921," (Ed.D. diss., Washington University, 1963), 50–135.

For more details on the members of the Mathematics Conference, see David Lindsay Roberts, "Mathematics and Pedagogy: Professional Mathematicians and American Educational Reform, 1893–1923," (Ph.D. diss., Johns Hopkins University, 1998), 89–133. Information on the schoolmen participating in the Conference is scant, but all evidence suggests that the Conference report largely reflected the opinions of the college and university members.

The emergence of the concepts of commutativity, associativity, and distributivity is sketched in Morris Kline, *Mathematical Thought from Ancient to Modern Times* (New York: Oxford University Press, 1972), 979–992 and Michael J. Crowe, *A History of Vector Analysis* (Mineola, N.Y.: Dover, 1994), 15–16. Such technicalities would again become bones of contention during the "new math" era of the 1950s and 1960s.

On the history of root extraction as a school subject, and the cube root block, see Peggy A. Kidwell, Amy Ackerberg-Hastings, and David Lindsay Roberts, *Tools of American Mathematics Teaching, 1800–2000* (Baltimore: Johns Hopkins University Press, 2008), 123–138.

B.5. E. H. Moore: Leader of a New Generation of American Mathematicians

The following standard histories of mathematics make no mention of mathematical astronomers Hansen, Delaunay, Le Verrier, or Newcomb: Dirk Struik, *A Concise History of Mathematics*, 4th revised ed. (Mineola, N.Y.: Dover, 1987); Carl B. Boyer, *A History of Mathematics* (Princeton: Princeton University Press, 1968); E. T. Bell, *The Development of Mathematics* (New York: McGraw-Hill, 1940); and Morris Kline. One historian who makes a concerted effort to avoid the traditional pure mathematical bias is Ivor Grattan-Guinness, and he does indeed list Hansen, Delaunay, and Le Verrier in his index; but not Newcomb. See Grattan-Guinness, Norton History. Poincaré's work, in celestial mechanics and much else, is discussed in all these books.

On Newcomb's relationship with Charles Sanders Peirce, see Carolyn Eisele, "The Correspondence with Simon Newcomb," in R. M. Martin, ed., *Studies in the Scientific and Mathematical Philosophy of Charles S. Peirce* (The Hague, Mouton Publishers, 1979); Moyer, 58–62; and Joseph Brent, *Charles Sanders Peirce: A Life* (Bloomington and Indianapolis: Indiana University Press, 1998), 150–55.

In 1924 Raymond Clare Archibald (1875–1957), a mathematician of the new generation, categorized Newcomb's publications and determined that of

541 titles 318 should be classified as belonging to astronomy, 35 to mathematics, 42 to economics, and 146 to miscellaneous. See Archibald, "Newcomb Bibliography," 69.

On William Howard Taft's warm recollections of his days at Woodward High School, see Henry F. Pringle, *The Life and Times of William Howard Taft* (New York: Farrar & Rinehart, 1939), 1:28–29. Joseph Ray was "doctor" by virtue of an M.D. from the Medical School of Ohio located in Cincinnati. Charles Carpenter, *History of American Schoolbooks* (Philadelphia: University of Pennsylvania Press, 1963), 145. On Ray see also Jerry Dennis, "Joseph Ray," *Ohio Archaeological and Historical Quarterly* 46 (1937), 43–49, and William J. Reese, *The Origins of the American High School* (New Haven: Yale University Press, 1995), 112–113.

The support of Chicagoans for the university is affirmed in Willard J. Pugh III, "The Beginnings of Research at the University of Chicago" (Ph.D. diss., University of Chicago, 1990), 749. On the contrary situation at Clark University, see Dorothy Ross, *G. Stanley Hall: The Psychologist as Prophet* (Chicago: University of Chicago Press, 1972), 209.

Julie Reuben argues that, contrary to the claim of James P. Wind, William Rainey Harper's understanding of science was not inductive or Baconian but progressivist, meaning that theories were continually being "tested and either improved or rejected." See Reuben, 44, 305. I hold that the difference was not significant for professionalizing mathematicians such as E. H. Moore, who were less interested in philosophical niceties than in choosing words to advance their goals.

For more on the collaborative character of early University of Chicago mathematics research, see R. L. Wilder, "The Mathematical Work of R. L. Moore: Its Background, Nature and Influence," in Peter Duren, 3:266–274; and Karen Hunger Parshall, "In Pursuit of the Finite Division Algebra Theorem and Beyond: Joseph H. M. Wedderburn, Leonard E. Dickson, and Oswald Veblen," *Archives Internationales d'Histoire des Sciences* 33 (1983): 274–99.

Additional details on mathematics at the Columbian Exposition can be found in Karen V. H. Parshall and David E. Rowe, "Embedded in the Culture: Mathematics at the World's Columbian Exposition of 1893," *The Mathematical Intelligencer* 15 (1993): 40–45. With occasional deviations because of global politics, the world's mathematicians have tried to gather every four years since 1893. See Donald J. Albers, G. L. Alexanderson, and Constance Reid, *International Mathematical Congresses: An Illustrated History 1893–1986* (New York: Springer-Verlag, 1987).

B.6. The Development of E. H. Moore's Pedagogic Program

Additional background on John Dewey's educational work at Chicago can be found in Katherine C. Mayhew and Anna C. Edwards, The Dewey School (New York: Appleton-Century-Crofts, 1936); Arthur G. Wirth, John Dewey As Educator: His Design for Work in Education (1894–1904) (New York: John Wiley & Sons, 1966); Laurel N. Tanner, "The Meaning of Curriculum in Dewey's Laboratory School," Journal of Curriculum Studies 23 (1991): 101–117; and Ida DePencier, The History of the Laboratory Schools: The University of Chicago 1896–1965 (Chicago: Quadrangle Books, 1967). On Johann Friedrich Herbart's influence in America, see Krug, *High School,* 98–107.

Moore's "Sufficient unto the day is the precision thereof" dictum was used much later by philosopher Imre Lakatos to help support his views on the nature of mathematical research, as expounded in his *Proofs and Refutations: The Logic of Mathematical Discovery* (Cambridge: Cambridge University Press, 1976), 54, although Lakatos replaced Moore's "precision" by "rigour" (British spelling). According to Joe Crosswhite, Moore's proposal to let high school students create their own axioms for geometry was the inspiration for a course taught by Harold P. Fawcett at the Ohio State University lab school in the 1930s. F. Joe Crosswhite, Oral History Interview with David L. Roberts, July 29, 2003, *National Council of Teachers of Mathematics Oral History Project Records,* 1992–1993, 2002–2004, Archives of American Mathematics, Center for American History, University of Texas at Austin. Fawcett described the results of his experiment in his doctoral dissertation at Teachers College, Columbia University, which was published as *The Nature of Proof, 13th Yearbook of the NCTM* (New York: Bureau of Publications, Teachers College, Columbia University, 1938). Fawcett found the experiment encouraging, but it is clear he was dealing with a rather select group of students in an unusually supportive setting. Neither he nor anyone else that I am aware of has pursued this approach on a consistent basis in secondary school geometry.

B.7. Moore's Pedagogy in Relation to Contemporary Educational Thinkers

Morton White has discussed the Henry Fine-John Dewey controversy, but solely for purposes of elucidating the development of Dewey's philosophical thought, with no allusion to professional issues. White judges Dewey's reply to be philosophically confusing and evasive of Fine's major points. See Morton White, *The Origins of Dewey's Instrumentalism* (New York: Columbia University Press, 1943), 126–133.

For an outline of Felix Klein's career, see Werner Burau and Bruno Schoeneberg, "Klein, Christian Felix," *DSB*, 7:396–400. Also see Parshall and Rowe, 147–260. Gert Schubring has stressed that through much of the nineteenth century there were important differences in the educational systems of Protestant north Germany and the Catholic south. He sees Klein as greatly influenced by his teaching experiences in Munich, so that Klein's subsequent pedagogical reform efforts were in part designed to bring the Bavarian formulation of mathematics education to all of Germany. Gert Schubring, "Pure and Applied Mathematics in Divergent Institutional Settings in Germany: The Role of Felix Klein," in David E. Rowe and John McCleary, eds., *The History of Modern Mathematics* (Boston: Academic Press, Inc., 1989), 2:175–78.

On the general disconnection between American physicists and mathematicians during E. H. Moore's era, see John Servos, "Mathematics and the Physical Sciences in America, 1880–1930," *Isis* 77 (1986): 611–629.

B.8. The Reception of Moore's Program

On the University of Chicago's incorporation of the Chicago Institute, see Richard J. Storr, *Harper's University: The Beginnings* (Chicago: University of Chicago Press, 1966), 300–302. On Francis Parker, see Cremin, *Transformation of the School*, 128–135.

The constitution of the AMS stipulated that new members must be proposed in writing by two current members, and at a subsequent meeting of the Society be recommended by the Council and receive a majority vote of the members present (RAMS, Box 12, Folder 4). A hasty survey of the Society minutes from its inception to the 1920s suggests that nominees for membership were rarely, if ever, turned down. See also Della Dumbaugh Fenster and Karen Hunger Parshall, "A Profile of the American Mathematical Research Community: 1891–1906," in Rowe and McCleary, 3:220.

The history of functional analysis is sketched in Morris Kline, 1076–1095, and Grattan-Guinness, *Norton History*, 686–87. On Moore's work in "general analysis," see Parshall and Rowe, 387; Saunders Mac Lane, "Mathematics at the University of Chicago: A Brief History," in Peter Duren, 2:134–35, 140; and especially Reinhard Siegmund-Schultze, "Eliakim Hastings Moore's 'General Analysis'," *Archive for History of Exact Sciences* 52 (1998): 51–89.

The engineering tradition of graphical analysis is touched on in Ronald R. Kline, *Steinmetz: Engineer and Socialist* (Baltimore: Johns Hopkins University Press, 1992), 37–39. On the expansion of engineering education, see Edwin T. Layton, *The Revolt of the Engineers: Social Responsibility and the American Engineering Profession* (Baltimore: Johns Hopkins University Press, 1986), 3–4.

The problems of ascribing knowledge of mathematical concepts to historical actors are emphasized in I. Grattan-Guinness, "A Residual Category: Some Reflections on the History of Mathematics and Its Status," *Mathematical Intelligencer* 15 (1993): 4–6. On the history of the function concept in research and teaching, see Morris Kline, 950; Israel Kleiner, "Evolution of the Function Concept: A Brief Survey," *College Mathematics Journal* 20 (1989): 282–300; and Jones, 47, 57, 73, 205–206.

The classic account of the American-indifference-to-basic-science thesis is Richard Shryock, "American Indifference to Basic Science," *Archives Internationales d'Histoire des Sciences* 2 (1948): 50–65. His student, Nathan Reingold, offers a critique in "American Indifference to Basic Science: A Reappraisal," in *Science, American Style* (New Brunswick: Rutgers University Press, 1991), 54–75.

B.9. School Mathematics on the Defensive

Further relevant commentary on the evolution of American education in this period can be found in Cremin, "Revolution in American Secondary Education"; David John Hogan, *Class and Reform: School and Society in Chicago, 1880–1930* (Philadelphia: University of Pennsylvania Press, 1985), 51–59, 123–135; and Krug, *High School*, 170–71, 215–16.

For further commentary on the Cardinal Principles, see Cremin, "Revolution in American Secondary Education," 305–307; Krug, *High School*, 378–406; Kliebard, *Struggle for the American Curriculum*, 113–115; Richard Hofstadter, *Anti-Intellectualism in American Life* (New York: Vintage Books, 1963), 332–341; and Ravitch, 123–129.

On the educational testing movement, see Joel Spring, *The American School 1642–1990* (New York: Longman, 1990), 238–245, and Kidwell, Ackerberg-Hastings, and Roberts, Chapter 3.

Franklin Bobbitt's views are discussed in Kliebard, *Struggle for the Curriculum*, 97–99, 115–118 and Ravitch, 163–69. On Harold Rugg, see Ravitch 190–196; Ellen Condliffe Lagemann, *An Elusive Science: The Troubling History of Education Research* (Chicago: University of Chicago Press, 2000), 125–29; and Peter F. Carbone, Jr., *The Social and Educational Thought of Harold Rugg* (Durham: Duke University Press, 1977).

B.10. Conclusion

For more on the trend toward stratification of mathematics in the twentieth century, see David L. Roberts, "Albert Harry Wheeler (1873–1950): A Case Study in the Stratification of American Mathematical Activity," *Historia Mathematica* 23 (August 1996): 269–287.

Index

American Journal of Mathematics, 47, 115, 117, 140
American Mathematical Monthly, 223, 237
Annals of Mathematics, 117
Cardinal Principles of Secondary Education, 242, 256, 257, 267, 272, 273, 294, 314
Mathematics Teacher, 263, 279
Nautical Almanac, 60–62, 64, 67
Reorganization of Mathematics in Secondary Education, 2, 4, 242, 267, 272, 280, 294
Report of the Committee on Secondary School Studies and see Committee of Ten and Committee of Ten Report, 1, 83

academies, 10, 11, 37, 38, 48, 74, 120, 146, 149, 306
accreditation of secondary schools, 38, 146, 163
Adams, Herbert Baxter, 46
Agassiz, Alexander, 64
Airy, George Biddell, 61, 307, 308
algebra (advanced), 45, 116, 130–133, 223, 233
algebra (elementary), 3, 6, 12, 18, 23–25, 28, 37, 40, 49–51, 68–71, 73–75, 92–94, 100, 102, 103, 105, 108, 160, 161, 209, 213, 220, 221, 229, 245, 249, 254, 262, 273, 281, 303
American Association for the Advancement of Science, 136, 171, 291
American Mathematical Society, 116, 133, 143, 160, 164, 170, 173, 202, 215, 238, 313
Bulletin, 213, 216
Transactions, 129, 140
analysis and synthesis, 50, 51, 306, 307
Archibald, R. C., 278
arithmetic, 3, 11–13, 21–25, 29, 37, 51, 53, 57–59, 61, 69, 74, 75, 88–92, 96, 101, 103, 106, 155, 156, 183, 249, 273, 303
articulation of school and college, 38, 298
Association of Colleges and Preparatory Schools of the Middle States and Maryland, 163
astronomers, 17, 60, 64, 112, 115, 129, 138, 233, 234, 310
astronomy, 16, 17, 30, 42, 43, 46, 55, 66–68, 88, 114, 116, 121, 127, 129, 136, 177, 197, 209
Austin, Charles M., 261, 263, 289

Bagley, William C., 290
Baker, James H., 85
Ball, Robert S., 67
Baltimore City College, 48

317

Barus, Carl, 120
Bass, Hyman, 7
Beman, Wooster W., 155
Betz, William, 245, 246
Bezout, Étienne, 59
Birkhoff, George David, 133, 233, 252
Blair, Vevia, 266, 274, 281
Bliss, G. A., 133
Bobbitt, Franklin, 281, 314
Bolza, Oskar, 123, 128, 135
Breslich, Ernst, 261, 269
Brown University, 134
Brown, E. W., 114, 234
Bryn Mawr College, 48
Butler, Nicholas Murray, 163, 194
Byerly, William E., 86, 88

Cajori, Florian, 47–54, 86–88, 250, 251
calculus, 28, 29, 49, 59, 66, 68, 70, 76,
  79–81, 118, 167, 173, 198, 209,
  225–227, 234, 236, 255, 307
Canadian education, 56–58, 307
Cantor, Georg, 118
Cauchy, Augustin, 81, 307, 309
Central Association of Science and
  Mathematics Teachers, 219, 248, 266
Chautauqua College of Liberal Arts, 124
Chautauqua Summer Schools, 124
Chicago Manual Training School, 169
Child, Francis J., 20
Clark University, 96, 147
Colburn, Warren, 12, 94
college domination, 53, 96, 107, 108, 203
College Entrance Examination Board,
  163
Columbia University, 116, 163, 192
  Teachers College, 193, 252, 269, 274
Columbian University, 36
commercial arithmetic, 2, 11, 12, 89–92,
  97–100, 105, 108, 109, 157
Commission on the Reorganization of
  Secondary Education, 256
Committee of Fifteen, 102, 251

Committee of Ten, 2, 20, 28, 37, 47, 55,
  67, 83, 100, 106, 156, 202, 256, 283,
  309
Committee of Ten Report, 83, 95
Committee on College Entrance
  Requirements, 159, 160, 251
common schools, 37, 306
commutative law, 92, 93, 100, 157, 273
Conant, Levi L., 36
correlation coefficient, 277
correlation of studies, 103, 161, 165, 170,
  174, 189, 211, 212, 219, 220, 228,
  250, 279
Craig, Thomas, 64, 255
Crane Technical High School, 271
Crathorne, A. R., 266, 277
cross-section paper, 169, 220–223
cube root, 57, 100–102, 106, 158, 249, 310
Cutler, Arthur, 88

d'Ocagne, Maurice, 222
D. C. Heath and Company, 77
Davies, Charles, 59, 74, 81
Day, Jeremiah, 9, 15, 18, 27, 28, 92, 304
deduction, 2, 30, 31, 35, 36, 45, 127, 164,
  172, 182, 204, 221
Delaunay, Charles-Eugène, 113, 310
Denison University, 124, 127
Dewey, John, 125, 151, 168, 176, 179,
  182, 247, 255, 276
Dexter, Edwin, 104, 105
Dickson, Leonard Eugene, 133, 159, 209,
  223

Educational Review, 165, 194
efficiency, 3, 4, 9, 24, 25, 27, 36, 90, 108,
  146, 151, 226, 232, 244, 245, 247,
  249–251, 253, 256, 276, 277, 281,
  285, 298, 299, 305
elective system/principle, 26, 27, 29, 96,
  98, 247, 260, 278, 298
elementary schools, 1, 29, 37, 74, 103,
  152, 172, 220
Eliot, Charles W., 1, 10, 14, 17, 18, 33,
  40, 55, 84, 93, 96, 107, 125, 163, 243

enrollments, school and college, 20, 56, 66, 242, 243, 305
equivalence of studies, 14, 21, 96
Euler, Leonard, 112
Evans, G. W., 266
Evanston Colloquium, 139, 161, 196

Fine, Henry, 92, 117, 160, 183
Finkel, Benjamin, 154, 158
Fisk University, 48
Fiske, Thomas, 116, 163, 173, 212, 232
Flexner, Abraham, 267, 271, 292
Foberg, J. A., 266, 271
Forsyth, A. R., 193
Fréchet, Maurice, 224
Franklin Female College, 48
Froebel, Friedrich, 31, 305
function concept, 4, 94, 138, 161, 198, 208, 218, 220, 221, 231, 232, 274–276, 279, 282, 288, 314

Göttingen, 121, 171, 197, 199–201
Gauss, Carl Friedrich, 112
general analysis, 130, 132, 207, 224, 225, 313
General Education Board, 267
geometry
  analytic, 22, 51, 52
  concrete, 93, 103
  constructive, 277
  deductive, 221
  demonstrative, 94, 97, 104, 188, 277
  descriptive, 52
  elementary, 18, 68, 287
  euclidean, 51, 177, 186
  experimental, 93, 277
  formal, 88
  four-dimensional, 112
  Greek, 22
  higher-dimensional, 115
  intuitional, 188
  intuitive, 187, 247, 270, 277
  inventional, 88
  non-euclidean, 45, 78, 103, 177, 186
  plane, 40, 186, 213
  projective, 95, 287

projective differential, 130
rational, 187
sensuous, 187
solid, 186
synthetic, 51, 52, 95
George Washington University, 36
Gibbs, J. Willard, 121, 229
Gildersleeve, Basil, 43
Gilman, Daniel Coit, 17, 31, 42, 55, 62, 123, 125
Gore, J. Howard, 36
grammar schools, 11, 28, 37, 157, 158, 219, 250, 306
graph paper, 179, 207, 220, 229, 275, 282
Greenwood, J. M., 99, 102
Greeson, W. A., 48

Hale, George Ellery, 136
Hall, G. Stanley, 96, 98, 147
Halsted, George Bruce, 45, 50, 186–188
Hamilton, William (philosopher), 30, 52
Hamilton, William Rowan (mathematician), 92
Hancock, Harris, 153, 159, 266
Hansen, Peter Andreas, 113, 310
Hanus, Paul, 29, 91
Harper, William Rainey, 122–124, 145, 172, 174, 179, 180, 226
Harris, William T., 23, 84, 102
Harvard College/University, 15, 18–20, 39, 40, 48, 60, 61, 65, 72, 86, 122, 133, 134, 145, 163, 213, 230, 234, 309
Harvard Observatory, 61
Hedrick, Earle R., 247, 250, 275
Henry Holt and Company, 68, 71–73, 308
Henry, Joseph, 60
Herbart, Johann Friedrich, 161, 174
Herbert, H. A., 67
Herrick, Robert, 131
Hilbert, David, 138, 171, 184, 186, 197, 199, 201, 246
Hilgard, J. E., 60
Hill, G. W., 234
Hill, Thomas, 18
Holden, E. S., 67

Holt, Henry, 74, 75, 308
Howard University, 48
Huling, Ray Greene, 20
Huntington, Edward V., 230
Huxley, T. H., 31, 127

induction and inductive teaching
    methods, 2, 9, 10, 12, 16, 30, 31, 33,
    36, 45, 84, 93, 108, 125–127, 150,
    156, 172, 182, 193, 204, 246, 283,
    305, 306, 311
Ingraham, Andrew, 48, 49
Institute for Advanced Study, 292
International Commission on the
    Teaching of Mathematics, 227,
    252–254, 271, 279
International Mathematical Congress,
    1–3, 55, 111, 117, 135–139, 198, 199,
    252, 311

James, William, 71, 244, 274
Johns Hopkins University, 10, 22, 30, 31,
    41, 44, 55, 62, 64, 76, 118, 140, 153,
    300, 304, 308
Johnson, William Woolsey, 117
Jordan, David Starr, 29
Judd, Charles, 245
Judson, Harry Pratt, 164, 226

Karpinski, L. C., 258
Keyser, C. J., 232
Kilpatrick Report, 273
Kilpatrick, William Heard, 242, 254–260
Kingsley, Clarence D., 256, 259
Klein, Felix, 63, 135, 172, 179, 180, 195,
    247, 252, 284, 300, 313
Kronecker, Leopold, 183

laboratory method, 3, 4, 9, 20, 31–33,
    143, 144, 156, 165–169, 177, 178,
    183, 185, 204, 207–212, 218, 219,
    223, 225–228, 246, 251, 259, 279,
    284–288, 305
Lagrange, Joseph-Louis, 112
Laplace, Pierre-Simon, 59, 112
Lawrence Scientific School, 20, 39, 61

Lawrenceville School, 48, 85
Le Verrier, Urbain Jean Joseph, 113
Lennes, N. J., 133, 223
limit concept, 49, 70, 80, 81, 118, 177,
    307, 309
Lincoln, Abraham, 11
Locke, John, 15
Long, Edith, 165, 178
Lunn, A. C., 167, 209

Mac Lane, Saunders, 224, 292
Mackenzie, James C., 85
manual training, 29, 31, 52, 96, 150, 168,
    256, 277, 306
Maschke, Heinrich, 128, 255
Massachusetts Institute of Technology,
    20, 75, 197
Mathematical Association of America, 2,
    4, 154, 207, 227, 236, 238, 242, 264,
    265, 294
Mathematics Conference of the
    Committee of Ten, 86, 88–95, 156,
    157, 202–204
McClelland, B. F., 100
McGee, Anita Newcomb, 68, 309
McLellan, James, 183
Men's Mathematics Club of Chicago,
    241, 260, 279
mental arithmetic, 94, 100
mental discipline, 2, 4, 6, 9, 13–15,
    25–28, 33, 34, 58, 59, 79, 81, 84, 89,
    91, 94, 97, 101, 106, 107, 131, 161,
    176, 204, 228, 241, 244, 249,
    252–254, 256, 259, 273, 274, 276,
    277, 279, 281, 283, 284, 288, 304
Mercer College, 255
Michelson, A. A., 130
Mill, John Stuart, 78
Miller, G. A., 218
Millikan, Robert, 130, 181, 201, 228
Millis, J. F., 248, 258
Mittag-Leffler, Gösta, 136
Moore, C. N., 266
Moore, E. H., 3, 55, 64, 75, 93, 111, 119,
    122, 133, 138, 143, 160, 167, 175,

188, 202, 207, 254, 265, 277, 279, 280, 284, 292
Moore, Robert Lee, 133
Morrill Act of 1862, 39
Morrison, Joseph R., 67, 308
Moulton, Forest Ray, 129–131
Mount Holyoke Female Seminary, 48
Muskingham College, 123
Myers, G. W., 209, 223, 227

National Academy of Sciences, 171
National Committee on Mathematical Requirements, 263–281, 283, 288, 289, 294
National Council of Teachers of Mathematics, 228, 241, 263, 279, 289, 294
National Education(al) Association, 1, 84, 159, 212
Naval Academy, 117
Naval Observatory, 61
New Hope Female Academy of the Choctaw Nation, 48
New York Mathematical Society, 116, 139, 202
Bulletin, 117
Newcomb's Mathematical Series, 68
Newcomb, Simon, 2, 23, 24, 39, 41, 46, 55, 67, 103, 112, 134, 138, 140, 179, 202, 234, 270, 283, 291
Newton, Hubert Anson, 121
Newton, Isaac, 11, 18, 59, 101, 112
Nightingale, A. F., 160
nomography, 222, 229
normal schools, 38, 39, 48–50, 53, 163, 178, 214, 280, 306
North Central Association of Colleges and Secondary Schools, 163
Northwestern University, 139
number line, 69

Olds, George, 48, 49, 87, 88
Osgood, W. F., 230, 234, 258, 264

Parker, Francis, 151, 210
Peano, Guiseppe, 171

Peirce, Benjamin, 19, 44, 55, 60, 80, 112, 115, 234
Peirce, Charles Sanders, 45, 114, 310
Peirce, James Mills, 19
Perry, John, 169, 172, 176, 179, 192, 229, 284
Pestalozzi, Johann, 31, 305
Phillips Andover, 48
Phillips Exeter, 48, 72
Phillips, D. E., 158, 159
physics, 30, 44, 47, 93, 130, 136, 165, 169, 170, 175–177, 181, 185, 194, 197, 199, 201, 228–230, 255, 284, 287, 288, 301–303
Plato, 15
Poincaré, Henri, 114, 153, 172, 182, 310
primary schools, 37, 102, 174, 175, 178, 306
Princeton University, 48, 72, 117, 133, 134, 160, 183, 226, 287, 292
professionalization and professional jurisdiction, 3, 37, 64, 93, 99, 108, 109, 143, 162, 168, 175, 176, 191, 198–200, 203, 204, 214, 219, 246–248, 264, 285, 286, 291, 301–303

quadratic equations, 40, 73
quantics, 184
quaternions, 22, 92

Ray, Joseph, 51, 69, 120, 157, 311
Richardson, R. G. D., 134
Robinson, Oscar D., 14, 85
Rockefeller, John D., 145, 267
Rowland, Henry, 30, 31, 46
Rugg, Harold, 281
Rule of Three, 11
Ryerson Physics Laboratory, 123

Safford, T. H., 48, 52, 77, 91
Schorling, Raleigh, 269, 277
scientific and technical schools, 21, 39, 40, 96, 192, 199
Seaver, E. P., 99, 102
secondary schools, 2–5, 10, 18, 20, 21, 23, 25, 27, 37, 38, 40, 49, 53, 70, 71, 85,

95–98, 107, 111, 120, 146, 151, 155, 162–164, 174, 175, 178, 193, 198, 203, 209, 210, 220, 233, 242, 243, 249, 250, 256, 264, 266, 269–271, 273, 281, 284, 285, 292, 298
Seerley, H. H., 100
Sheffield Scientific School, 39, 42
Sinclair, Upton, 124
Slaught, Herbert, 149, 154, 209, 218, 227, 237, 242, 250, 261, 263, 289
Smith College, 48
Smith, David Eugene, 39, 155, 193, 212, 214, 227, 250, 252, 266, 276, 289, 306
Smithsonian Institution, 60
Snedden, David, 254
social utility, 3, 249
South Side Academy, 169
Spencer, Herbert, 33
square root, 57, 69, 73, 100, 249
standardized tests, 278, 314
State Colored Normal School of North Carolina, 48
Story, William E., 45, 46, 64, 112
student differentiation, 84, 85, 97–99, 105, 107, 243, 251, 257, 259, 298
student interest, 9, 27, 59, 84, 106, 107, 190, 283
Study, Eduard, 186
Swain Free School of New Bedford, 48, 49
Sykes, Mabel, 249
Sylvester, James Joseph, 31, 42, 44–46, 62–64, 66

Taft, William Howard, 120, 311
Tappan, Henry Philip, 28
textbooks, 11, 17, 18, 57, 59, 67–77, 79, 90, 106, 120, 126, 150, 155, 157–159, 227, 229, 282, 307–309
Thomson, James, 192
Thorndike, E. L., 244, 256, 274
Tyler Report, 213
Tyler, H. W., 213, 258, 264

University of Chicago, 122, 147, 153, 167, 180, 209, 220, 255

University of Chicago Mathematical Club, 133, 159, 165, 211
University of Cincinnati, 153, 266
University of Colorado, 85
University of Illinois, 104
University of Michigan, 28
University of Rochester, 48
University of Virginia, 44, 153
University Secondary School, University of Chicago, 169
Upton, Clifford Brewster, 278

Van Vleck, E. B., 236, 264
Veblen, Oswald, 133, 134, 140, 169, 185, 208, 266, 287, 292
Veblen, Thorstein, 124
vocational education, 243, 244, 251, 254, 298

Walker, Francis Amasa, 30, 75, 101
Weil, André, 6
Wentworth, G. A., 69, 70, 72, 309
Whewell, William, 78
White, H. S., 139
Whitney, William Dwight, 128
Wilczynski, E. J., 215, 227
Williams College, 48
Wilson, Edwin Bidwell, 231
Woodward High School, 120
Worcester Polytechnic Institute, 36, 40
World's Columbian Exposition, 1, 86, 111, 135, 137, 311

Yale College/University, 9, 17, 39, 40, 42, 43, 48, 72, 121, 123, 124, 127, 128, 134, 304, 308
Yale Report, 9, 15–18, 28, 90, 274, 304
Youmans, Edward Livingston, 33
Young, Jacob William Albert, 147–149, 152, 159, 160, 165, 185, 209, 212, 219, 258, 280, 294
Young, John Wesley, 134, 148, 228, 242, 268, 271, 287, 289, 294

Ziwet, Alexander, 202, 213

www.ingramcontent.com/pod-product-compliance
Lightning Source LLC
Chambersburg PA
CBHW070554100426
42744CB00006B/267